Operationism in Psychology

Operationism in Psychology
AN EPISTEMOLOGY OF EXPLORATION

Uljana Feest

The University of Chicago Press   Chicago and London

The University of Chicago Press, Chicago 60637
The University of Chicago Press, Ltd., London
© 2025 by The University of Chicago
All rights reserved. No part of this book may be used or reproduced in any manner whatsoever without written permission, except in the case of brief quotations in critical articles and reviews. For more information, contact the University of Chicago Press, 1427 E. 60th St., Chicago, IL 60637.
Published 2025

34 33 32 31 30 29 28 27 26 25   1 2 3 4 5

ISBN-13: 978-0-226-83837-3 (cloth)
ISBN-13: 978-0-226-83839-7 (paper)
ISBN-13: 978-0-226-83838-0 (e-book)
DOI: https://doi.org/10.7208/chicago/9780226838380.001.0001

Library of Congress Cataloging-in-Publication Data

Names: Feest, Uljana, author.
Title: Operationism in psychology : an epistemology of exploration / Uljana Feest.
Description: Chicago : The University of Chicago Press, 2025. | Includes bibliographical references and index.
Identifiers: LCCN 2024036147 | ISBN 9780226838373 (cloth) | ISBN 9780226838397 (paperback) | ISBN 9780226838380 (ebook)
Subjects: LCSH: Psychology, Experimental. | Psychology—Research.
Classification: LCC BF181 .F44 2025 | DDC 150.72—dc23/eng/20240905
LC record available at https://lccn.loc.gov/2024036147

CONTENTS

INTRODUCTION: TOWARD AN EPISTEMOLOGY OF EXPLORATION IN PSYCHOLOGY · 1

1. Topic and Main Theses of the Book 1
2. Contexts of Discovery and Justification 7
3. Integrating Philosophy of Science and History of Science 12
4. Objects of Research: Moving Targets of Scientific Investigation 18
5. Addressing the Crisis of Confidence in Psychology 20
6. Does My Analysis Generalize beyond Psychology? 22
7. A Quick Overview of the Chapters 24

1. OPERATIONISM IN PSYCHOLOGY: (SOME) HISTORICAL BEGINNINGS · 30

1.1. Introduction 30
1.2. Stanley Smith Stevens and the Operational Treatment of Sensations 35
1.3. Of Rats and Psychologists: An Analysis of E. C. Tolman's Operationism 46
1.4. Clark Hull and the Role of Operationism in Theory Construction 56
1.5. Conclusion 68

2. OPERATIONISM: THE SECOND GENERATION · 71

2.1. Introduction 71
2.2. Early Debates (1930s/1940s) 74
2.3. Some Midcentury Developments (Interlude) 81
2.4. The Construct Validity of Psychological Tests (1955) 87
2.5. Converging Operations 97
2.6. Conclusion 100

3. OPERATIONAL DEFINITIONS AS TOOLS · 102

3.1. Introduction 102
3.2. Operational Definitions and Research Designs in Memory Research 104
3.3. Operational Definitions as Tools: What Do They Do? 113
3.4. Conceptual Development and Reference: Another Look at the Case Studies 119
3.5. Operational Definitions vis-à-vis Philosophical Analyses of Concepts 124
3.6. Scientific Concepts and Investigative Practice 129
3.7. Conclusion 132

4. OBJECTS OF RESEARCH AS TARGETS OF EXPLORATION · 135

4.1. Introduction 135
4.2. Delineating and Describing Objects of Research: A First Approximation 138
4.3. Describing Empirical Features of Objects of Research 141
4.4. Exploratory Research 152
4.5. Conclusion 160

5. PHENOMENA AND OBJECTS OF RESEARCH · 163

5.1. Introduction 163
5.2. Phenomena vs. Data and vs. Objects of Research? Conceptual Groundwork 165
5.3. Objects of Psychological Research as Explanandum Phenomena? 174
5.4. Psychological Discovery as Phenomenal Decomposition? 182
5.5. Toward an Analysis of Norms of Exploration in Psychology 189
5.6. Conclusion 192

6. WHAT KINDS OF THINGS ARE PSYCHOLOGICAL KINDS? · 194

6.1. Introduction 194
6.2. (Natural) Kinds: Setting the Stage 197
6.3. Pluralism, Mechanisms, and the Whole Organism 204
6.4. Similarity Judgments at the Whole-Organism Level: Echoes from Ecological Psychology 211
6.5. Psychological Kinds and Cognitive Ontology 215
6.6. Conclusion 223

7. OPERATIONAL ANALYSIS AND CONVERGING OPERATIONS · 226

7.1. Introduction 226
7.2. Inferences in Psychological Experiments 228
7.3. Experimental Inferences as Constrained by Operational Analysis 236
7.4. Converging Operations 243
7.5. So What Do Converging Operations Converge On? 250
7.6. Conclusion 260

CONCLUDING REMARKS · 262
  1. Introduction 262
  2. Main Points 262
  3. Current Relevance and Future Directions 264

ACKNOWLEDGMENTS · 283
REFERENCES · 287
INDEX · 317

INTRODUCTION

Toward an Epistemology of Exploration in Psychology

1. Topic and Main Theses of the Book

In the 1980s, psychologists and neuroscientists found a functional dissociation between the results on two different types of memory tests: (*a*) tests in which subjects were instructed to report items they recalled or recognized from a previous study phase and (*b*) tests in which the subjects' memories of such items were measured by means of priming tests, where no conscious effort of recollection was required. This experimental dissociation was one of the factors that led scientists to posit the existence of a type of memory they dubbed "implicit memory," which does not rely on processes of conscious recollection. Subsequent decades saw a lot of experimental research into this purported type of memory. In trying to understand how scientists conducted this research, however, we face a puzzle: How did they go about investigating implicit memory given that little was known about it? Surely, in order to subject implicit memory to experimental research, scientists had to rely on an empirical concept of it. Yet, as long as both the existence and the central features of the concept's referent were still debated, the concept itself remained in flux. Indeed, it was continually revised, modified, and questioned as a result of the very research that it made possible. But what precisely was the concept informed by? What kinds of assumptions did it contain? And how did it enable researchers to grasp the objects presumed to be in its extension? In other words, in practical terms, how did scientists proceed in their empirical investigations of implicit memory given, on the one hand, the intriguing empirical results they attributed to it and, on the other, their relative ignorance of it?

This example (and the questions it raises) draws our attention to the issue of how to analyze the investigative processes in areas of research that are characterized by a high degree of conceptual openness and epistemic uncertainty. The aim of this book is to present an answer to this question. Accordingly, I seek to contribute to a growing literature that concerns itself with the analysis of experimental research processes, understood as dynamic, ongoing endeavors that are aimed at delineating and refining our very understanding of the subject matter under investigation (see Feest and Steinle 2016, sec. 3). More specifically, the book provides a philosophical analysis of the dynamic research practices in experimental cognitive psychology.[1] This analysis is informed by the contention that the experimental practice in this field give rise to philosophical questions about the ways in which conceptual work and experimental work are intertwined.

With the agenda pursued in this book, I take inspiration from important work about scientific discovery (esp. Bechtel and Richardson 1993) as well as related work in philosophy of neuroscience (e.g., Craver and Darden 2001, 2013). However, while previous work has focused on the discovery of explanatory mechanisms, I argue that a lot of experimental research in cognitive psychology should be understood as an attempt to gain a better descriptive and taxonomic understanding of what I term "objects of research," that is, the targets of scientific investigations, such as memory. I argue that research aimed at exploring objects of psychological research includes but is not limited to the discovery of mechanisms.

In addition, this book builds on a rich body of literature about the history and philosophy of experimentation, such as can be found in work about exploratory experiments (e.g., Burian 1997; and Steinle 1997), epistemic things (Rheinberger 1997), and the iterative nature of scientific practice (Chang 2004; Elliott 2012). This and other work was instrumental in forging a process-oriented approach to the philosophy of experimentation as a continuation of the new experimentalism that emerged in the late 1970s (Ackermann 1989). Prominent figures of this movement, such as Ian Hacking, Allan Franklin, and Deborah Mayo, rightly rejected a theory-centric approach in philosophy of science, pointing out that many experiments are designed not only to test theories but to perform other epistemic functions as well, such as detecting entities or calibrating instruments. Even so, the first wave of experimentalist philosophy of science was not primarily interested in the dynamic research process as such. Likewise, while there has long been an interest in conceptual change in science, few have explicitly linked this interest

---

1. Note that, while my main focus is on psychology, there is not always a sharp line between cognitive psychology and cognitive neuroscience, as will become apparent at various points in the book.

to the specific role of experiments, though this has begun to change in recent years.[2]

Given this first rough sketch of the topic and agenda of the book, the question is how to approach it. The method of this book represents a particular answer to this question. Keeping in mind the predicament of the conceptual openness and epistemic uncertainty of research in cognitive psychology and neuroscience, a philosophical analysis should, I claim, start out by paying close attention to the strategies the scientists themselves pursue in dealing with this uncertainty. In other words, I argue that an important source to which to turn for our normative philosophical analysis of experimental practices is scientific methodology. This approach is not new, of course, as other philosophers of science have drawn inspiration from scientific methods as they are applied in scientific practice. Along these lines, Deborah Mayo (1996) has drawn attention to "methodological rules," focusing in particular on the methods of statistical analysis that scientists use in their attempts to eliminate experimental errors. Likewise, Allan Franklin (1999) has singled out an open-ended list of "epistemological strategies" used by experimental scientists in their attempts to justify particular experimental results. The approach taken in this book is compatible with theirs, but, whereas Mayo has focused her attention on the severe testing of hypotheses, I am interested in the methods governing the interplay between conceptual work and empirical work. In turn, while Franklin's focus has been largely on physics, my interest is in the methodologies that shape the material practices of experimentation in psychology.

In approaching the issues just outlined, I place particular emphasis on the question of how scientists come empirically to delineate and study particular objects of research in the widest sense of this concept (i.e., including phenomena, effects, mechanisms). With this I highlight that it is not at all obvious how to capture philosophically what constitutes an object of research in psychology and cognitive neuroscience (and, I would argue, in other areas of the social and behavioral sciences as well). While there is some suggestion in the literature that the targets of psychological research should be conceived as *phenomena* (e.g., Bechtel 2008b), this notion itself, for all its seemingly evident meaning, is

---

2. Within philosophy of neuroscience, Jacqueline Sullivan (e.g., Sullivan 2009, 2010, 2016b, 2017a, 2017b) has also pointed to the complexities and uncertainties of experimental practice, and some recent work has started looking into the issues of concept formation and exploratory experimentation in neuroscience and psychology (e.g., Colaço 2018a, 2018b, 2020; and Haueis 2017, 2018). I draw connections to these works at various points throughout this book.

notoriously difficult to pin down (as evidenced in the debates about Bogen and Woodward [1988]). This is especially true when it comes to the phenomena studied by cognitive psychology and cognitive neuroscience, such as memory and learning. To return to the example of implicit memory, how should we describe the phenomenon? In terms of a particular experimental effect? In terms of some neural mechanism underlying the effect? In terms of a combination of the two? And, if so, how? In addition, as I will show, the word "phenomenon" is often used ambiguously, referring sometimes to open-ended research targets, such as implicit memory, and sometimes to specific regularities, such as the priming effect. This book offers some novel terminological distinctions that will shed light on these questions. To put it briefly, I argue that objects of research in psychology are often conceptualized as clusters of phenomena.

I contend that the philosophical difficulties of coming to grips with the question of what objects of psychological research are correspond to very real scientific difficulties in saying precisely what constitutes an empirical description of the relevant target of research and how to arrive at one. This insight prompts me to explore the question of why this is the case. Developing an idea first formulated by Hans-Jörg Rheinberger (1997), I argue that the objects of psychological research are *epistemically blurry*. By this I mean (*a*) that scientists find themselves in a position of epistemic and conceptual uncertainty vis-à-vis their research objects and (*b*) that this is due, in part, to the fact that research objects are typically composed of multiple, more confined phenomena. The challenge for researchers is to provide descriptive accounts of specific objects as composed of multiple phenomena. For example, when we think of something like spatial memory, we typically take this object to include any number of instances of navigational behavior *as well as* the multitude of cognitive and neural processes that make such behavior possible (Sullivan 2010). This complexity of the psychological subject matter partially accounts for the epistemic situation of uncertainty vis-à-vis their research targets. In addition, I argue that the phenomena that comprise a specific research object like spatial memory do not neatly determine the object's identity condition.

I show that this is true of objects investigated by psychologists in general and that the "distributed" nature of such objects corresponds to a rather holistic set of interrelated conceptual assumptions with which cognitive psychologists approach their disorderly and complex field of study, though I argue that folk psychology provides basic structuring categories that researchers use. Owing to the lack of detailed knowledge, these assumptions are often ill-formed. The question I address in this book is how they can nonetheless inform productive experimental research. Or, to rephrase this in the terminology used above, the question is what kinds of methods scientists draw on in their attempts to pursue research about a particular object of research and how these practices are underwritten by normative methodological considerations.

The main thesis of this book is that in experimental psychology and cognitive neuroscience we find a methodological attitude at work, which (following scientific usage) I refer to with the term "operationism."[3] This attitude can be characterized as a package of interrelated informal maxims designed to guide the ways in which scientific concepts inform and are informed by experimental operations. Hence, operationism, as I construe it, has several components. In the most general terms, operationism recognizes that whenever researchers design an experiment to investigate a specific question, they need to *operationalize* the question, that is, they need to specify the empirical manipulations and measurement procedures that they will use in pursuit of their epistemic aims. One of the issues that needs to be addressed in the course of this procedure is that of how to measure the object of research. Given the reality of a situation wherein researchers often do not have a fully articulated conceptual grasp of their objects of investigation, operationism proposes (*a*) that researchers pin down preexisting concepts in terms of specific, tentative measurement procedures and (*b*) that novel concepts can be introduced into the scientific discourse by appeal to a novel measurement procedure that is presumed to reveal a previously unknown object. In both these cases, scientists talk of providing "operational definitions" of their research objects.

With this twofold understanding of operationism, I capture Percy Bridgman's (1927) contention that researchers should be careful not to assume that an existing concept can be extended to a novel domain or a different scale unless they have specific measurement operations to back this up. Conversely, sometimes caution demands that a new research object is posited in accordance with a new measurement technique. At the same time, I also hope to capture the insight that existing concepts often do get utilized, modified, and extended in the research process (Wilson 2006) as well as highlight the iterative dynamics of measurement operations and conceptual development (Chang 2004). As should be clear at this point, I follow the way in which scientists understand the concept of an *operational definition*, which differs from the way it

---

3. Throughout this book, I use the term "operationism" unless I am citing from a source that uses "operationalism." For the purposes of my analysis, the two terms are synonymous.

is usually understood within philosophy. On my analysis, operational definitions specify what at a given point in time experimental scientists take to be typical conditions of application for a term thought to denote their object of research. But there is more to the operationist attitude that I describe and champion in this book. It has to do with the fact that there is no guarantee that any given operational definition—or its implementation in a specific experiment—in fact captures a genuine object of interest. The reason for this is not only that an (operationally defined) concept might not have a referent but also, and more importantly, that a given implementation of an operational definition in a specific experiment relies on countless background assumptions about the subject domain. Specifically, in treating the data of an experiment as indicative of their research objects, psychologists have to assume that the data are not contaminated by any uncontrolled features of the experiment. I argue that this calls for another aspect of the operationist attitude, one corresponding to what Bridgman (1938, 1945) called "operational analysis," which is essentially a form of experimental critique. It involves the explication and the critical evaluation of the material presuppositions required for specific experimental inference. I argue that this activity of operational analysis is not merely an act of error probing (Mayo 1996) but in fact a vital and constructive component of the process whereby the scope of a given concept is determined and its referent described.

A fundamental contention of this book is that, when psychologists and cognitive neuroscientists talk about operational definitions, they typically have in mind paradigmatic measurement procedures (or tests) thought to reveal the presumed object in especially clear form. Those paradigmatic procedures, in turn, figure as parts of experimental paradigms for the empirical investigation of the phenomenon in question. In the case of implicit memory, paradigmatic measurement procedures might be instances of a specific kind of memory test, that is, tests that do not require subjects to draw on conscious recollections. Experimental paradigms, more generally, include typical experimental interventions that are performed prior to the measurement procedure being administered. Experimental paradigms embody general conceptions of the subject matter (e.g., the assumption that memory consists of learning, storage and retrieval), whereas paradigmatic measurement procedures specify operational definitions, which embody specific conceptual presuppositions about the subject matter, pertaining to typical behavioral effects that are indicative of the object of research. Operational definitions do not (and are not intended to) exhaust a term's meaning. I argue that scientists are well aware that such definitions are tentative, hypothetical, and in important ways theoretically underdetermined. I show that it is for this reason that the methodological maxim of operational analysis is such a powerful research tool as it encourages scientists to explicate, question, and test the conceptual assumptions built into their research designs as well as the implementations of those research designs.

The connection between operational definitions and experiments is no coincidence. When psychologists conduct experimental tests, they proceed by performing experimental operations, such as manipulating an independent variable and instructing research subjects to respond to the independent variable in a particular way (Feest 2011a). Just as those manipulations are informed by preexisting conceptual assumptions about the object under investigation, so, too, are the measurement procedures used to record the outcomes of the experimental manipulations. In this book, I argue that this close relation between scientific concepts and experimental operations allows us to forge a link between two equally important aspects of science: that of representation and that of intervention. Ian Hacking (1983) has famously argued that traditional philosophy of science—coming out of the analysis of language—has focused too much on the representational nature of scientific theories, thereby creating skeptical problems that do not seem so relevant to scientific practice once we start taking a closer look at experimental intervention. I hope to show that to understand experimental practice we need to bring scientific language back into the picture and that we can do so while retaining Hacking's important insight that experiments have many lives of their own. By picking concepts—as opposed to theories or experimental systems (e.g., Rheinberger 1997)—as the units of analysis, my account of the dynamics of research is related to some of the work on conceptual change in the history and philosophy of science (e.g., Andersen, Barker, and Chen 2006; and Nersessian 1984). Unlike those works, however, my account explicitly links scientific concepts and experimental intervention. Therefore, I highlight the ways in which objects of research are conceptualized by scientists and how such conceptualizations shape *and are shaped by* the experimental interventions scientists perform.

2. Contexts of Discovery and Justification

One way of summarizing the project described thus far is to say that it is concerned with the discovery process in psychology. While there is a sense in which this is true, the terminology of "discovery" is highly ambiguous and can easily give rise to misunderstandings. Let me begin by distinguishing between two notions of discovery: one pertaining to a process and one pertaining to a product. When I talk about discovery in this book, I have in mind the former of these two senses. That is to say, my focus is not on reconstructing how successful discoveries come about but on how scientists proceed when they do not yet know what the result of a successful discovery might look like.

A number of objections can be raised against this kind of project. Let me consider some of these objections by looking at issues surrounding discovery and justification in twentieth-century philosophy of science. In doing so, I hope to clarify my own approach. Though he did not use this terminology, Karl Popper (1935/1992) presented one version of the distinction between discovery and justification. His vision still informs a particular approach to doing philosophy of science according to which it is entirely irrelevant for philosophical purposes how a scientist arrived at a hypothesis—all that matters is how that hypothesis can be scientifically tested. In Popper's case, of course, this view was informed by his radical rejection of inductivism. Consequently, his critique of philosophical analyses of discovery had two strands. On the one hand, he was inclined to view the processes that lead up to a discovery as irrational (as illustrated by famous examples of scientific findings that were inspired by dreams). On the other hand, he would have said that, even in cases where a hypothesis was arrived at by means of rational argumentation, this was irrelevant to philosophical analyses of scientific knowledge since, by virtue of being inductive arguments, the arguments invoked when developing a hypothesis could not add any justification to the hypothesis in question. Such justification could—according to Popper—be granted to a hypothesis only by virtue of not being empirically refuted. (As is well-known, Popper was careful to emphasize that this did not amount to a confirmation of the hypothesis.)

Many philosophers of science today reject Popper's extreme anti-inductivist stance. However, his approach still exemplifies a particular vision of philosophy of science, namely, as concerned with purely formal relations between theory and evidence that can be analyzed regardless of the temporal dynamics by which both theory and evidence are constructed in scientific research (see Nickles 2009). Even Hanson's attempt to formulate a logic of discovery (Hanson 1958, 1961), which describes the generation of hypotheses (as opposed to the logic of testing hypotheses), for all its genuine insights, still bought into the idea that discovery and justification can be distinguished in terms of purely formal characteristics. It is this vision of philosophy of science that I wish to challenge in this book since it systematically ignores two important features of scientific practice: (a) the fact that scientific reasoning practices (especially in an experimental context) are typically constrained by assumptions about the physical world (Norton 2003) and (b) the fact that scientific practices more generally have a temporal character (Rouse 2015). In this book, I aim to take both these features seriously. By emphasizing the material and temporal nature of scientific practice, this book has some affinities with treatments of scientific practice coming from the history of science and science studies, especially after the turn to material culture (e.g., Galison 1996). The question therefore is what characterizes my project as a philosophical one. Very broadly speaking, my response is that the project pursued in this book is philosophical insofar as it seeks to provide an account of the rationality of the investigative process, where this rationality is conceived not in terms of formal and universal rules of reasoning but in terms of informal strategies, rules, and maxims that are applicable in specific types of contexts, given specific problem situations.

This turn to informal and context-specific kinds of scientific reasoning raises far-reaching questions regarding the scope and method of philosophical analyses. For example, does my interest in specific contexts and material practices imply a turning away from the aim of providing a universal analysis of scientific rationality? In evaluating this question, we need to distinguish between two issues: one concerning formality and one concerning context specificity. Regarding formality, it seems to me that it is possible to abandon the search for a formal notion of scientific reasoning without thereby giving up the aim of constructing a general theory. John Norton's (2003) material theory of induction can be regarded as an example of such a general theory that is not formal. Regarding context specificity, I argue that it is, ultimately, an empirical question whether different context-specific accounts of scientific rationality can be subsumed under a more general account (formal or informal). Thus, I view informal and context-specific approaches as at least in principle compatible with more universalizing analyses of scientific rationality. For the purposes of this book, however, I restrict my attention to an analysis of reasoning practices in psychology. While I do claim some generality for my account within this realm, it is a question for future work to determine whether my analysis also has insights to offer with regard to other scientific domains. More importantly, I suggest a shift of focus. While much of the debate about scientific rationality has concentrated on the logic of theory testing, this is not the topic I pursue here. The simple reason for this is that I do not view theory testing as the only, or the most important, function of scientific experiments. The process I hope to get into clear view, instead, is that of concept formation and theory development.

With the assessment that there is more to scientific experimentation than theory testing, I situate myself in the tradition of new experimentalism (Ackermann 1989), which was instrumental in directing our attention to other epistemic functions of experiments, such as the exploration of phenomena, the calibration of instruments, or the formation of concepts (see also Feest and Steinle 2016). One of the key aims of this literature is to take seriously the philosophical questions that arise when we look at what experimenters *in fact* do. Applied to the topic of this book, the question thus is how they *in fact* reason and how their de facto reasoning strategies contribute to the ways in which they generate scientific knowledge. This sentiment is expressed nicely by Ian Hacking's critique of well-known twentieth-century accounts of scientific rationality, in which he concludes: "Normal science . . . is not in the confirmation, verification, falsification or conjecture-and-refutation business at all." Hacking goes on to say that normal science "does, on the other hand, constructively accumulate a body of knowledge and concepts in some domain" (1983, 7). The insight that science constructively accumulates bodies of knowledge in ways that are not adequately captured by traditional models of scientific rationality raises the question of how such "constructive accumulation" proceeds. The analysis of operationism that I lay out and develop in this book is intended to provide an answer to this question, at least as regards the process of knowledge accumulation in psychology.

There is a second worry about the philosophical nature of this project that also needs to be addressed here. If my aim is to address the question of knowledge generation by turning to the methodologies actually employed by scientists, I might run the risk of blurring another aspect of the distinction between the contexts of discovery and justification dear to some philosophers of science, namely, that between the normative aims of philosophy of science and the descriptive aims of the history, sociology, and psychology of science (see Hoyningen-Huene 1987). This aspect was addressed by Hans Reichenbach (1938) when he distinguished between three tasks of epistemology: a descriptive task, a critical task, and an advisory task. Reichenbach argued that, for the purposes of epistemological analysis, a description of scientific reasoning should already be "cleaned up" or rationally reconstructed. His notion of the descriptive task of epistemology overlaps with that of the critical task. He acknowledges this but argues: "The tendency to remain in correspondence with actual thinking must be separated from the tendency to obtain valid thinking, and so we have to distinguish between the descriptive and the critical task" (1938, 7). With this remark, Reichenbach puts his finger on a problem that arises if we want our philosophical theory of scientific reasoning to "remain in correspondence with" actual scientific thinking while also offering critical but constructive evaluations of scientific practice. What is an appropriate vantage point from which we can do this?

It would seem that the aim of being in correspondence with what scientists really do requires that we have an accurate empirical description of it. This raises the question at what level such a description should be pitched. For example, we could pitch our description at the level of the cognitive processes of individual scientists or at causal processes of social interaction in the lab, to name just two. In this book, I pitch my description at the normative methodological standards that scientists themselves endorse. This, then, raises the questions of (a) how to arrive at a descriptively accurate account of norms of scientific reasoning if we mainly work with scientists' own accounts of those norms or (even worse) if we do not even have any such accounts to work with and (b) what the status of such an account is vis-à-vis both what scientists really do and our philosophical aim of providing normative evaluations of what they do.

The relevance of the first problem is highlighted if we look at philosophical accounts of reasoning strategies in science, such as the one provided by Lindley Darden (1991). Darden explicitly does not claim that her reconstructions of discoveries in early molecular biology capture the reasoning strategies that scientists actually used, stating instead that she aims to provide hypotheses about strategies that *could* have been used to arrive at certain findings (thereby demonstrating that rational belief change was at least in principle possible). Now, historical evidence regarding the reasoning strategies used by classical geneticists is, of course, often hard to come by (Weber 2004, 70). Nonetheless, I hold that we should aspire for our philosophical analyses to be closely in touch with actual scientific practice. However, it is not so easy to say what this amounts to. When we are concerned with reasoning practices, it is certainly helpful to look at cases in which reasoning strategies are explicated in scientific writings and are, therefore, more easily accessible to philosophical reconstruction. However, reasoning strategies are not always clearly explicated. This creates some pitfalls, as evidenced by the fact that it is possible to cite one and the same historical episode of scientific reasoning in support of competing philosophical accounts of scientific reasoning (for an instructive analysis of this, see Tulodziecki [2013]), which shows that there is a danger of reconstructing specific scientific episodes in accordance with one's favorite philosophical theory of reasoning.

I take up this issue again in the following section. For now, let's assume that an adequate reconstruction of scientific methodology is possible. This brings me to my second question, regarding the philosophical status or significance of this account. Given my contention that we should turn to scientific methodology to gain insights into the process of knowledge construction by experimental means, what is the relationship between such scientific norms of inquiry and the normative evaluations of philosophy? My response is that it is one of critical engagement. By this I mean that we should treat scientific methodologies as frameworks that express the methods and commitments to which scientists in a given community hold themselves and others normatively accountable. By taking seriously the normative frameworks that scientists themselves endorse, we do not necessarily have to endorse them ourselves. Nor do we have to assume that scientists always act in accordance with those norms. What I attempt to do in this book, thus, is to provide an explication of a central methodological maxim of psychological research, *operationism*. This explication is a rational reconstruction in Reichenbach's sense, but it is one that aims to capture what scientists are actually committed to. It has the status of an ideal type, which we can use as a foil to understand and normatively evaluate scientific practice on two levels: that of evaluating the normative commitments of scientists in their own right and that of evaluating the ways in which those commitments are implemented in specific cases.

I hope that this work will speak not only to philosophers of science but also to practicing scientists. Reichenbach's talk of the advisory task of epistemology sounds a little too condescending to my ears. Still, this book aims for an analysis that attempts to offer an original perspective on scientific research without telling scientists how to conduct said research. Accordingly, I hope to engage psychologists in a critical and at times perhaps provocative conversation about their methodological and material commitments.

3. Integrating Philosophy of Science and History of Science

This book is committed to integrating history of science and philosophy of science in two respects. First, it takes methodological discussions in twentieth-century psychology as a point of departure, thereby aiming to develop a philosophical question by way of an engagement with historical and contemporary case studies. Second, it draws attention to scientific research as an ongoing historical process, thereby shifting the focus of philosophy of science and its concern with "static," formal problems.[4]

---

4. I am aware, of course, that other (e.g., Bayesian) approaches also model processes of rational belief change. In that sense, it is not entirely fair to refer to them as *static*. Yet I hold that, even though they are interested in change, their formal methods require them to look at the relationship between theory and evidence at specific time slices rather than analyzing what drives the *generation* of evidence.

I look at these two aspects in turn. There are some parallels between this outlook and Hasok Chang's notion of a "complementary science" (as laid out in chap. 6 of his *Inventing Temperature* [2004]), though I try to recover not so much lost knowledge as lost contexts of methodological analyses. There are also parallels with Jutta Schickore's research on the history of methodological reflections (Schickore 2017), but, whereas her focus is on the history of accounts of the experimental method, my own approach here is better described as using a historical approach to explore the ways in which method and ontology are intertwined in psychology.

The analysis begins by looking at several prominent figures in psychology in the 1930s, focusing specifically on how they reasoned for their endorsement of operationism, what they meant by this concept, and how this endorsement figured in their research projects. While I show that operationism has often been misunderstood by both philosophers of psychology and historians of psychology, I do not deny that, as formulated by its early proponents, the position had significant problems. These problems were addressed in the subsequent methodological literature in psychology, a fact that we miss if we direct our historical attention only to particular time slices. For this reason, the exposition of the historical emergence of operationism is presented in chapters 2 and 3, following the historical actors in the ways in which the position was explicated and developed as part of a methodological conversation within psychology. In other chapters, I draw on case studies from more recent history of psychology, specifically focusing on the investigation of memory phenomena. There is no direct historical link between the reconstruction of the origins and development of operationism provided in the earlier chapters and the case studies presented in later chapters. Nonetheless, the more recent cases reveal the extent to which the various elements of operationism (outlined above) are still relevant and/or scientists are still committed to them. My reconstruction of operationism thus provides an analytic framework that is used for an account of research practices in contemporary cognitive psychology.

While a lot of my reconstruction of operationism (especially in chap. 1) is based on historical research, it does not claim to provide a thick historical description. Rather, it is self-consciously selective and geared toward affording insights into a particular set of philosophical issues. Needless to say, this raises some methodological challenges that I have already touched on briefly in the previous section. In particular, the following two questions arise. First, what is the status of historical case studies for philosophical arguments, and are they really needed to make the point? Second, can we draw on historical material with the aim of making a philosophical point without violating historical sensibilities? The two worries are related. If historical cases are used as mere illustrations of philosophical arguments, there is a concern that philosophers will draw on historical material selectively and tell the story in a way that fits their needs. This kind of anachronistic history is rightly rejected by historians of science, but the question is whether it is possible to practice a nonanachronistic history that is still fruitful for systematic philosophical purposes. These and related issues have been discussed on and off over the past fifty or so years. Well-known catchphrases are whether the marriage between history of science and philosophy of science is one of love or mere convenience, and what a truly integrated history and philosophy of science might look like (see Chang 2012a; and Schickore 2011).

I believe that there is a wider issue at stake here and that the integration of historical material in a philosophical analysis is merely a special case of the more general question of how to integrate *any* empirical description of real science into a philosophical analysis. There are two aspects to this question. First, what is the role of (past or present) case studies in philosophy of science? Second, how should such case studies be constructed? In this book, I take a stance on both issues. With respect to the first, my approach is informed by the contention that philosophy of science (like science itself) is an ongoing and dynamic endeavor with not only changing answers but also changing questions. Therefore, I hold that new philosophical questions can emerge as a result of the emergence of new kinds of sciences and/or methods (e.g., the science of complexity, nanotechnology, brain imaging, and machine learning) but also as a result of original insights about past or present science. While those insights are, in turn, inevitably constructed from a particular vantage point, they can result in a shifting of that vantage point (Schickore [2011] refers to this as a "historicist" conception of the integrated history and philosophy of science).

My reconstruction of the status and significance of operationism in American psychology of the 1930s reveals the historical actors involved to have been concerned with an issue that is not widely recognized within contemporary philosophy of science, namely, how to proceed with one's research in the face of both the complexity and the very limited understanding of the subject matter. According to the analysis presented in this book, the significance of operationism with regard to this problem was that it urged scientists to start out with tentative specifications of the paradigmatic procedures expected to provide insights into a given topic. More importantly, however, it urged that these specifications and the ways in which they were implemented in specific experiments be explicated so as to allow for maximal transparency concerning the circumstances and assumptions that underlie a given experimental finding. In turn, I show that such transparency allows for specific critiques and tests of experiments, thereby contributing to an iterative process of knowledge generation. Hence, I argue that attention to the methodological maxim of operationism as it emerged in the early twentieth century and is still practiced today opens up a field of philosophical analysis in which the dynamic and ongoing processes of experimental research in psychology (and conceivably also other fields) can be understood.

Now, with respect to the question of what it takes to give an adequate empirical description of (past or present) science, my brief answer is that, while we do not approach our case studies from a neutral perspective, the things we can assert about them are still going to be constrained by the historical record. Clearly, the account provided here is—in some sense—fairly internalistic in that my interest is in strategies of reasoning and experimental practice. This means that my analysis is bound to be incomplete, but it does not mean that it has to be inaccurate. There are a number of different kinds of historical questions that we can ask about operationism in psychology, and the adequacy of the answers will be determined not only by whether they address the question but also by whether those answers are backed by the evidence. In this book, the historical question I ask about operationism as it was formulated in the 1930s is what problems it was supposed to address and how it was thought to address them. I argue that to answer these questions it is not sufficient to look at methodological writings by themselves as the methodological pronouncements of scientists are often misleading and sanitized. Instead, I base my reconstructions on methodological *and* scientific writings in order to show how specific methodological strategies in fact played out in scientific research. In addition, I provide descriptions of the types of research problems individual advocates of operationism were concerned with, in turn placing these in the wider contexts of psychological debates at the time. While, like any historical account, my historical account can be challenged on empirical grounds, I maintain that the philosophical nature of my question does not present a special problem.

I now turn to the second sense in which my approach is historical, namely, insofar as it aims to provide a theory of a temporal process. As mentioned at the outset, I am not alone in my contention that the iterative process of scientific research is worthy of philosophical analysis. If we agree with the statement that "scientific discovery should be viewed as an extended process that occurs in cycles of generation, evaluation, and revision" (Darden 2009, 44), the question is how best to approach the study of such processes. Lindley Darden focuses her analysis on two features of research, (*a*) the generation, evaluation, and revision of hypotheses and (*b*) the discovery of mechanisms (see also Craver and Darden 2013). While this may well be a useful way to think about discovery in other scientific domains, I believe that both features just pointed to are too narrow to capture the nature of the investigative process in cognitive psychology. Spelling out what I mean by this brings my own analysis into clearer view. With regard to the first point, I contend that, while there is an important sense in which every tentative assumption is hypothetical, it does not do justice to the process to think of all such tentative assumptions as *hypotheses*. For example, if we consider the case mentioned at the outset, in which a particular type of test is tentatively treated as providing data about a particular object of research, such assumptions are not hypotheses in the sense that they are necessarily targets of subsequent scientific tests (though they can be). Rather, they are preconditions for the very possibility of conducting experiments about the presumed object of interest, whether those experiments test hypotheses or not (e.g., the experiments may also be geared toward providing phenomenological descriptions of the phenomenon of interest [cf. Steinle 1997]). In highlighting this point, I do not mean to deny that there are cycles of hypothesis generation, evaluation, and revision as envisioned by Darden. But what is needed here is a richer analysis of the iterative nature of the scientific process, one that includes the evaluation and reevaluation not only of scientific hypotheses but also of the instruments and methods of investigation.

Kevin Elliott (2012) pushes in a similar direction by pointing out that, while the notion of *iteration* has become ubiquitous in both philosophy and science, different accounts have focused on different aspects. Specifically, he distinguishes between two types of iterations, one epistemic and one methodological. Taking inspiration from Chang (2004), Elliott defines epistemic iteration as "a process by which scientific *knowledge claims* are progressively altered and refined via self-correction or enrichment." By contrast, he characterizes methodological iteration as "a process by which scientists move back and forth between different modes of research practice," such as hypothesis-driven, exploratory, question-driven, or technology-oriented modes (Elliott 2012, 3). Elliott goes on to present an analysis of the ways in which those two kinds of iteration are, in fact, integrated, focusing specifically on the question of how methodological iteration can contribute to epistemic iteration by initiating, equipping, or stimulating it. For example, an iteration between exploratory, hypothesis-driven, and technology-oriented research can help scientists set up a preliminary framework for studying nanoparticle toxicity, which in turn initiates future processes of epistemic iteration.

In applying Elliott's framework to the case of implicit memory mentioned at the beginning of this introduction, we might say that a functional dissociation between the results of two memory tests can give rise to an iterative process whereby a particular type of measurement procedure (priming tests) is used for exploratory studies (mode of research) of a purported target object (implicit memory). Such exploratory studies might then give rise to a hypothesis about the object, which in turn gets tested (mode of research), which in turn leads to a formulation of results (epistemic iteration), which in turn leads to the reevaluation of the experimental setup (technology-oriented iteration), etc. While this captures aspects of the analysis I provide in this book, I highlight a key element of this process as it plays out in psychology, that is, the extent to which it is organized around specific *objects of research* (e.g., implicit memory) and relies on conceptual assumptions about these objects. Researchers engaged in such an iterative research process, thus, share not only an ontological commitment (e.g., that implicit memory exists) but also an epistemic presupposition (e.g., that implicit memory can be accessed via priming effects). Objects of research are therefore conceptualized in such a way that it makes specific measurement procedures seem naturally suited to the purpose of producing data about them. However vague these scientific preconceptions of their objects of study may be, I argue that they play an important role in structuring research and that they do so by virtue of being ingrained in research designs via operational definitions. In this way, my analysis draws attention to the ways in which epistemic iteration and methodological iteration are integrated by being directed at the same objects of research.

This brings me to the question of what scientists are looking for in their attempts to explore objects of research in psychology and the behavioral sciences (including cognitive neuroscience). One prominent answer we find in the philosophical literature is that these sciences are concerned with the search for mechanisms (Bechtel 2008a, 2008b; Bechtel and Richardson 1993; Craver and Darden 2001, 2013; Darden 2009). I contend that while the search for mechanisms is an important feature of scientific investigations in psychology, mechanisms are not the sole objects of psychological research, as is evident when we consider that the search for mechanisms typically already relies on some prior understanding of what the mechanism is supposed to explain (a point acknowledged by the authors referenced above). In this book, I argue that much of research in psychology can be characterized as taxonomic and descriptive, in the sense that it attempts to delineate and describe its very objects. An epistemology of exploration must analyze the rationales by which preexisting, and often partial, conceptual assumptions about research objects aid the construction and development of objects of research as a whole.

4. Objects of Research: Moving Targets of Scientific Investigation

This formulation—"the object of research as a whole is constructed"— takes us back to the beginning of this introductory chapter, which I began by raising the question of how scientists go about studying a research object that they do not understand yet. As I have indicated (and as I spell out in the course of this book), a central thesis of this work is that, in psychology, operational definitions play an important role in empirically delineating the presumed objects of research for the purposes of empirical research. While anything that is not well understood can become an object of research, I argue in this book that the domain of psychological research is, ultimately, structured by way of the categories provided to researchers by folk-psychological vocabulary, which, I argue, individuates the units of analysis at the level of the whole organism (chap. 6). With this I do not mean to suggest that our folk-psychological categories (and the scientific terms derived from them) automatically get it right. By conceding this, I would like to capture the intuition that there is a material reality that puts constraints on the ways in which scientific concepts get refined and fine-tuned in the course of research; though, as I argue, material reality cannot fully account for the identity conditions of psychological kinds.

In trying to do research about an object that is outside our cognitive grasp, we face the challenge of figuring out how to grasp it. Operational definitions, I argue, address this challenge. Needless to say, the mere existence of an operational definition does not guarantee that there really is a corresponding object; nor does it guarantee that this object is unchanging. In other words, a given operational definition might not successfully latch on to something real and robust, and it might not pass the test of time (either because the operational definition is discarded in favor or a more accurate one or because the object changes). Hence, we need to differentiate clearly between (*a*) the assertion that such definitions function as tools in pursuit of presumed objects of research and (*b*) the assertion that the tools in question can be regarded as valid from the outset or that they succeed in individuating an object of lasting significance. While my analysis endorses the first of these two statements, I certainly do not endorse the second, as I explain in chapters 4–7 and my concluding remarks.

Now, the notion that a definition can be a tool for the investigative pursuit of an object that might turn out not to exist may seem like a contradiction. With this formulation, I want to make room for the fact that, as scientists learn more about their domain (with the help of operational definitions and experimental paradigms), they can attain novel insights about how the domain is structured. Putting this in the more traditional language of philosophy of science, I want not only to acknowledge the reality of conceptual change in scientific research but also to provide an analysis of this process, an analysis that highlights the productive role of the concepts themselves in bringing about their own change via the concrete physical manipulations they enable. They can play that productive role not by virtue of being operationally defined alone but also by virtue of a wealth of other assumptions concerning the concept's meaning. With this account, I reject as too rigid the traditional causal theory of reference, arguing instead that a given term's reference can shift in the course of scientific investigation. However, I am also critical of description-theoretical approaches to concepts, which have typically not paid much attention to the question of what one can do with scientific concepts and, specifically, to the material constraints imposed on concept use.[5]

With the basic contention of this work—that the objects of psychological research are not very well understood (and, hence, that our understanding even of what are the relevant objects can shift over time)—I echo a well-known account of the discovery process in molecular biology, namely, that of Hans-Jörg Rheinberger (1997). Rheinberger introduces the notion of an *epistemic thing* to describe objects of research that attract our scientific curiosity. He emphasizes that they do so because of what we do not know about them and argues that it is precisely this "irreducible vagueness" of epistemic things that makes it possible for the creative dynamics of the investigative process to unfold. I find the notion of an *epistemic thing* very helpful insofar as it highlights an aspect of scientific research that was, for a long time, neglected in philosophy of science, that is, the fact that research is often conducted precisely because of our limited knowledge of the objects of research. But my analysis departs from Rheinberger's in two important ways. First, I prefer to speak of epistemic things as "blurry" (rather than "vague"), thereby stressing the epistemic predicament of scientists as one of trying to get

---

5. My account thus has some affinities with pragmatic approaches to concepts, such as Wittgenstein-inspired versions of inferentialism (e.g., Brandom 1998). More importantly, there are also affinities to Mark Wilson's (2006) patchwork account of concepts, but these will not be explored in great detail here (but, for forays into this territory, see Bursten [2018] and Haueis [2018, 2024]).

their presumed objects of research *into clear focus* (Feest 2011a). Second, Rheinberger deliberately uses the term "epistemic thing" sometimes to refer to the object under investigation and sometimes to refer to the ways in which such objects are conceptualized by scientists. His focus is on the material culture (as embodied by experimental systems) in which research objects are constituted, both physically and conceptually. My account shares with Rheinberger's an emphasis on the material dimension of experimentation, and I also agree with him that there is a sense in which our only access to objects of research is via our own conceptualizations of them. Nonetheless, I emphasize that, if we want to understand the research process, we must recognize the basic grammatical distinction between objects as the targets of research, on the one hand, and concepts of those objects as playing an essential role in the process of exploring the (presumed) objects, on the other. In stating this, I am not committing to the notion that scientific conceptualizations of research objects correspond to mind-independent natural kinds (for a discussion of this, see chap. 6). However, I argue that experimental practice in psychology is underwritten by a commonsense realism that we can capture only by making a clear distinction between concepts and objects that have their own kind of materiality. I suggest that my account of the role of operational definitions provides some insights into the ways in which the links between the material and the conceptual are forged.

5. Addressing the Crisis of Confidence in Psychology

The analysis presented in this book is optimistic that psychology has the resources to make real and legitimate empirical and conceptual progress. I am not committing to the claim that psychologists always (or even mostly) utilize these resources appropriately. However, I think that my account, which is based on historical and contemporary reflections from within psychology, offers some insights and suggestions as to how they might proceed. With this, I respond to current debates about the crisis of confidence in psychology, debates that have specifically focused on the fact that many psychological results cannot be replicated.[6] Now, it seems intuitively obvious that this is a problem. If scientific findings were reliable and robust, it should be easy to replicate them. This not

---

6. Fidler and Wilcox (2018) provide a comprehensive overview of the philosophical debates raised by the replication crisis, which covers but is not restricted to psychology. Sturm and Mülberger (2012) point out that, since the late nineteenth century, the history of psychology has been littered with crisis discussions and crisis diagnoses. In light of this, it seems that the replication crisis is only the latest in a history of crises.

being the case, the question is what it is about psychological research practices (and the more general context of research and publication cultures) that stands in the way of replicable results. One obvious answer to the low replication rates in psychology is to question the initial results, that is, to raise the suspicion that a study cannot be replicated because there were flaws in it. In turn, this raises two kinds of questions. First, what is it in the nature of the flaws that misleads researchers into accepting false results? Second, what is it about the way psychological research is organized that stands in the way of flaws being caught and corrected as the traditional picture of science as self-correcting might have it (e.g., Romero 2019). With regard to the first question, a lot of attention has been focused on practices of inflating the statistical significance of findings, also known as "p-hacking," prompting some to diagnose this crisis as one that is at heart about statistics (e.g., Gelman and Loken 2014). To be sure, both questionable research practices and the reasons that encourage them are deeply problematic. However, the analysis developed in this book has a different focus in that I address questions like what we mean by concepts like *robust findings* and *experimental result* to begin with and what specific challenges psychology faces in achieving robust experimental results.

My overall contention is that the focus on formal and procedural mistakes in scientific reasoning sometimes detracts from the fact that such methods are applied to data that are already deeply entrenched in conceptual/material assumptions about the subject matter at hand. While it may be a truism that scientific data are human creations, my analysis of operationism highlights the importance of considering the purposes for which data are created. Data are created with the assumption that they are legitimate evidence for questions about specific research objects.[7] They are invested with this assumption by virtue of the operational definition built into the experimental design (Sullivan 2009) as well as the design's implementation in a specific experiment. But, of course, the question of whether these assumptions are accurate is very much at stake. By this, I do not mean to question the legitimacy of using operational definitions in experimental research. Rather, my point is that a crucial aspect of the conceptual and experimental work in psychology consists in a constant process of probing the meaning of the experimental data generated in an experiment and (simultaneously) the scope of the conceptual claims to which the data are assumed to speak (chap. 7). This process is what I describe with the concept of *operational analysis*.

---

7. A similar point is emphasized by Sabina Leonelli's (2015, 2016) relational theory of data (see also chap. 5 below).

Now, what are the implications of this for debates about the replicability of results? I argue that, whenever researchers do (or redo) an experiment, they make an inference to an experimental result. This inference is sound relative to the background assumption that the data do indeed speak to the question at hand, which in turn hinges on the adequacy of the operational definition and its implementation. My main claim is that there will be very little progress in psychology if we become too fixated on whether a given effect can be replicated rather than asking what it shows and whether our conceptual apparatus for describing the result is adequate and pitched at the right level of generality (see also Yarkoni 2022). Addressing this question, I argue, involves probing the background assumptions in play in specific experiments. This was recognized by advocates of operationism in psychology already in the 1940s, as chapter 2 lays out.

6. Does My Analysis Generalize beyond Psychology?

One question that can obviously be raised with regard to my analysis is whether the arguments I am making are specific to the investigative process in psychology or whether they generalize to other areas of research. A possible way to attack this question is to ask what the similarities and differences are between psychology and other disciplines, such that my analysis might be applicable to psychology but not to other disciplines. There are two possible (mutually compatible) answers to this that I consider here.

The first reason psychology might be considered to require a special analysis is that it is a much younger discipline than, for example, the physical sciences. Thus, it is not surprising that there are still many underdeveloped concepts, thereby calling for a philosophical analysis of how researchers deal with the resulting uncertainty. This suggests that my analysis is relevant to psychology and other younger sciences rather than to more mature disciplines. It also suggests that my approach might provide insights into earlier moments in the histories of other disciplines. There is a certain sense in which this answer seems obviously right. Indeed, I situate my analysis in the spirit of previous works in the history of physics, such as Hasok Chang's work on temperature and Friedrich Steinle's work on electricity. Chang (2004, 2019), in particular, has explicitly connected his account of the interrelated dynamics of the conceptualization and the measurement of temperature with an original reading of Percy Bridgman's operationism, reconstructing how a sophisticated concept of *temperature* was developed on the basis of our ability to differentiate between warm and cold. Thus, my account incorporates and adapts elements of such previous work, capitalizing on notions like *iteration* and *exploration* as they apply to psychological research.

However, I do not want to suggest that the analysis I am developing here applies only to the early stages of conceptual development. Whenever a given object is treated as an object of research, there are conceptual questions at stake and, hence, conceptual dynamics in play, regardless of how mature a given concept or discipline is. In addition, and as we know from Harry Collins's work (e.g., Collins 1985), even when we have a very well-developed theoretical concept (such as that of *gravitational waves*), the issue of how to operationalize questions pertaining to the object in question and in particular, how to know that one has successfully measured the object is far from trivial. Accordingly, I suggest that many of the questions addressed here are highly relevant to scientific experimentation in different disciplines and that this is the case regardless of which historical point in time one looks at. So, while the fact that psychology is a young discipline can bring out some issues particularly clearly (in part because young disciplines tend to have more vigorous methodological discussions than more established fields), the issues as such are not confined to the early stages of concept formation and conceptual development.[8]

But there is a second question that can be raised here, namely, whether there is anything special about the discipline of psychology, such that insights about its research practices or dynamics of concept formation cannot be generalized to other fields. This suggestion echoes a long history of thinking about whether psychology requires a special kind of method because of the special nature of its subject matter, such as the fact that it deals with mentality, which is inherently meaningful and accessible to a privileged first-person perspective in a way that is not the case in other disciplines. Responding to this briefly, let me emphasize that, even though I acknowledge that there are aspects of the psychological subject matter that might require unique methods, those methods, too, will have to make use of operational definitions in some shape or form. In other words, even if psychological measures are measures of subjective or meaningful mental states, they are always going to rely on observable behaviors (including verbal reports), which are elicited by an external prompt. This means that, in spite of differences in subject matter, there is in principle no difference

---

8. This is not to say that my analysis does not shed light on the ways in which new concepts get introduced, not only in psychology but also in other disciplines (e.g., see Bloch-Mullins 2012; and Boon 2012).

between psychology and other fields as far as the need to formulate operational definitions goes.

There is an additional important point here, which is that, even if the psychological subject matter includes subjective mental states, it cannot be reduced to those states. On the analysis I develop in this book, objects of psychological research are complex capacities of organisms to exhibit experiential, cognitive, and behavioral phenomena that are individuated at the level of the whole organism (see also Feest 2023). To be sure, such "cognitive-experiential-behavioral capacities" (as I call them) include intentional mental states that orient the organisms toward the environment (Sullivan 2014). But the overarching categories in which the psychological research field is structured refer to capacities of the behaving organisms as a whole that are psychology's objects of research. I argue (in chap. 4) that the dominant focus in philosophy of neuroscience on (mechanistic) explanation has detracted from the fact that objects of psychological research are located at a different spatiotemporal scale than are those of neuroscience and that the investigation and exploration of psychological objects of research are valuable and important research aims in their own right (see also Feest 2014a, 2017a), though this does not preclude neuroscience from contributing to this task.

7. A Quick Overview of the Chapters

Chapter 1 goes back to the origins of the usage of the term "operationism" in psychology and offers a construal of operationism that emphasizes both its emergence from and its relevance to issues of psychological experimentation. The term gained prominence in American psychology around the mid-1930s. While the relevant writings make reference to the physicist Percy Bridgman and his 1927 *The Logic of Modern Physics*, I argue that the development of what came to be called "operationism" (or "operationalism") in psychology was historically largely independent of Bridgman's ideas. Furthermore, while operationism is frequently equated—both historically and conceptually—with varieties of logical positivist theories of meaning and/or knowledge, the psychologists who adopted it were motivated by different concerns than those of the philosophers of science at the time. I substantiate these claims by way of an analysis of the methodological and scientific writings of three prominent operationists in the 1930s: Stanley Smith Stevens, Edward Tolman, and Clark Hull. In particular, I argue that these figures recognized that any experimental investigation of a purported psychological entity presupposed a conceptualization of that entity in a way that tied it firmly to specific experimental designs (I refer to such designs as "paradigmatic conditions of application") and that they tried to explicate this understanding by giving what they called "operational definitions" (see Feest 2005b). In addition, at least Clark Hull was motivated by questions about the nature of *theory construction*, which he wanted to be tied closely to experimental intervention and observation, and in which central concepts—however rudimentary—early on have empirical content. I argue that this point in particular already highlights two systematic issues that I emphasize throughout this book: (*a*) the dynamic nature of scientific research and (*b*) the material (as opposed to merely the formal) modes of reasoning in this process.

In chapter 2, I continue my historical overview of operationism in psychology by examining developments that took place in the 1940s and 1950s. The chapter has two main aims. First, I use my historical reconstruction of the midcentury debates about operationism to underwrite further and elaborate on my analysis, according to which operationism in psychology was intended to address practical epistemological concerns that arise in experimental research. Second, I use the resulting narrative to highlight a handful of systematic questions as well as potential tools with which to address these questions. As I show, advocates of operationism during those years highlighted the conceptual openness and epistemic uncertainty of experimental research in psychology. This gave rise to an emphasis on operational definitions as (*a*) allowing for an explication of hidden assumptions in an experimental design (this was tied to the idea of operational analysis) and (*b*) conceivably playing a positive role for purposes of research and teaching, insofar as—by virtue of being tied to specific experimental designs—such definitions created common reference points for empirical work. I also cover three influential approaches to the validation of experimentally defined concepts. The first (Cronbach and Meehl 1955) sought to validate concepts by embedding them in "nomological nets," the second (Campbell and Fiske 1959) argued that validation could be accomplished by multiple determination, and the third (Garner, Hake, and Eriksen 1956) argued that conceptual development required the systematic testing of competing explanations of experimental data. The main point of the chapter is that, in the 1940s and 1950s, even where they engaged with philosophy of science of the time, methodological writings all highlighted the dynamic nature of experimental work and emphasized the conceptual openness of the domain of research. This, in turn, allows me to outline some systematic questions that I analyze in the remainder of the book. I suggest that Garner, Hake, and Eriksen's (1956) notion of *converging operations* has valuable insights to offer the philosophy of psychological experimentation (a suggestion to which I return in chap. 7).

Chapter 3 is the first of several chapters that argue for the applicability of my analysis of operationism to more recent psychological research, specifically memory research. At the same time, it also continues to show how both the questions and the answers presented in this book compare with other approaches within philosophy of science. I argue that concepts are not simply end results of a scientific process; rather, they contribute to the process itself in that scientific investigations are typically intended to be *about something* and that *something* has to be conceptualized, however tentatively. Picking up from my characterization of operational definitions in the previous chapter, I show that this characterization is also applicable to the cases of implicit memory and short-term memory and provides a more general account of the way in which operationism is relevant to these two areas of research. I argue that concepts are crucial on two levels: (*a*) by schematically suggesting types of experimental designs for the study of any given object of research and (*b*) by virtue of paradigmatic experimental procedures that have operational definitions ingrained in them. This argument underwrites the main thesis of the chapter, namely, that, in the context of experimental psychology, operational definitions are *research tools*, thus shifting the focus away from questions about the representational functions of concepts, asking instead how concepts contribute to experimental interventions and measurements aimed at exploring the purported referents. The chapter concludes by situating my approach vis-à-vis both prominent twentieth-century philosophical theories of concepts (the causal theory of reference and the description theory of meaning) and more recent attempts to think about the role of concepts in investigative practice.

In chapter 4, I provide a more in-depth discussion of my notion of an *object of research* in psychology. My main contention is that (as spelled out in chap. 3) there is a sense in which scientists take their research to be about something real out there. Yet, at the same time, speaking from the epistemic perspective of the researchers involved, it is an open question whether any specific psychological concept will pan out to be a projectable taxonomic category. Therefore, I suggest that the objects of psychology are epistemically blurry and that an important objective of psychological research is to empirically delineate and describe such objects by experimental means. Once again focusing my attention on research about implicit and working memory, I argue that this research aims to provide experimentally grounded *descriptions* of the objects of research in question. I argue that the mere description of the objects of research in psychology is by no means trivial, and I illustrate this point by way of a discussion of descriptive features of implicit and working memories, such as the (supposedly unproblematic) fact that implicit memory involves no conscious processes and that working memory has a storage capacity of 4 ± 1. I situate my discussion within the wider context of literature about exploratory experimentation (e.g., Steinle 1997), which has rightly pointed out that experiments often aim at providing phenomenological descriptions, that is, descriptions of observable phenomena. While I agree with this contention, I emphasize (a) that in psychology and cognitive science researchers typically make assumptions about what such descriptions are descriptions *of* and (b) that we need a more sophisticated discussion of the meaning of "phenomenological." With regard to the former issue, I argue that operational definitions importantly specify preliminary taxonomic assumptions about objects in the psychological domain. With regard to the latter issue, I argue that the (presumed) descriptive features of objects of psychological research often go well beyond the actual data generated by the definition/paradigm in question, being derived instead from features of the measurement tools employed.

Chapter 5 picks up on the insight that objects under investigation are ill understood and that research efforts in psychology are typically directed at attempting to delineate and describe the object of research. In this chapter, I further clarify my notion of an *object of research* by contrasting this notion with that of a *phenomenon*. Along the way, I also situate my usage of the term "phenomenon" in the existing literature. The chapter makes two overarching points. First, objects of psychological research are clusters of phenomena that include both explanatory mechanisms and observable regularities. Second, operational definitions draw on existing, albeit preliminary, concepts to specify what kinds of experimentally generated phenomena constitute relevant evidence for specific descriptive claims about objects of psychological research. This analysis complements an important strand in philosophy of science (most prominently, philosophy of neuroscience), one that has focused on phenomena as the explananda of mechanistic explanations. I argue that my focus on objects of research (as opposed to explanandum phenomena) highlights the intricacies of establishing what one is even trying to describe and explain. My analysis also goes counter to Bogen and Woodward's (1988) analysis of phenomena, which understands them as the stable and general explananda of scientific theories and juxtaposes them with data, which are local and idiosyncratic. By contrast, and reading Bogen and Woodward against the grain, I argue that phenomena can occur on a continuum between the local and the context specific, on the one hand, and the general and the context transcendent, on the other (Feest 2011b). This allows me to argue that, when it comes to research practices in cognitive science, local and context-specific phenomena (or effects) typically function as evidence rather than as explananda. They are treated as evidence pertaining to given objects of research, which are intuitively understood as clusters of phenomena, some of which are more easily empirically accessible (or creatable) than others. The practice of treating certain phenomena as the (defeasible) evidence for (defeasible) claims about objects of research is made possible by virtue of their being conceptually tied to objects of research by way of operational definitions.

Now, given my contention (laid out in chaps. 4 and 5) that objects of research are (*a*) epistemically blurry and (*b*) composed of clusters of phenomena, some of which are closely tied to specific experimental contexts of data generation and, thereby, take on the role of evidence, the question arises as to the ontological status of those objects. This question is addressed in chapter 6, which touches on debates about natural kinds while also picking up, more specifically, on the recent literature about cognitive ontology. I develop and defend an account of psychological kinds according to which such kinds depend for their existence on our folk-psychological conceptual and causal practices. According to this account, which I call a "relational theory of kinds," psychological kinds come into existence at the interface between our sensory-conceptual apparatus and the physical world. I argue that it is precisely those conceptual and causal practices that also inform our operational definitions and the experimental designs that rely on them. With this I insert myself into the debate about whether natural kinds have an objective basis or are merely conventional. While I come down squarely in favor of the view that there is no objective fact of the matter as to what is the one correct way of taxonomizing psychological kinds, I resist the assumption that this makes the psychological kinds studied by psychologists unreal or arbitrary. Psychological kinds have, I argue, a robustly real status relative to the specific contexts in which they matter to us, the folk-psychological context being particularly salient. Accordingly, I argue that scientific psychology is well-advised to individuate psychological kinds in exactly the same fashion that folk psychology individuates them, that is, as whole-organism cognitive-experiential-behavioral capacities. This does not preclude the possibility of criticizing entrenched psychological kinds and the folk practices that stabilize them.

Chapter 7 returns to the experimental practices concerned with delineating and exploring the objects of psychological research. This concern is, of course, the flip side of the task of constructing and developing the corresponding concepts. I argue that this task is intimately tied to the process of evaluating the quality of the experimental data. This thesis, in turn, takes us back to the need of conducting operational analysis, first introduced in chapter 2, that is, the need for an analysis of the hidden background assumptions that figure in experiments. I make this more precise by highlighting that the assumptions in question are required for specific experimental inferences about the objects of psychological research. This prompts me to present a schematic account of experimental inferences in psychology and to identify three sets of background assumptions that give rise to three challenges that have to be met if the experimental inferences in question are to be regarded as sound: the individuation challenge, the manipulation challenge, and the measurement challenge. Turning to the question of how the accuracy of these assumptions can be investigated, I argue that the method of converging operations (again, first encountered in chap. 2) provides valuable insights. This method can be construed as probing the reliability of a given set of data by testing specific hypotheses about possible confounders. I argue, however, that it effectively does more than that. The method of converging operations creates novel data in pursuit of an evolving understanding of the objects in question. I distinguish between converging operations designed to interrogate the manipulation and the measurement challenges, on the one hand, and those designed to interrogate the individuation challenge, on the other. The latter, in particular, is highlighted as it pertains to questions about the scope of psychological concepts as well as corresponding questions about the shape of psychological kinds. The chapter concludes with an analysis of the unique epistemic challenges raised by the possibility that psychological kinds are characterized by a high degree of context sensitivity.

My concluding remarks reflect on the ongoing relevance of the analyses presented in the preceding chapters. Specifically, I argue that my analysis of psychological kinds calls for ecological theorizing and research designs, and I discuss the implications of my account for recent debates about the credibility crisis in psychology.

CHAPTER ONE

Operationism in Psychology

(Some) Historical Beginnings

1.1. Introduction

The term *operationism* (or *operationalism*) is usually associated with the Harvard physicist Percy Bridgman (1882–1961), who famously claimed: "In general, we mean by a concept nothing more than a set of operations; the concept is synonymous with the corresponding sets of operations" (1927, 5). In twentieth-century philosophy of science, this quote, or the sentiment expressed by it, has frequently been interpreted as voicing the semantic and epistemological tenets of a crude logical positivism or empiricism. According to the former position, the meaning of a statement is synonymous with the conditions that would constitute an empirical test for it. According to the latter position, the justification of statements about a given phenomenon is provided by pointing to the empirical conditions that would verify it. Accordingly, Hempel stated: "Operationism, in its fundamental tenets, is closely akin to logical empiricism" (1954/1956, 215).

As is well-known, the heyday of positivism, in both its semantic and its epistemological forms, has long since passed, and for good reasons. Briefly, it was recognized (*a*) that there are statements about objects that we intuitively recognize as *meaningful* despite the fact that the concepts that occur in those statements cannot be exhaustively defined in terms of operations and resulting observations and (*b*) that there are statements that scientists take to be *justified* even though it is impossible to exhaustively rephrase them in terms of observation sentences.[1] Now, given the decline of the core tenets of logical positivism, and given furthermore the assumption that operationism was

---

1. Some of the arguments for these changes are reviewed in the following chapters.

closely related to positivism, one might wonder why we should be interested in this doctrine and whether there is anything worth salvaging here. Indeed, until recently, there was a long-standing consensus in much of the philosophical literature to the effect that operationism was a deeply misguided program and that, regrettably, some scientists still failed to notice this. For example, the philosopher Fred Suppe writes: "It seems to be characteristic, but unfortunate, of science to continue holding philosophical positions, long after they are discredited" (1974/1977, 19). In the past twenty or so years, however, some historians and philosophers of science have taken a renewed interest in operationism as it was formulated by Bridgman (e.g., Chang 2004; and Humphreys 2004), arguing that it was a reasonable position in the scientific context in which it emerged. Inevitably, this has also led to reevaluations of what the position actually stated. For example, as Chang (2019) points out, Bridgman's main motivation was cautioning physicists to recognize their limited knowledge about the scope of their concepts and measurement operations. This concern was pertinent in light of the breakdown of Newtonian physics at the time. And Vessonen (2020) has recently defended a moderate version of ("respectful") operationism according to which it can be legitimate and fruitful to "operationally define" one's concepts by means of tests, so long as one recognizes that these concepts have "extraoperational" meanings as well.

In this book, I follow these leads in arguing that our evaluations of the methodological and practical choices of scientists ought to be informed by a focus on their actual questions and concerns. Rather than passing philosophical judgment on scientists' failure to notice changes in philosophical discourse, we should aim for a philosophy of science that takes the actual concerns of scientists as its starting point, basing it on a careful reconstruction of scientific practices that is sensitive to the historical circumstances in which those practices were adopted. My goal, then, is to provide a charitable reconstruction of operationism and its continuing relevance.

With regard to the context of operationism in psychology, it will be helpful to juxtapose it not only with positivist philosophy of science but also with another intellectual current in early twentieth-century American thought: *behaviorism*. It is customary to distinguish between *logical*, or *analytical*, behaviorism, on the one hand, and *methodological* behaviorism, on the other. The former is a philosophical movement that aims to define mentalistic concepts in terms of behavioral dispositions, whereas the latter is a psychological movement that argues against the use of introspective methods and mentalistic concepts altogether, replacing them with behavioral observations and laws.[2] The former kind of position has come to be associated with the work of Gilbert Ryle (1949/1983), but versions of it can also be found in logical positivists' work, such as Rudolf Carnap (1932) and Carl Hempel (1935/1980) (though Crawford [2014] has argued that Carnap was not, in fact, a logical behaviorist). The latter position has most prominently come to be associated with John B. Watson (1913) and B. F. Skinner (1938).[3] I argue in this chapter that the positions advocated by early operationists, such as S. S. Stevens (1906–72), Edward C. Tolman (1886–1959), and Clark Hull (1884–52), were influenced by and show echoes of both types of behaviorisms but ultimately do not sit comfortably with either of them. For example, Stevens investigated auditory experience (i.e., a conscious mental state), and both Tolman and Hull posited the existence of explanatory variables that intervene between stimuli and behaviors. None of them tried to eliminate mentalistic/intervening variable concepts from their vocabulary, and none of them claimed that the meanings of those concepts can be reduced to operations and observable behaviors. Instead, as I argue in this chapter, for these psychologists, operationism held the promise of allowing for the introduction of mentalistic terms and intervening variables in a rigorous and controlled fashion. As I show, their positions were, ultimately, not compatible with strict semantic reductionism about their concepts (i.e., with the logical behaviorism commonly attributed to early logical positivists).[4]

Roughly speaking, this chapter pursues two aims, one negative and one positive. The negative aim is to show that, as it was practiced by the three experimental psychologists whom I have singled out (Stevens, Tolman, Hull), operationism was not committed to the tenets of logical positivist philosophy of science and analytical behaviorism. First, it was not primarily intended as a theory of meaning, and, second, its advocates were, for the most part, not committed to an antirealism about the referents of theoretical concepts (as has been suggested by, e.g., Leahey

---

2. I agree with Strapasson and Araujo (2020), who argue that *methodological behaviorism* is not a very clearly defined concept, but, for my purposes, this distinction, which runs together Graham's (2019) *psychological* behaviorism and *methodological* behaviorism, will suffice.

3. It should be noted that, while Carnap (1932) explicitly aligns his position with that of American (methodological) behaviorism, the two projects are, in a certain sense, antithetical: logical behaviorism tries to make mentalistic concepts respectable by "physicalizing" them, whereas methodological behaviorism tries to eliminate them (see Feest 2017b, 2020b).

4. Section 2.3.1 below provides a slightly more detailed analysis of the context of logical positivism and, in particular, Carnap's views about psychology.

[1980]). Consequently, third, they were not committed to the view that no two operations can be tied to the same concept and, thus, refer to the same entity in the world (as suggested by Hempel [1954/1956]). The positive aim of this chapter is to show that the above-mentioned figures have in common the fact that their experimental investigations include what philosophers might call "unobservable entities," such as mental states or cognitive functions. The point of their operationist positions was, thus, to specify their usage of their central concepts in terms of experimental operations and the resulting data, thereby making the corresponding (putative) objects amenable to experimental investigation.

More specifically, I argue that appeals to definitional considerations entered their research practices on two levels. First, and often drawing on folk-psychological concepts, researchers broadly conceptualized their objects of research in terms of functional relationships between types of stimuli and responses. Second, and drawing on these broad conceptualizations, they attempted to formulate unique experimental conditions of application for specific instances of these broadly defined concepts. I call the former "wide operational definitions" (operational definitions^{wide}) and the latter "narrow operational definitions" (operational definitions^{narrow}). Strictly speaking, only these latter kinds of definitions make reference to specific measurement operations. Nonetheless, I argue that there is a continuity between them in that the folk-psychological origins of operational definitions^{wide} suggest and lend credibility to the construction of operational definitions^{narrow}. As we will see, both levels of definitions pave the way toward an experimental investigation of the corresponding (putative) entities. But only the latter provides concrete instructions by formulating paradigmatic experimental conditions of application for the concept. I argue that the intention behind both definitions was simultaneously to enable empirical research on an object of research (essentially functioning as a kind of pragmatic a priori) and to reflect empirical knowledge about the object of research (which means that these definitions were seen as revisable in principle).

Before delving into the details of the three case studies, let me briefly address two potential worries about my approach. The first worry is that I am cherry-picking my case studies so that they fit my thesis. In particular, I am conspicuously leaving B. F. Skinner out of my story even though (as has recently been pointed out in the literature) he formulated an operationist position even prior to Stevens (Verhaegh 2021). Skinner is known for a version of (operant) behaviorism that radically eschewed any talk of intervening variables and, thus, might be viewed as contradicting my thesis. I argue that a closer look at Skinner's position actually underscores my reading of the way operationism was construed by figures like Stevens, Tolman, and Hull. Indeed, Skinner came to be critical of the way in which some of his contemporaries appealed to operationism when introducing theoretical terms, arguing that he had long held that "the reinterpretation of an established set of explanatory fictions was not the way to secure the tools they needed for a scientific description of behavior" (Skinner 1945/1972, 381). The difference between the operationisms of Skinner, on the one hand, and Tolman and Hull, on the other, is also noted by Green (1992, 301), who reports that already by 1938 Skinner attacked the usage of (operationally defined) vernacular concepts he saw in some of his neobehaviorist contemporaries.[5] Thus, I should emphasize that, while grounded in historical case studies, my analysis of operationism does not claim to be true of all historical cases. However, as this book progresses, it will become clear that the operationist position we see emerging on the basis of an analysis of Stevens's, Tolman's, and Hull's work offers valuable insights into the epistemic predicament of more current psychological research precisely because it does not fit the radical behaviorist and positivist mold often associated with operationism.

The second worry is that the methodological positions of practicing scientists (then and now) are often ambiguous and philosophically hard to pin down. Moreover, there were some genuine philosophical disagreements among operationist psychologists, especially with regard to the realism/antirealism issue. However, as I argue in this chapter, once we read the methodological writings of these three authors in conjunction with their scientific papers and research reports, it is hard to construe their positions as verificationist, antirealist, or committed to the idea that every new operation requires a new concept. Moreover, I show that each of my three characters was responding to methodological questions inherent in their research that preceded any exposure to logical positivism. Therefore, this chapter proposes a *methodological* reading of early operationism, according to which its advocates were interested not primarily in the question of what a given concept meant but rather in how best to investigate its presumed referent: auditory experience in the case of Stevens, protocognitive processes in the case of Tolman, and habits in the case of Hull. To be sure, there is a sense in which their specifications of their research objects were, indeed, *definitional*, but the definitions in question had a pragmatic character and, thus, were not

---

5. For Green, this is evidence that (unlike Tolman and Hull) Skinner got the spirit of Bridgman's operationism right. By contrast, I offer an analysis of Tolman's and Hull's (and Stevens's) operationism on their own terms.

assumed to capture the full meaning of a concept.[6] Rather, they were intended to capture a preliminary understanding of the object under investigation, making it amenable to empirical investigation, and playing an essential role in the construction of a theory of the object and of the domain in which it is thought to be located.

1.2. Stanley Smith Stevens and the Operational Treatment of Sensations

S. S. Stevens, a Harvard psychologist of Mormon heritage (Stevens 1974b), is today known mainly for his writings on scales of measurement (Stevens 1939a, 1946, 1951). His main area of scientific expertise, however, was in the field of psychophysics, more specifically, in the psychophysical investigation of hearing. In this field, his greatest claim to fame is having modified Fechner's psychophysical law. The psychophysical law plots the subjective intensity of a sense experience against the intensity of the stimulus. While Fechner had argued that this was a logarithmic law, Stevens argued that it was a power law (Stevens 1974a). His earlier and less well-known work was concerned with the experimental investigation and individuation of perceived attributes of tones (Stevens 1934a, 1934b, 1934c, 1935c; Stevens and Davis 1936). This was part of his dissertation work, which he conducted with Edwin Boring (1886–1968) between 1931 and 1933. As he was publishing the results of these investigations, he also published the first in a series of papers in which he argued for an operationist position in psychology (Stevens 1935a, 1935b, 1936a, 1939b). I show that Stevens's views about operationism were informed by his research about tonal attributes. For him, this research not only necessitated conceptual reflections about the notion of experience but also prompted him to think about ways in which specific *types* of experience might be detectable by experimental means. (As we will see below, he subsumed both these concerns under the label *operationism*.)

To understand where Stevens was coming from in the early 1930s, his dissertation adviser, Edwin Boring, is of special importance. As the only psychologist in the joint department of philosophy and psychology at Harvard, Boring was interested in the history of experimental

---

6. Hibbert (2019) argues that, if we construe operational definitions that way, they are not really definitions. Her argument is informed by an essentialist notion of *definition* of which I am skeptical (see chaps. 4 and 7 below). More importantly, my main purpose is not to evaluate whether psychologists use the term "definition" correctly but rather to provide a sympathetic reconstruction of their investigative practice.

psychology (see Boring 1929, 1942) and considered it his task to firmly establish psychology as an experimental scientific discipline, outside the reach of the other five full professors (see Stevens 1968, 592).[7] As documented by a huge collection of letters and comments on papers, which can be found in the S. S. Stevens Archives at Harvard University, the intellectual relationship between Boring and Stevens was extraordinarily close.[8] Thus, Boring described his relationship to Stevens as "paternal" and "genuinely rewarding" (Boring 1961, 68). Boring himself—having initially trained as an engineer and a physicist before switching to psychology—appears to have enjoyed an equally close relationship with his own teacher, Edward Titchener (1876–1927), with whom he had studied at Cornell, earning his PhD in 1914. Titchener, in turn a student of Wilhelm Wundt's (1832–1920), is generally considered to have introduced a version of Wundtian, structuralist psychology to the United States (see Boring 1927; and O'Donnell 1985). Stevens himself viewed the lineage as extraordinarily tight (Stevens 1968, 594), and, as we will see in a moment, this is quite apparent in the operationist project he pursued.

1.2.1. Stevens's Operationism: Specifying the Subject Matter in Experimental Terms

In his first paper on operationism, Stevens remarked on the "lack of rigor in the formulation of fundamental concepts," stating that what was needed was "a straightforward procedure for the definition and validation of concepts": "Such a procedure is one which tests the meaning of concepts by appealing to the concrete operations by which the concept is determined. We may call it *operationism*" (Stevens 1935a, 323). At first sight, this statement is puzzling. Nowadays, when psychologists

---

7. The empirical study of mental phenomena at Harvard dates back to 1875, when William James established the first psychological laboratory. In the following decades, psychology remained institutionally affiliated with the Philosophy Department. It was not until 1936 that, owing to the efforts of Edwin Boring, psychologists were allowed to form a completely separate department at Harvard. Before that time, the Department of Philosophy had several famous psychology professors (as we will see below, Tolman studied with Langfeldt and Yerkes in the 1910s). However, when Stevens first arrived there, Boring was the only full professor.

8. The unpublished material by Stevens and Boring employed in this chapter resides in the Harvard University Archives (shelf number HUG (FP)-2.45, boxes 1 and 2). Box 1 was made available to me by the Harvard University Archives; box 2 was made available to me by Gary Hardcastle. I refer to this material hereafter by box number.

talk about the validation of a concept, they mean the procedure of establishing that the concept really picks out a phenomenon of scientific interest. But why would a concept need to be validated if it has already been *defined* in terms of a particular procedure? And what does Stevens mean by the notion of testing the meanings of scientific concepts? Scientists do not test meanings. They might test (*a*) whether a given object falls into the extension of a particular concept or (*b*) hypotheses about descriptive features and identity conditions of the object. These two points are related to the issues of classification and description. Notice that there is a sense in which the two issues mutually presuppose each other. (I can classify a specific object as falling in the extension of a concept only if I have a clear empirical understanding of the circumstances under which the concept applies, and I can arrive at such a clear understanding only if I can investigate instances of the objects in the extension of the concept.) I argue that this is precisely the issue Stevens points to here, suggesting that operationism can provide a procedure for breaking out of this apparent circularity.[9]

On my construal, then, the question at stake was how to conduct scientific investigations of a research object when its very existence, nature, and identity conditions are controversial. Stevens's own research highlights this question nicely. When he was writing his articles on operationism, his research project consisted in showing that humans not only perceive the loudness and pitch of a tone but in addition experience two distinctly different types of tonal attributes, namely, the *density* and the *volume* of a tone (see Stevens 1934a, 1934b, 1934c, 1935c). As I explain below, his aim was to formulate what he called "operational definitions" of these concepts, cast in terms of paradigmatic procedures used to detect their referents. First, however, he required a broader understanding of his subject matter. Perceptions of tonal volume and density are, he suggested, types of experiences. This prompted him to inquire into the notion of *experience* as such. This concept, he wrote, "denotes the sum total of the discriminatory reactions performed by human beings, for *to experience* is, for the purpose of science, *to react discriminatively*" (Stevens 1935b, 521). While this statement makes it look as if Stevens was equating the meaning of the term "experience" with "discriminative reaction," I argue that the scope and the significance of this assertion are more restricted to a particular context and aim. Surely, Stevens was aware of the

---

9. Notice that this circle is very similar to the one Hasok Chang (2004, 2016) makes out in the study of temperature. Like Chang, I argue that a historically sensitive reconstruction of operation(al)ism can provide insight into how researchers attempt to address it.

fact that not every experience is triggered by an external stimulus. Nor is every experience accompanied by outward behavior. Moreover, it is an essential part of the ordinary meaning of the term "experience" that experience has a phenomenal feel and that it is introspectively accessible. Indeed, the whole field of psychophysics—which tries to correlate phenomenal and physical conditions of experience—presupposes this (cf., e.g., Feest 2017b). This point is explicitly acknowledged by Stevens when he explains his research rationale: "We have, on the one hand, adequate introspective evidence that volume is phenomenally a separate and distinct attribute of tonal stimuli; but, on the other hand, experiments in which volume has been used as the basis of judgment have not yielded stable quantitative results. Consequently, there remain grounds for the divergence of opinion which exists among theorists as to the existence, the nature, and the possible physiological basis of the volumic attribute" (Stevens 1934c, 398). His point, then, was not to deny the phenomenal nature of experience but (*a*) to highlight the simple insight that the scientific investigation of experience relies on discriminatory behavior and (*b*) to emphasize the need for experiments that would yield stable quantitative results about such discriminatory behavior. Both these assertions express fairly basic premises of psychophysics as it has been practiced in psychology since the mid-nineteenth century. These premises demand that—in the context of scientific research on sensory experience—one needs to control the stimulus material and cannot assume that subjects are having a sensory experience unless they can be prompted to exhibit robust discriminative behavior in response to the stimulus material.

I argue that Stevens's specification of the meaning of the term "experience" is an instance of operational definition[wide] as roughly capturing a folk-psychological understanding of the conditions under which we are likely to attribute experiences to other creatures. Stevens then goes on to argue that, in order to investigate specific types of experience, one needs to define the corresponding concepts in terms of specific types of stimulus material and corresponding instructions to elicit specific types of discriminatory reactions. Thus, it is here that operational definitions more narrowly conceived (operational definitions[narrow]) come into play. Specific types of experience (e.g., the experience of tonal density) are defined in terms of particular experimental operations and responses. Accordingly, Stevens sought to define the expression "tonal attribute" in terms of a differential reaction to a tone by a listener under a particular set (of stimuli and instructions), or "Aufgabe" (Stevens and Davis 1938, 449–50). So, here he is saying that, insofar as they exist at all, specific kinds of experience can be shown to have a unique quantitative profile that is defined in terms of functional relationships between stimuli and responses, as determined experimentally. We need to be clear to distinguish two issues here. On the one hand, we can say that "tonal volume" is operationally defined in terms of the relationship between specific experimental stimuli and responses, that is, it is a summary of the experimental result. On the other hand, we can say that the experimental result is possible only by virtue of having set up the experiment this way to begin with, that is, with the conceptual assumption that the resulting data are going to be treated as indicative of the experience of tonal volume. So operational definitions can be either experimental findings or the conceptual presuppositions required for such findings. In chapter 3, I argue that, in this latter sense, operational definitions should be considered tools of research. Right now, I simply wish to emphasize that conceptual and empirical issues are clearly closely related here. And, since empirical findings are fallible, it seems highly unlikely that Stevens wanted to say something definitive or exhaustive about the meaning of terms like "tonal volume."

Before pursuing this further, let me briefly lay out some of the theoretical and historical context of Stevens's research project. The question Stevens pursued was how to devise experiments that would reliably elicit a specific experiential effect—as reflected in a stable behavioral response—distinct from different experiential effects that might be triggered by the same stimulus. Specifically, Stevens posed the question of whether auditory stimuli had other experiential effects apart from loudness and pitch. In doing so, Stevens was challenging what Boring (1935, 238) referred to as the "classic view," according to which there are only two experiential qualities that directly correspond to the two physical properties of auditory stimuli—frequency and energy—and that can be varied independently (Külpe 1893).[10] By the time of Stevens's work, the assumption of independent variability had been called into doubt by two things. First, it had been found that the simple one-to-one correspondence between stimulus dimensions and experienced attributes did not always hold. Thus, Boring reports that "it has been known for some years that loudness varies with frequency, when energy

---

10. It appears that the terms "energy" and "intensity" are used interchangeably in the literature. It also seems to me that Stevens (1934b) is not always careful to distinguish between intensity and loudness. Occasionally, the terms "amplitude" and "pressure" are used to describe the physical aspect of sound that complements frequency. For this reason, Stevens's psychophysical articles from the mid-1930s are hard to follow because they presuppose some knowledge of the fundamental terms and magnitudes of acoustics. In *Hearing* (1938), Stevens and Davis acknowledge that a precise use of language is necessary and give a fairly clear account of the terminology.

is constant," and that "pitch varies with energy when frequency is constant" (1935, 238). Second, Titchener (1908) had contemplated the possibility of there being more than two experienced attributes of tones by postulating the existence of the tonal attribute of volume. Both these questions stimulated some research in the 1910s and 1920s.[11] Given that only two physical stimulus dimensions were known, the question was whether a particular tonal attribute could be a function of more than one feature of the stimulus. In his "Attributes of Sensation," Boring answered this question by saying that, *theoretically*, there is no upper limit to the number of attributes a tone might have (1935, 245; see also Isaac 2017, 105). The question of what are the *practical* limits of our ability to differentiate between attributes of tones is, however, clearly an empirical one. It is this empirical question that Stevens (1934a, 1934b, 1934c) took on when asking whether there could be experimental evidence for the existence of the experienced tonal attributes of volume and density. Answering this question required, among other things, coming up with a standardized experimental arrangement that would allow for the determination of the tonal attribute in question.

1.2.2. The Experimental Operation of Individuating Tonal Attributes: Volume and Density

In asking about the tonal attribute of "volume," "largeness," or "massiveness" (see Stevens 1934c, 397), Stevens was picking up on some existing work. For example, Halverson (1924) had found that soft tones (low energy) as well as high tones (high frequency) tend to be perceived as "small" (low volume), suggesting that, just like experienced loudness and pitch, tonal volume is a function of *both* stimulus dimensions. However, findings as to the exact quantitative characterization of this attribute were inconsistent (Boring 1933). What Stevens thus aimed for was a functional characterization of volume in which *sameness* of experienced volume was to be plotted against relative *changes* in the two stimulus dimensions of frequency and intensity. The challenge was to find an experimental paradigm that would "tune" the subjects so that they would be able to make the relevant discriminations: "This tuning of the observer is one of the fundamental operations underlying the concept of attribute" (Stevens 1935b, 525). The challenge facing the psychologist, then, was to formulate experimental instructions in such a way that subjects were enabled to report their discriminations without

---

11. References to this research can be found in Boring (1935), Gundlach and Bentley (1930), Rich (1916, 1919), Stevens (1934b, 1934c), and Zoll (1934).

any distorting factors. Thus, it is not surprising that Stevens explained the lack of consistent experimental findings as a result of the fact that researchers had previously used inadequate experimental designs. To explain this, let us turn to the types of discriminations experimental subjects are usually required to make and the types of experimental paradigms typically used by psychophysicists.

Traditionally, psychophysical research has tried to establish the smallest units of sensation, also known as "just-noticeable-differences," that is, the differences between stimuli below which no differences can be perceived. (The origins of this approach are usually attributed to Ernst Weber and Gustav Fechner.) The first systematic account of available experimental paradigms in psychophysics was given by Fechner (1860) (see Boring 1942). Stevens (1934c) distinguishes between two main approaches, the "method of constant stimuli" and the "method of average error." The main difference between these two methods is that the former presents subjects with fixed stimuli and lets them report their sensations and the latter presents subjects with a standard stimulus and instructs them to change a comparison stimulus so that the two sensations seem equal. The average difference between the standard stimulus and the comparison stimulus (i.e., the average mistake) is taken as the physical unit that distinguishes two sensations. It is this second method that Stevens proposed to use in his study of volume, stating that this method was most likely to overcome the reasons why volume is difficult to judge (the relative unfamiliarity with this aspect of tonal stimuli, the fact that the subjects' criteria for judgment are very unstable, and the fact that the difference in pitch is more conspicuous than the difference in volume) (Stevens 1934c, 399).

Stevens's experiments proceeded by giving the subject two tones with different frequencies and asking them to regulate the intensity of the lower one until the perceived volumes of both tones were equal. It turned out that subjects increased the intensity of the stimulus with the lower frequency in order to make its volume equal to the stimulus with the higher frequency and that the required increase in intensity was smaller for pairs of tones with high frequency. Thus, Stevens provided evidence for the existence of the tonal attribute volume by showing that it depends on both stimulus dimensions in a unique way: "Finally, we might ask what tonal volume is. We are justified in saying that volume is a phenomenal dimension of tones. It is a dimension in terms of which an O [organism] is able consistently to make discriminable judgments, and in terms of which he is able to equate two tones which differ in respect of every other phenomenal *dimension*" (1934c, 407). Stevens's work on the tonal attribute of density resulted in the opposite

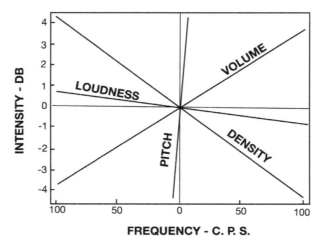

Figure 1: The experience of equal loudness, pitch, volume, and density as a function of the two stimulus dimensions, energy/intensity and frequency, measured in decibels and cycles per second (hertz), respectively. Reproduced from Stevens 1934a, 458.

finding: "In order . . . to make [two tones] equal in density, however, the lower tone must be made louder than the higher" (1934b, 589). His findings on tonal attributes (see Stevens 1934a) are illustrated in a graph in which all four tonal attributes (loudness, pitch, volume, and density) are plotted against the two stimulus dimensions (see fig. 1).

Now, we might ask whether and in what sense Stevens viewed either these quantitative results or the measurement procedures that gave rise to them as definitions of the terms "tonal density" and "tonal volume." As we saw above, he clearly regarded the experiential aspect of tonal attributes as central to the meanings of the terms. For this reason, it seems clear that he did not intend to provide an exhaustive characterization of those meanings. Rather, the experimental setup in question, in conjunction with the behavioral response, was treated as providing a paradigmatic condition of application for the terms because he took the experiments to produce a paradigmatic instance of the phenomenon under investigation.

This raises two additional questions. First, if central terms are defined in terms of paradigmatic conditions of application, the question is what makes these conditions paradigmatic? More pointedly, paradigmatic of what? Presumably, the defining experiment individuates the research object in a particularly clear form. This means (*a*) that there is an assumption of the reality and independent existence of the research object, which in turn implies (*b*) that there are potentially different ways of getting at the research object. The second assumption brings to the fore the question of how it is determined that different measuring procedures indeed do get at the same thing. As we will see in the remaining two case studies, and as I spell out in chapter 2, early operationists were quite aware of this question, and it would continue to be of concern for subsequent generations of psychologists. A second question is whether the practice of *defining* a given concept rules out whether the concept might subsequently change, and, if not, whether the analysis provided here gives us some inkling of how conceptual change might come about. This, too, is a question to which I return later in this book. For now, let me highlight that Stevens did not adhere to a rigid understanding of the nature of definitions, arguing, for example: "[Definitions] take into account the state of factual knowledge at a given time. It is for good reason that the discovery of new related facts may make a revision of the criteria necessary so that we may include or exclude the new observation from the class denoted by the original definition" (Stevens 1935b, 519).[12] The idea that definitional assumptions can shift as a result of new findings is very nicely illustrated in Stevens's own research. In the case at hand, he concludes that Oswald Külpe's definition of the term "sensory attribute" has to be rejected. Külpe had assumed that there was a one-to-one correspondence between stimuli and sensory experience, leading him to assume that it was possible to vary types of sensory experience while holding others constant. But, as Stevens states, the relationship turns out to be such that no single experiential attribute can be varied independently while all others remain constant (Stevens 1935b, 524).

1.2.3. Contextualizing Stevens's Research and Methodological Views

Now, given that the characterization offered above paints an innocuous picture of what Stevens intended with his operationism, one might wonder how operationism got such a bad reputation. Part of the answer, surely, is that Stevens and other psychological advocates of operationism did not always formulate their views precisely. Another part is that the origins of this position are not widely known. In this section, I provide a little more background about the intellectual and institutional context in which Stevens developed and articulated his views.

Stevens's research on tonal attributes has to be seen as originating in the structuralist tradition of research on sensory attributes (going back to Titchener and Wundt) at a time when consciousness as an object

---

12. Also: "No concept can be defined once and for all: every concept requires constant purging to keep it operationally healthy" (Stevens 1935b, 527).

of scientific investigation had come under attack. This gave rise to the demand to put the psychophysics of auditory sensations on a more rigorous footing. Both these themes come out nicely in Edwin Boring's 1933 *The Physical Dimensions of Consciousness*, which he wrote while supervising Stevens's dissertation. In this book, Boring argued for a mind/brain identity theory and tried to find a middle ground between the *eliminativist* attitude of behaviorism (according to which consciousness cannot be an object of scientific study) and what he considered to be the *dualism* of the tradition of introspectivist psychology.[13] The main thrust of Boring's argument here was (*a*) that conscious experience has dimensions, (*b*) that each dimension of consciousness is identical with a physiological state of the brain, and, therefore, (*c*) that experiential dimensions were amenable to rigorous experimental investigations. This last point was also in line with Boring's conviction that "historically science is physical science": "Psychology, if it is to be a science, must be like physics" (Boring 1933, 6). With this battle cry, Boring sought to distance himself from what he perceived as a methodological dualism on the part of his own teacher, Edward Titchener. At the same time, he saw himself as carrying on Titchener's project of identifying basic attributes of sensations, thereby responding to the behaviorist agenda of banning consciousness as an object of scientific study altogether: "My book was a move away from Titchener, but it also served to make his dimensionalism clear" (Boring 1961, 53).

Several sources indicate that Stevens was closely involved in the process of writing *The Physical Dimensions of Consciousness* (Boring 1933, viii; Stevens 1968, 596; Stevens Private Notebook, 1933).[14] Both Boring and Stevens retrospectively report that what the book was attempting to do (if not quite satisfactorily) was to formulate the methodological position Stevens would shortly thereafter call "operationism" and, as Stevens noted, that "an operational restatement of psychology's basic concepts was Boring's real target" (Stevens 1968, 597). However, neither Stevens nor Boring was entirely satisfied with the result, and the book was no great success. Stevens later attributed this to the fact that Boring was not yet in possession of the empirical results that would back up his dimensionalism, suggesting that his own research on attributes of tonal sensations would clarify Boring's point (Stevens 1968, 597).

---

13. Whether the structuralist psychophysical tradition is adequately described as dualist and whether, indeed, there really was such a thing as introspectionism are different questions that will not be pursued here (but see Danziger 1980; Hatfield 2005; and Feest 2012b, 2021).

14. Harvard University Archives, box 1.

By drawing out the close relationship between Stevens's operationism and the project pursued in Boring's *Physical Dimensions of Consciousness*, we can now also better appreciate why Stevens was not an antirealist about the referents of the concepts he defined operationally. Assuming that he shared Boring's mind/brain identity theory, he would presumably have held that conscious states are brain states and hence have an ontologically robust status. In turn, this implies that he would have had to be open to the possibility that a particular attribute of consciousness could in principle be accessed in more than one way.

Now, we may ask to what extent Boring's and Stevens's positions were similar to or perhaps even influenced by those of Percy Bridgman and the logical positivists. Bridgman's role is particularly interesting because (*a*) he was at Harvard at the time Stevens wrote his papers on operationism, and it is conceivable that they might have met, and (*b*) Boring's ambition to model the science of psychology on physics would have attracted him to a methodological doctrine put forth by an eminent physicist. Green (1992) has suggested that Herbert Feigl brought Bridgman's operationism to Boring's attention in 1930. However, if we follow Boring's own retrospective account, then he was unaware of Bridgman even in 1932: "P. W. Bridgman . . . had expounded 'operationism' in 1927. . . . Later he had great influence on American psychologists, but neither they nor I knew about his book then. In an unclear way my book was using this new 'operationist' logic" (Boring 1961, 52–53). On the other hand, Verhaegh (2021) shows convincingly not only that B. F. Skinner, who was also a graduate student of Boring's, was aware of Bridgman's work by the late 1920s but also that he had already tried to "operationally define" the concept of *reflex* in a chapter of his dissertation in 1930 (i.e., before Stevens was Boring's student). In turn, this work was harshly criticized by Boring (see Verhaegh 2021). We can speculate that Boring's apparent change of heart only a few years later was due to the fact that the Boring/Stevens type of operationism was very different from Skinner's in that the former tried to tie talk of conscious dimensions to behavioral indicators while the latter wanted to eschew talk of consciousness altogether. Be that as it may, while it is possible that (through Skinner) Boring encountered Bridgman already in 1930, there is evidence of Stevens's operationist position in an unpublished 1933 paper entitled "Materialism" in which neither Bridgman nor the term "operationism" are mentioned (Stevens 1933). The bottom line of my analysis of Stevens's relationship to Bridgman is, thus, that Stevens's operationism was motivated by and articulated in response to very specific questions in the history of psychophysics and not significantly influenced by Bridgman (or Skinner, for that matter).

With respect to Stevens's relation to logical positivism and in particular to the analytical behaviorism associated with the project of physicalizing psychological concepts (Carnap 1932), I argue that, while there are certainly affinities between the two projects, Stevens's methodological focus on experimental research as well as his dynamic conception of the notion of *definition* also made his position significantly different from that of logical positivism.[15] Stevens encountered members of the Vienna Circle at Harvard in the second half of the 1930s, got involved with the Unity of Science movement in the late 1930s, and explicitly related his operationism to this movement in Stevens (1936a).[16] However, there is no mention of logical positivism in Stevens's early notebooks of 1932 and 1933. Moreover, Hardcastle (1995) has made the case that Stevens's knowledge of German was limited and that he was, therefore, unlikely to have been exposed to logical positivism before Carnap's "On the Character of Philosophical Problems" (1934) appeared in English.

1.3. Of Rats and Psychologists: An Analysis of E. C. Tolman's Operationism

The psychologist Edward C. Tolman was the son of a liberal, upper-middle-class family.[17] He graduated from MIT in 1911 (with a major in electrochemistry) and went to graduate school at Harvard in the joint department of philosophy and psychology, where he finished his PhD in psychology in 1915. Some of his professors were the new realists Edwin B. Holt (1873–1946) and Ralph Barton Perry (1876–1957), the psychologist Herbert Langfeldt (1879–1958), and the comparative psychologist Robert Yerkes (1876–1956). After finishing his PhD, Tolman taught at Northwestern University for three years and then assumed a position at the University of California, Berkeley, in 1918, where he stayed until the end of his life. In 1923, he spent a one-year sabbatical in Giessen, Germany, where Kurt Koffka (1886–1941), one of the main proponents of Gestalt psychology, was a professor. Ten years later (upon completion of *Purposive Behavior in Animals and Men* [1932b]), in 1933, he spent another sabbatical in Europe, this time in Vienna. There he formed a close

---

15. For an analysis of the relationship between physicalism and psychophysics, see Feest (2017b).

16. Stevens presented the paper "On the Problem of Scales for the Measurement of Psychological Attitudes" at the Fifth International Congress for the Unity of Science, which was held at Harvard University in September 1939 (see Stadler 2001, 389).

17. For more details about Tolman's life, the reader is referred to David Carroll's excellent 2017 biography, which also covers Tolman's lifelong political activism.

personal and professional relationship with Egon Brunswik (1903–55) (Smith 1986).

Tolman is particularly well-known for his experimental work on the behavior of rats in mazes (Tolman 1948), which he took to indicate that rats are able to form mental maps of their environments. This work earned him the reputation of having been a protocognitivist in a behaviorist era, though Tolman himself always insisted on calling himself a behaviorist. This may appear surprising given today's understanding of behaviorism as involving a radical rejection of all reference to mental states. However, in the early 1920s, behaviorism was by no means a unified position.[18] In the mid- to late 1930s, Tolman also published several papers in which he explicated his notion of *operationism* (Tolman 1936, 1937, 1938). In the present section, I provide an analysis of this notion. I show that, even though there are some superficial similarities to a verificationist semantics, analytical behaviorism, and empiricist epistemology, Tolman's operationism was, like Stevens's, in fact firmly anchored in his scientific work. Briefly, my argument is that Tolman's operationism needs to be related to his aim of developing a theory of purposive behavior. An essential component of this theory was the notion that behavior can be causally explained as resulting from what Tolman called "drives/demands" and "cognitive postulations." I argue that Tolman's operationism was an attempt to salvage the mentalistic concept of *desire* (drive, demand, instinct) as a legitimate scientific concept by showing that its referent could be empirically grounded. Furthermore, as in the case of Stevens's articulation of operationism, we find a distinction between operational definitions^{wide} and operational definitions^{narrow}. And, finally, Tolman's concept of *cognitive postulation* figured as part of a theory of cognition in rats that bears a surprising resemblance to his theory of scientific cognition as articulated in his views about operationism.

1.3.1. Essentials of Tolman's Operationism

Tolman first used the term "operationalism" in "Operational Behaviorism and Current Trends in Psychology" (1936). He characterized the position as asserting (a) the existence of a number of intervening variables, (b) certain functions whereby these intervening variables are related to particular independent variables, and (c) further functions whereby an organism's final behavior results from combinations of these intervening and independent variables (Tolman 1936, 118). The expression

---

18. For a sketch of the different varieties of behaviorism to be found at the time, see Roback (1923/1937).

"intervening variable," in turn, had first been used by Tolman in an article published the previous year (Tolman 1935) in which he had proposed that any given behavior of an organism should be regarded as a function of such variables and that the aim of a complete psychological theory was to identify them.

Tolman is not always clear in his usage of the expression "intervening variable." For example, when he says that theories are sets of intervening variables, he appears to mean the terms or concepts of a theory, but in other places he seems to have in mind the entities or processes denoted by those terms. Moreover, when he says that intervening variable terms are to be defined operationally, he sometimes appears to suggest that intervening variable terms are perhaps only shorthands for particular input-output functions or experimental laws: "Mental processes . . . will figure only in the guise of objectively definable intervening variables" (Tolman 1936, 116). However, other statements by him strongly suggest a more realist understanding, to the effect that intervening variables are detected or inferred by means of experimental data: "We . . . assume that, when we carefully select our experimental setups, these intervening variables will be betrayed directly in the selected quantified features of the resulting behavior" (Tolman 1936, 125). In a similar way he refers to an experiment as "indicating and defining [the intervening variable's] immanent expectations" (Tolman 1932b, 79).[19]

In their much-cited "On a Distinction between Hypothetical Constructs and Intervening Variables," Kenneth MacCorquodale and Paul E. Meehl famously argued that Tolman's notion of *intervening variables* "involves nothing which is not in the empirical laws that support them" (1948, 100). In contrast, and despite the inconsistencies in Tolman's work, I suggest that we take his more realist formulations seriously as this accords well with the nature of his research as well as with explicit commitments he voices in other places. Therefore, I argue that, when Tolman used the expression "operational definition," he did not intend to reduce the meanings of the terms to input-output functions but rather took them to refer to processes that causally intervene between stimulus and response. Tolman—like Stevens—formulated his operational definitions on the basis of the relevant folk-psychological concepts (operational definitions[wide]). And—again like Stevens—he viewed operational definitions (more narrowly conceived) as having the dual status of making experimental research possible to begin with

19. This latter statement is, of course, somewhat confused since, presumably, we detect *things* and we define *concepts* or *terms*. However, I argue that talk of "detecting" something experimentally betrays a realist sentiment.

but also as reflecting the state of empirical knowledge that results from such research (operational definitions[narrow]). Operational definitions (in both senses) make experimental research possible by virtue of being ingrained in the experimental designs (experimental manipulations and expected behaviors) that are expected to provide insights into specific objects under investigation. In turn, some experiments conducted on the basis of that prior conceptualization might then turn out to be what Tolman called "defining experiments" (see below), which delineate the entity in question in particularly clear form. Tolman uses the concept of "operational definitions" for both, but, as I show below, there is an important sense in which, as with Stevens, the latter is the result of experimental research that presupposes and can perhaps also correct the former type of operational definition.

Let me start with an example of the broad way in which Tolman's experimental work was informed by explications of folk-psychological presuppositions. I do so by looking at an intervening variable that figures importantly in Tolman's system, namely, *demand* (see Tolman 1937, 1938). Demands are desires for particular things (food, sex, etc.). On a very general level, Tolman worked with the assumption that the behavioral effects of a demand can be brought about by singling out the environmental variable that is thought to affect the demand in question and observing how changes in that variable affect behavior. For example, in research on the *hunger* demand, the relevant environmental variable is the availability of *food*. Hunger is, then, operationally defined as the functional relationship between "time since last feeding" (the independent variable) and "food-searching behavior" (the dependent variable). Now, on the basis of this broad understanding of hunger as an instance of *demand*, Tolman proposed to devise a "defining experiment," that is, an experiment that individuates the variable of interest (in this case, hunger) in a particularly clear, quantitative form. In Tolman's words: "In certain carefully chosen, controlled and 'standard' experimental setups, one tries to hold all but one, or one small group, of the independent variables constant and studies the functional connection between the variations in this one independent variable ... and the correlated variations in some quantifiable feature of the final behavior" (1936, 122). In my terminology, the "definitions" provided by such standardized experimental setups are operational definitions[narrow], but their plausibility relies on a broader and more colloquial understanding of the phenomenon under investigation (operational definition[wide]). Now, with respect to the question of what motivated Tolman to devise such standard experiments, I make out two answers. The first is that frequently, for Tolman, the point of appealing to an operational definition of a specific demand was not to do research on that particular demand per se but to get a grip on this variable in order to be able to control for it when testing hypotheses pertaining to *other* variables. For example, in his *Purposive Behavior in Animals and Men* (1932b) (i.e., before the publication of his operationism papers), Tolman draws on his definition of *hunger* as "time since last feeding" in order to conduct experiments about the question of whether different kinds of food are equally attractive to rats given the same level of hunger. (*Purposive Behavior* contains many examples of this strategy.) When Tolman (1937) later provides a "defining experiment" for the notion of *hunger*, he is simultaneously presenting an experimental result (the functional relationship between food deprivation and search behavior) and attempting to legitimate the scientific practice in which he is already engaged, that is, attempting to control for the effects of intervening variables by controlling the independent variable of which the effect is a function.

Tolman's second motivation for wanting to define intervening variables was the recognition that, if theories were sets of intervening variable terms, it was important to find a way to establish their legitimacy independently of one another. As mentioned above, Tolman posited two types of intervening variables as central to psychological theory: *demands* and *cognitions* (or cognitive postulations). For him, the point of designing defining experiments for individual intervening variables was to tease them apart. Both demands and cognitions made their appearances early on in Tolman's theorizing (i.e., before he explicitly stated his views on operationism), albeit under different names, as early as, for example, "Purpose and Cognition: The Determiners of Animal Learning" (Tolman 1925b). While he believed that any given action is always going to be determined by both types of intervening variables, it is interesting to note that, already at this early point, he recognized that the two variables needed to be experimentally distinguishable in order to figure in a respectable scientific theory: "Previous work . . . has made no distinction between an animal's mere *knowledge* of the behavior possibilities which will get him to food, and his *desire* for those behavior possibilities. . . . Experimental work must be done to tease these two factors apart" (Tolman 1925b, 47). In subsequent years, Tolman simultaneously (*a*) elaborated on the question of how to "tease apart" these factors and (*b*) formulated a theory of the behavioral system that specified the nature of the variables and their impact on behavior. Overall, however, it is not clear that he succeeded in this aim of experimentally prying apart the two determiners of goal-directed behavior he posited. Indeed, as I argue in chapter 3, singling out the individual contributions that different intervening variables make to a behavioral end product is one of the central problems with which subsequent generations of operationists had to deal (and with which psychologists deal to this day). For the purposes of the present chapter, however, I merely wish to highlight that the distinction between two kinds of intervening variables was a central component of Tolman's theoretical outlook and that he was quite aware that this posed methodological problems that he needed to tackle (and that he hoped to tackle by means of his operationism).

1.3.2. Demands and Cognitive Postulations as Crucial Intervening Variables

According to Tolman's theory of purposive behavior, environmental features are cognitively represented in terms of how they can be used or manipulated in order to attain certain goals. Or, to use Tolman's own terminology, environmental features are represented as "means-objects," which figure in "cognitive postulations" (or hypotheses) as to what would happen if the object were to be manipulated in a certain way (Tolman 1932b; Tolman and Krechevsky 1933). In the case of mental maps, the rat cognitively represents the maze with respect to features that would be helpful if the rat were trying to get to the food. Thus, Tolman's notion of *cognitive representation* is not that of a mere neutral description of the environment. Rather, such representations are always formed relative to certain goals (e.g., the goal to get to the food), and they involve expectations as to the outcomes of hypothetical actions (e.g., "if I were to take the left lane, this would get me to the food").

In what follows, I cover several formative influences as they pertain to (*a*) Tolman's notion of *demand* and (*b*) his notion of *cognitive postulation*. His thinking about this can be traced back to two of his philosophy professors at Harvard, Edwin Bissel Holt and Ralph Barton Perry, both of whom belonged to the philosophical movement of New Realism. This movement emerged at the turn of the twentieth century, mainly in opposition to idealist tendencies in American philosophy at the time (Holt 1912; Holt et al. 1910; Robischon 1967). While Holt was instrumental in shaping Tolman's views about the *description* of behavior, Perry laid some foundations for Tolman's views about both purposes and cognitions as *explanatory* variables. Tolman's continuing concern with both these questions is evident into the 1920s. I begin with the former issue.

Tolman described his brand of behaviorism as "molar," distinguishing it from what he termed J. B. Watson's "molecular" behaviorism (or "muscle-twitchism") (Tolman 1925a). Essentially, to describe behavior in molar terms means describing it teleologically, that is, as directed toward a particular goal or purpose (e.g., food-seeking behavior or reproductive behavior) (Tolman 1925a). It is very likely that Tolman's views in this regard were influenced by Holt, who had argued that under certain circumstances the synthesis of many known physical processes results in the occurrence of new phenomena and laws, thereby requiring a descriptive terminology that adequately captures such (emergent) phenomena. One such phenomenon, he argued, is the appearance of behavior as going beyond the sum of all reflex actions. Rather than breaking down the behavior into its successive states, he proposed a "functional view," which studies someone's behavior "until we have found that object, situation, process . . . of which his behavior is a *constant function*" (Holt 1915b, 163), that is, the object, situation, or process upon the attainment of which the behavior stops. Smith (1986) suggests that Holt's ideas about molar behavior laid the foundation for Tolman's operationism, understood (by Smith) as an attempt to reduce the mentalistic terms to input/output functions. I agree that, by emphasizing the molar, goal-directed character of behavior, Holt importantly influenced Tolman's outlook. However, given my contention that Tolman also wanted to *explain* behavior by appeal to internal states (demands and postulations), this cannot be the whole story. It is, therefore, important to recognize another important formative figure during Tolman's early education.

During his time as an undergraduate, Tolman was also exposed to arguments that emphasized not only the need to describe behavior in teleological terms but also the need for internal variables as explanatory of the behavior in question (and hence as more than mere shorthands for empirical laws). As Tolman recounts, while in college he took a class with Ralph Barton Perry that "laid the basis for my later interest in motivation and, indeed, gave me the main concepts . . . which I have retained ever since" (1952, 325). These "main concepts" were precisely the motivational and the cognitive determinants of behavior that later got consolidated in his notions of *demands* and *cognitive postulations*. In the late 1910s, Perry published a series of papers (Perry 1917a, 1917b, 1918) that dealt with the notion of *purpose* and critiqued existing accounts (such as Thorndike's notion of *trial and error*). He argued that associationist theories of learning can account only for learning—given that the animal keeps trying—but they do not have an explanation for *why* the animal keeps trying. Accounting for this, he argued, requires (*a*) the postulation of drives toward particular kinds of goals (he called this the "selective propensity") and (*b*) some kind of representation of the means by which the goal might be achieved that can be adjusted as a result of learning. Perry's general line of thought was further explicated in another series of papers (Perry 1921a, 1921b, 1921c). There, he stated that behaviorists "regard the mind as something that intervenes as an arc or circuit or general causal nexus" (Perry 1921a, 87) but added that purposive behavior requires interests and beliefs, which can be independently varied.[20] By his own account, Tolman was "tremendously influenced by Perry, who demonstrated that both cognition and motivation could be treated as perfectly objective facts, without taint of teleology or of subjectivism" (Tolman 1959, 94).

In the decades that followed, Tolman elaborated on his conceptions of *demand* and *cognitive postulations*. With regard to the former, in particular, the debate about the nature of instincts proved to be fruitful. Tolman returned to this concept again and again (Tolman 1920, 1922a, 1923b, 1925b, 1926b), gradually elaborating his position. Later, it reemerged, virtually unchanged, in his writings of the 1930s (Tolman 1932b, 1936, 1937) as he situated himself vis-à-vis two broad notions of *instinct* at the time: one that equated instinct with "inherited reflex pattern" (e.g., Watson 1913, 1914) and one that equated it with some kind of subjectively experienced urge to behave in certain ways toward particular kinds of objects (e.g., McDougall 1908/1914),[21] ultimately distinguishing between first- and second-order drives (Tolman 1926b).[22] Tolman aligned his own approach with that of William McDougall, whose list of instincts he adopted (e.g., curiosity, gregariousness, self-assertion, self-abasement, and imitativeness as examples of the latter type of instinct), though he took his own approach to be more rigorously empirically based. This idea remained basically unchanged in his main work, *Purposive Behavior in Animals and Men* (1932b, e.g., 305). I argue, therefore, that we see here evidence not only of his ongoing interest in demands as causally explanatory intervening variables but also of his grappling with the issue of how to individuate them empirically.

With respect to Tolman's evolving notion of *cognitive postulations*, it is likely that another of Tolman's Harvard teachers, the animal psychologist Robert Yerkes, had an early impact on Tolman. In 1916, Yerkes published *The Mental Life of Monkeys and Apes*, which was based on research carried out in the spring of 1915. Tolman took a class in comparative psychology with him in which J. B. Watson's 1914 *Behavior* was used as

---

20. Perry's work culminated in *General Theory of Value* (1926), which attempted to offer a naturalistic treatment of purposeful behavior.

21. For a fuller discussion, see Feest (2005a).

22. It can be speculated that we see here a shift in terminology, from "instinct" to "drive" or "demand," because of the increasing unpopularity of the concept of *instinct* among enlightened psychologists. However, it seems clear to me that Tolman is essentially talking about the same thing as before.

a textbook (Tolman 1952). Yerkes reported several experiments that he conducted with orangutans in which he observed their ability to reach bananas by means of poles or by climbing on boxes stacked on top of one another. Though the results were mixed, he concluded (with respect to one particular individual): "'Trial and error' had no obvious part in the development of the really essential features of the behavior. The ape had an idea and upon it depended for guidance" (Yerkes 1916, 97).

Yerkes's experiments are reminiscent of the well-known experiments carried out by the Gestalt psychologist Wolfgang Köhler (1887–1967), whose work on the problem-solving abilities of chimpanzees Tolman references (Köhler 1917; Tolman 1926a).[23] Two other Gestalt psychologist had an impact on Tolman. The first was Kurt Lewin (Tolman 1952; but see also Tolman 1932a, 1932b) and in particular his conception of objects having a certain "valence" (Lewin 1931). It is thus surely no coincidence that Tolman referred to the mental representations of means-end relations as "sign-gestalts" (hypotheses about the outcomes of hypothetical manipulations) (e.g., Tolman 1933). The second Gestalt psychologist explicitly singled out by Tolman was Egon Brunswik (Tolman 1952). As mentioned above, the two met during Toman's sabbatical in Vienna in 1933–34, and they subsequently published a paper together (Tolman and Brunswik 1935).[24] At the time of their first meeting, Brunswik was in the process of finishing a book on object permanency (Brunswik 1934) in which he introduced the notion that perception consists in an "intentionalistic attainment" of environmental entities. As Tolman states in various places, his and Brunswik's positions were very similar even though Brunswik focused more on perception and Tolman focused more on behavior. There was also a major difference between their approaches, however. Brunswik was opposed to positing internal processes, thinking that it detracted too much from the influence of the environment and resulted in nonrepresentative experimental designs. (I return to this issue in chap. 9.) Consequently, he was a proponent of what he called the "psychology of the empty organism" (Hammond 1966), a notion that became especially prominent in James Gibson's ecological theory of perception (e.g., Gibson 1979). Nonetheless, Brunswik underscored two elements

23. This is probably no coincidence as Yerkes was actually engaged in correspondence with Köhler, who was the director of the anthropoid research station of the Prussian Academy of Science on Tenerife. For an excellent account of Köhler's activities on Tenerife as well as the correspondence between Köhler and Yerkes, see Ash (1995).

24. In the academic year of 1935–36, Brunswik had a fellowship to come to Berkeley. The following year, he returned to Berkeley as an assistant professor, eventually becoming a full professor in 1947.

that were already present in Tolman's thinking, that is, the notion that cognition is to be described as functionally embedded in the environment and the notion that all behavior is based on fallible hypotheses about the nature of the environment, the latter having been suggested by Brunswik's probabilism (the precursor of his later "probabilistic functionalism" [see Brunswik 1943; and Tolman 1956]).

The notion that sign-gestalts are hypotheses had already begun to appear in the context of Tolman's experimental work shortly before his encounter with Brunswik. In particular, it was tied to work that Tolman conducted with his student Isadore Krechevsky (Krechevsky 1932a, 1932b; Tolman and Krechevsky 1933).[25] This (and more) work was later summarized in Tolman's well-known article on cognitive maps in rats (Tolman 1948). In 1931, Krechevsky (who had previously been at New York University) decided to continue his graduate studies in Berkeley. In his autobiographical article, Krech describes his first meeting with Tolman (who had asked him about his master's research): "I told him, but not at length, for I had no sooner given the bare outline of my findings than he caught fire and jumped out of his seat and practically shouted out 'Hypotheses in rats!'" (Krech 1974, 229). Krech also mentions that Tolman later insisted that the term "hypothesis" had been Krech's own, which Krech denies (Krech 1974, 229). For our purposes, the important point here is that hypotheses are what Tolman in other places called "cognitive postulations." Finally, let me briefly mention that Tolman was also well aware of some then current epistemological theories, such as C. I. Lewis's conceptualistic pragmatism and Stephen Pepper's contextualism, specifically Pepper's 1942 *World Hypotheses*. Both philosophers had also been students of Perry's at Harvard, both remained friends with Tolman, and both were, for a while, professors at Berkeley (Lewis was there from 1918 to 1920; Pepper was there for much of Tolman's tenure at Berkeley) (Smith 1986).

Tolman's idea that environmental objects are cognitively represented in terms of hypotheses about how objects can be manipulated in order to satisfy demands finds an intriguing parallel in Tolman's reasoning about scientific cognition and, thus, in his operationism. If the aim is to satisfy the demand for knowledge, the environmental objects have to be treated as "means-objects," which, if manipulated in the right way, will reveal something about the subject matter under investigation. The environmental objects are, in this case, the factors that go into the

---

25. Today, Krechevsky is better known by the name "David Krech," which he adopted in 1943 owing to the intense anti-Semitism he encountered throughout his professional life. The decision to change his name was prompted by the fact that he got married and wanted to spare his son the name Krechevsky (Krech 1974).

experiments: rats, food, mazes. Obviously, those objects are not going to be manipulated randomly. Rather, the manipulation in question is going to be guided by hypotheses about the subject matter. Given his sense of naturalism—shared by many post-Darwinian psychologists at that time—it is not surprising that Tolman took the difference between humans and other animals to be only one of degree, not one of kind (Hatfield 2002a). His (half-joking) references to the similarities between human beings and rats in mazes are scattered throughout his writings. For example: "The world for philosophers, as for rats, is, in the last analysis, nothing but a maze of discrimination-manipulation possibilities, extended or narrow, complex or simple, universal or particular" (Tolman 1926a, 61). Hence, I suspect that he would have endorsed my analysis of his operationism as a normative extension of a descriptive account of problem-solving in rats (for a similar point, see Smith [1986]).

It is clear from what I have outlined above that Tolman was exceptionally open-minded and willing to take on and acknowledge the ideas he encountered in his psychologist and philosopher friends, so it is hardly surprising that he had adopted the language of operationism by the mid- to late 1930s. (As he confirmed [see Tolman 1937], he had come across Stevens's and Bridgman's formulations of operationism by then.) It is also likely that, during his visit to Vienna (where he met Brunswik), he read Carnap's "Psychology in a Physical Language" (1932) and may have attended meetings of the Vienna Circle (Smith 1986), though Stadler (2001) does not have any record of this. Tolman was approached to present a paper at the 1936 Unity of Science congress but was unable to attend.[26] However, in light of my historical analysis of his position, it seems clear that he was only adopting a label for a position that was already fairly consolidated and that differed from the kind of position that is usually associated with operationism (for an example, see Leahey [1980]).

1.4. Clark Hull and the Role of Operationism in Theory Construction

Clark Leonard Hull, the son of farmers in rural Michigan, was the first person in his family to pursue a higher education. He enrolled at Alma College, where he studied mathematics and physics in preparation for a career in engineering. While there, he first contracted typhoid fever and then polio. During the lengthy period of recovery from the latter, he decided to switch fields and study psychology, an area he perceived to

---

26. The paper he was planning to present, "An Operational Analysis of 'Demand,'" is listed in the program (see Stadler 2001, 376).

be rapidly growing and full of opportunities for a young scientist with an interest in theoretical work. After recovering, Hull got a bachelor's degree from the University of Michigan. He then pursued a graduate degree in psychology at the University of Wisconsin–Madison, where he earned his PhD in 1918 with an experimental study on memory (see Hull 1920). He stayed in Wisconsin until 1929, gradually working his way up from assistant professor to full professor. In 1929, he became a research professor at Yale University. (For an autobiographical account, see Hull [1952b].)

During his time at the University of Wisconsin, where he did work on aptitude testing and hypnosis (Hull 1928, 1933), Hull became a dedicated behaviorist (see Smith 1986). Unlike the two psychologists discussed above, Hull never actually published any paper that specifically dealt with operationism. He referred to operationism once in his *Principles of Behavior: An Introduction to Behavior Theory* (Hull 1943a, 30), and he wrote an article in the same year dealing with the status of intervening variables. He mentioned the need for operational definitions once in an article (Hull 1937, 5), but he distanced himself from that notion at a later point (Hull et al. 1940). However, I argue that his methodological concerns were very similar to the ones that prompted Tolman and Stevens to argue for operationism, that is, a concern as to the criteria that were to govern the introduction of theoretical terms referring, at least partly, to entities and mechanisms that are assumed to take place inside the organism. Like Tolman, he aimed at the construction of a theory of behavior, which involved intervening variable terms, and he thought of intervening variable terms as referring to causally efficacious organismal entities and processes, though, as we will see, Hull's theory was much more formal and explicitly mechanistic. Unlike both Tolman and Stevens, however, Hull was very cautious to avoid mentalistic vocabulary such as "experience," "beliefs," and "demands" as this would have been incongruent with his behaviorist commitments. His theory was concerned with principles that govern the ways in which simple learning mechanisms result in the emergence of complex adaptive behavior.

Probably the most distinctive feature of Hull's theorizing was his ambition to develop an axiomatic theory of behavior. For example, in *Principles of Behavior*, Hull presented a number of fundamental axioms from which hypotheses about mammalian behavior were to be deduced. A second volume, *A Behavior System: An Introduction to Behavior Theory concerning the Individual Organism* (1952a), appeared a decade later.[27] Between his arrival at Yale and the appearance of his first book, Hull published a series of theoretical papers in which he laid down what he

---

27. A third and last volume (intended to be an application of Hull's theory to social phenomena) was never completed.

termed "miniature systems." These systems laid the foundations of the ideas that were to appear in his books (the papers are reprinted in Amsel and Rashotte 1984). He also coauthored *Mathematico-Deductive Theory of Rote Learning* (Hull et al. 1940), in which he and his coauthors tried to cast their theory in a rigorously logical form.

1.4.1. (In What Sense) Was Hull an Operationist?

Within twentieth-century philosophy of science, the notion of *axiom systems* has usually been discussed in terms of the question of how to provide an empirical interpretation of an already existing formal system (e.g., Nagel 1961). In contrast, Hull wanted to *develop* an axiomatic theory on the basis of empirical data. Thus, we find in his work a curious mix of a commitment to a very formal notion of theory and the idea that such a theory ought to be developed from the bottom up, its central terms linked to experimental operations (see also Feest 2012a). I argue that, like Stevens and Tolman, Hull put his finger on the question of how to ground his efforts at concept formation and theory construction empirically. And, again like Stevens and Tolman, he did so by reflecting methodologically about the relationship between the definitional elements and the factual elements of experimental work.[28]

Before looking more closely at Hull's methodological views, let me briefly summarize his general theoretical and epistemological outlook. As revealed by excerpts from his multivolume "idea book" (published posthumously in 1962), Hull wanted to model *Principles of Behavior* (1943) on David Hume's 1739 *Treatise on Human Nature* (see Hull 1962, 835). His ambition of updating classical empiricism in light of recent advances in behaviorist learning theories dated back to a 1925 seminar he taught on behaviorism. We can, thus, appreciate that the philosophical background of this approach (other authors whose work he stressed were Auguste Comte, Frederick Woodbridge, George Santayana, Edwin Holt, Ralph Barton Perry, John Dewey, and William James) (see Smith 1986). Hull saw the conditioned reflex as the behaviorist analog of Hume's simple impressions and laws of association, and he envisioned his own project as a form of associationism, with the conditioned reflex (which he subsequently called "habit") as the basic unit of knowledge (see Hull 1930). He also posited "symbolic habits," which he took to be more abstract than others and to "logically entail" other habits (Hull

---

28. In the following chapter, I elaborate on this question and how its treatment by psychologists was received by philosophers of science at the time.

1931a).²⁹ By the mid-1930s, Hull conceived of habits as forming a hierarchy that enabled one habit to override another (Hull 1934, 1935b). Both the notion of *abstract habits* and the related idea of *habit hierarchies* were expressed in 1926 in his idea book (see Hull 1962, 821). However, as early as 1916, he also laid out in his idea book his plan of writing a book about "*the psychology of abstraction and concept formation*, and perhaps, ultimately, of reasoning." In the same entry, he writes: "I must also someday work out my idea of *hierarchies*" (Hull 1962, 814, 815). Thus, his interest in reasoning processes preceded his conversion to behaviorism by almost ten years, suggesting that in his later works he attempted to cast his early interest in cognition in a behavioristic mold.

By the 1930s, the driving force of Hull's theorizing was giving a behaviorist account of what cognitive psychology might refer to as "reasoning processes." In aiming for such an account, Hull was keenly aware of the limits of traditional behaviorism in that it did not seem to be able to explain the apparently purposeful nature of behavior, which often seems to require intermediary steps not all of which are immediately displayed in behavior. His basic idea, then, was that behaviorist laws of learning can be used to provide an account of an internal chain of reasoning by providing a mechanistic account of how internal habits are formed. Put differently, Hull posited that while the concept of *internal habits* could not be reductively defined in terms of overt behavioral regularities, the formation of such habits could be explained behavioristically. This was first expressed in Hull (1930b) and further elaborated in subsequent papers (Hull 1931a, 1931b, 1932). Hull posited the existence of several kinds of internal stimuli that could jointly explain why organisms are able to behave in an apparently purposeful manner, that is, as if they were able to anticipate future events (Hull 1930a). Those internal stimuli included (*a*) the proprioception of responses to external stimuli, (*b*) "pure stimulus acts" (s), and (*c*) persistent stimulus acts (s_p).³⁰ The first stimuli were essential as they allowed "the world" to stamp "the pattern of its actions upon a physical object" (Hull 1930b, 514). The second stimuli were the nervous trace of an external stimulus S, and the third were proprioceptively perceived bodily needs, for example, in the form of

---

29. As we will see in a moment, these are supposed to capture those elements of human functioning that we might today refer to as "cognitive."

30. Thus, we see here that already in the 1930s Hull tried to capture the distinction between an internal representation (s) of an external stimulus (S), on the one hand, and an internal drive, on the other. In his *Principles of Behavior*, Hull uses the symbol D (drive) and includes a discussion of the problem of isolating this intervening variable from others (1943a, 66).

stomach contractions. The basic idea was that the process of learning to respond to a particular external stimulus is accompanied by physiological changes that subsequently are able to function as internal stimuli and hence evoke the next response.

What is striking here is that, like Tolman, Hull found himself positing internal variables as explanatory of behavior, though, unlike Tolman, he took pains to avoid any suggestion that these variables had a mentalistic flavor to them. His commitment to behaviorism played a dual role here. First, as we just saw, he drew on behaviorism as a theory of learning in order to account for the internal structures (intervening variables) that—qua steps in the chain of reasoning—enable organisms to adjust to the external environment. Second, he saw behaviorism as a methodological doctrine, requiring him to formulate behavioral criteria for positing and experimentally investigating the intervening variables in question. With regard to this issue, references to operational definitions are pertinent insofar as those definitions both guide and constrain the ways in which intervening variables can be introduced in the process of theory construction.

Below, I say more about Hull's notion of *theory*. For now, it will suffice to keep in mind that, for Hull, deductive reasoning was the hallmark of scientific method. His ambition was to formulate a formal theory that would enable him to derive explanandum phenomena (such as the phenomenon of rote learning) deductively from a small number of theoretical axioms (Hull et al. 1940). His commitment to construct such an axiomatic theory from the bottom up required that he specify his theoretical variables in experimental terms. This comes out in the miniature systems articles of the 1930s (see also Amsel and Rashotte 1984) that led up to the books of the 1940s. In these papers, he formulated hypotheses and tried to derive empirical predictions in support of these hypotheses from a number of definitions and postulates (or axioms).

In pursuing this project, Hull took himself to be following the method Newton had used in the *Principia* (1687), which, Hull stated, had started with eight definitions (of concepts such as *mass*, *inertia*, and *centripetal force*) and three postulates (the laws of motion). But what were the corresponding concepts and postulates within Hull's system? For the most part, the (eleven) definitions Hull offers in "The Conflicting Psychologies of Learning" (1935a) provide an inventory of his basic experimental material. For example, he defines "rote series" and "massed"/"distributed" practice (two types of training procedures). However, he also gives definitions of terms that appear to pick out behavioral dispositions, such as "excitatory tendency," which he defined as follows: "An 'excitatory tendency,' as emanating from a stimulus, is a tendency for a reaction to take place more … vigorously other things being equal, soon after the organism has received said stimulus than at other times" (Hull 1935a, 501). By contrast, his postulates are intended as (presumptive) factual statements about the referents of some of the terms defined in the list of definitions. Examples include the postulate "that the remote excitatory tendencies of Ebbinghaus exist; that remote excitatory tendencies of Ebbinghaus possess the same behavior characteristics as do the trace conditioned reflexes of Pavlov, [etc.]" (Hull 1935a, 498; see also Hull et al. 1940). It seems that, on the one hand, the empirical/factual statements that constitute the postulates are possible only on the basis of a prior conceptual understanding of the subject matter at hand. But, on the other hand, this prior conceptual understanding (Hull's definitions) already makes empirical assumptions, such as that excitatory tendencies exist and can be measured in terms of reactions to stimuli. This means that here—as with Stevens and Tolman—the definitions enable empirical research while also presupposing empirical assumptions about the world.

By the 1940s, Hull himself was grappling with the question of how to distinguish definitional from factual questions, noting that many of his theoretical terms refer to (putative) unobservable entities such as stimulus trace, excitatory potential, etc. and that attempts to define the corresponding terms inevitably blur the lines between definitional and factual statements (Hull et al. 1940, 3): "It thus appears that even definitions must ultimately be validated by observations" (4). Hull and his coauthors then suggest cautiously introducing theoretical terms along with empirical postulates and mathematical laws and deriving testable theorems from these elements, explicitly acknowledging the underdetermination of the resulting inferences. They nonetheless express some confidence that theory construction can and should proceed in this way.

But this still leaves the question of how the terms referring to unobservable entities ought to be introduced to begin with. I argue that in this respect Hull grapples with a question not so dissimilar from one Stevens and Tolman faced as well, that is, how to introduce novel theoretical terms into the language of psychology in a way that is both fruitful and empirically grounded. And it is clear that by the early 1940s Hull explicitly aimed at providing quantitative formulations of the conditions under which such grounding could be accomplished. I suggest that, if we understand the term "operational definition" in the liberal sense that I have attributed to Stevens and Tolman (i.e., as specifying paradigmatic conditions of application for concepts that are based on previous experimental research, which in turn relies on conceptual presuppositions about the subject matter), Hull's position is not unlike theirs. Moreover, like them, he believed that intervening variable terms were not just convenient fictions or mere shorthands for experimental laws. Thus, while he was wary of the term "operational definition," Hull endorsed what he took to be the substance of Bridgman's operationism: "The moral of Bridgman's treatise is that the intervening variable (X) is never directly observed but is an inference based on the observation of something else, and that the inference is critically dependent upon the experimental manipulations (operations) which lead to the observations" (Hull 1943a, 30). Like Stevens's and Tolman's, then, Hull's experimental operations were grounded in prior assumptions about the variable/phenomenon in question, assumptions that he took to be revisable in light of further research. Accordingly, he stated: "If it is discovered that some of the individual phenomena included within a definition behave according to a different set of principles or laws, the concept must either be abandoned or redefined" (Hull 1943a, 4).

To sum up, I argue that Hull's brand of neobehaviorism called for a bottom-up and (in some sense) hypothesis-guided approach to theory construction. It was hypothesis guided in the sense that it posited particular kinds of intervening variables as following from prior theoretical assumptions (Pavlov's behaviorism in conjunction with ideas about internal stimulus-response couplings). Yet it was also intended to be bottom-up in the sense that there were supposed to be empirical constraints on the kinds of intervening variables posited. Hull expressed this by arguing that intervening variables had to be "securely anchored." By that he meant the formulation of mathematical functions of the relationships between (a) stimuli and intervening variables and (b) intervening variables and responses. He optimistically asserted:

> Once (1) the dynamic relationship existing between the amount of the hypothetical entity (X) and some antecedent determining condition (A) which can be directly observed, and (2) the dynamic relationship of the hypothetical entity to some third consequent phenomenon or event (B) which also can be directly observed, become fairly well known, the scientific hazard largely disappears. . . . When a hypothetical dynamic entity, or even a chain of such entities each functionally related to the one logically preceding and following it, is thus securely anchored on both sides to observable and measurable conditions or events (A and B), the main theoretical danger vanishes. (Hull 1943a, 22)

It must be noted that, in taking such a theoretical approach at all, Hull faced criticism by some of his contemporaries. Upon introducing some intervening variable terms in *Principles of Behavior* (1943a), he repeatedly acknowledged that this went against certain "antitheoretical" or "positivist" convictions that were held by some of his colleagues.[31] He justified his approach by pointing to the heuristic value of theoretical assumptions. For example, in a paper presented at a meeting of the Southern Society for the Philosophy of Psychology (at Duke University in 1939), one speaker had argued (against Hull) that the job of science is description and that the construction of theoretical systems was too speculative.[32] In a letter to the author, Hull replied: "Your chief objection to my systematic theory was that it tended to blind people to certain possibilities which were not included in the theoretical structure. This is, of course, quite true. On the other hand, you seem to forget that a genuine theoretical structure may raise exceedingly significant questions which a person with no theoretical background will never see" (quoted in Amsel and Rashotte 1984, 60). Hull was also aware of the epistemological problems with his approach; that is, he knew that, simply by positing the existence of certain intervening variables on the basis of empirical data, he had not thereby provided independent evidence for their existence. As he put it: "It is relatively easy to find a single empirical equation expressing vigor of reaction as a function of the number of hour's food privation or the strength of an electric shock, but it is exceedingly difficult to break such an equation down into the two really meaningful component equations involving hunger drive (D) or motivation as an intervening variable" (Hull 1943a, 66).

Like Tolman, Hull ultimately had no principled answer to this problem, indicating that he was quite aware of the problem of the underdetermination of theoretical posits by experimental data: "In case postulational error is suspected, the suspicion falls more or less over the entire group of postulates involved in the derivation of the 'sour' theorem." However, he was optimistic that it would be possible to establish the validity of individual postulates "[in] a gradual manner—one of successive approximation" (Hull et al. 1940, 6). Such successive approximation meant that the "secure anchoring" of theoretical concepts and the development of theory should go hand in hand.[33] I believe this to be an important insight, one to which I return when discussing the method of converging operations in chapter 3 and chapter 7.

31. For example: "It may confidently be predicted that many writers with a positivist or antitheoretical inclination will reject such a procedure [of positing intervening variables] as both futile and unsound" (Hull 1943a, 66; see also Hull 1943b).
32. The author's name was Gelhard. I have not been able to find out more about him.
33. Smith (1986) cites a letter Hull wrote to him: "The only thing which I would be inclined to use is that operationalism should be combined with parallel attempts at theoretical systematization" (220).

1.4.2. Some Intellectual and Institutional Background

Hull's focus on theory construction has also been recognized by other commentators. For example, Koch (1941a, 1941b) contrasts two notions of theory construction, one that he calls the "interpretive" and one that he calls the "telescopic," the former starting out with an uninterpreted axiom system and then attempting to provide an empirical interpretation by way of coordinating definitions, and the latter elaborating the formal and the empirical parts of the theory simultaneously. He correctly asserts that the latter approach is the one espoused by Hull (and explicitly argued for in Hull [1938]). A related, though slightly misleading, distinction was made by Bergmann and Spence (1941), who distinguished between a "hypothetico-deductive" and a "postulational" approach to theorizing and put Hull in the latter camp because he started with postulates, which were informed guesses about intervening variables. This characterization is misleading because it seems to suggest that hypothetico-deductive and postulational are mutually exclusive. However, as we just saw, Hull not only aimed to introduce his postulations from the bottom up (attempting to anchor them empirically) but also did so with the explicit aim of deriving theorems and predictions from them in a hypothetico-deductive manner. As MacCorquodale and Meehl (1948) have rightly observed, Hull's "intervening variable" terms were indeed hypothetical constructs, though they did not come in a package with an already completed (and uninterpreted) axiomatic theory.

But what was the status of Hull's hypothetical constructs? Did he take them to be mere logical fictions, designed to make sense of observations and reducible to empirical laws, or did he think of them as referring to causally efficacious internal processes? Hull himself at times suggests the former, for example, when he says: "The use of logical constructs thus probably in all cases comes down to a matter of convenience in thinking, i.e., an economy in the manipulation of symbols" (Hull 1943a, 111). However, I suggest the following (contrasting) interpretation. The term "habit" was, for Hull, a logical fiction, not because he was an agnostic or an antirealist about habits, but because he did not think they existed apart from their neural basis.[34] Habits, for him, were hypothetical because, even though he had an empirically adequate psychological model of how they are formed, his ontology was ultimately reductionist, and, thus, he took his model to be only a placeholder for future neurophysiological explanations.

---

34. In this respect, Hull's position is similar to Boring's identity-theoretical commitments regarding dimensions of consciousness.

This interpretation is backed up by Smith (1986), who reports that early drafts of Hull's 1943 *Principles of Behavior* contained many more biological references, some of which were edited out by Spence because they did not fit in with the positivism of the time. When Hull talked of the hypothetical character of psychological constructs, he meant, therefore, not that they were incapable of picking out real causal processes but merely that psychology could and should proceed in its theoretical work even though the physical/biological basis of behavior had not yet been worked out. Like Tolman, Hull distinguished between molar and molecular behavior theory, presenting the situation for psychologists as a choice between two possibilities. One was to wait "until the physicochemical problems of neurophysiology have been adequately solved before beginning the elaboration of behavior theory," and the other was to proceed "in a provisional manner with certain reasonably stable principles of the coarse, macroscopic or molar action of the nervous system whereby movements are evoked by stimuli" (Hull 1943a, 20). Suggesting that the former option would be like asking Newton and Galileo to wait "until the micro-mechanics of the atomic and subatomic world had been satisfactorily elaborated" (Hull 1943a, 19–20), he firmly opted for the second option, that is, to formulate a molar theory of behavior. He accorded a dual status to such a theory's theoretical (or intervening variable) terms. On the one hand. he argued that they had the useful function of facilitating our thinking. On the other hand, he suggested that at least in principle they might identify unobservable entities or processes that would account for observations: "The habit presumably exists as an invisible condition of the nervous system quite as much when it is not mediating action as when habitual action is occurring" (Hull 1943a, 21). Thus, even though *habit* was, for Hull, a hypothetical construct in the sense that it referred to an unobservable process, he assumed habits to be real by virtue of being identical with particular conditions of the nervous system.

Let me take a step back here and explore the origins and development of Hull's reductionist sentiment as well as his interest in deductivism. Hull's early training in engineering and his deep appreciation of Darwinism made it natural for him to think of biological organisms as complex biological machines that can be understood in functional terms. In fact, constructing a theory of behavior was much like constructing a machine as well as being—at least early on—closely tied to the actual physical construction of a model that simulated the behavior Hull wanted to explain (e.g., Hull and Baernstein 1929; Krueger and Hull 1931; and Rashevsky 1931). In Krueger and Hull (1931), for example, an electrochemical model of classical conditioning is presented. In subsequent years, Hull shifted his attention away from such physical simulation models, but he still thought of the deductive structure of an axiomatic theory as machine-like and therefore ensuring rigor and objectivity. For Hull, biological organisms and machines have in common the fact that both represent knowledge about the world in a systematic way, with the result that novelty can be generated. In the case of organisms, what is generated is adaptive behavior. Such behavior is novel because it involves adaptation to novel situations. In the case of a theory of organisms, what is generated are predictions about behavior. Such predictions are novel because they are applications of the theory to novel situations. Hull reasoned that, if the theory in question is a theory about the behavior of biological organisms, then the principles that govern the prediction of the phenomena (i.e., of behavior) ought to be similar to the principles that govern the actual phenomenon (i.e., behavior).

According to Smith (1986), Hull became convinced of the importance of deductive reasoning sometime around 1930 when he spent a few months at Harvard, where he apparently became familiar with Newton's *Principia* (which henceforth served as a model of scientific method) and with Bertrand Russell and Alfred Whitehead's *Principia Mathematica* (1910–13/1962), which he took to be a model of how to formalize an axiom system. This was also right around the time Hull arrived at Yale, which in 1929 had founded the Institute of Human Relations. This institute, which was funded by the Rockefeller Foundation, had an explicitly interdisciplinary focus. It was supposed to integrate the newly emerging social sciences, attend to social problems, and be cooperatively managed by scientists from a variety of departments (see Morawski 1986). However, the early years of the institute did not yield any cooperative projects or attempts at an integrated theory. After the five-year review, Mark May (who had already—in 1932—drafted a proposal for systematizing research projects) became the director of the institute and tried to introduce policies that would force individual members to work together more closely. The notion of an integrated theory of all social sciences struck a chord with Hull, and he became deeply involved in restructuring the institute (see Morawski 1986), developing blueprints for collective studies of various behavioral phenomena (food seeking, hunger, frustration, conflict). The institute members ultimately decided on the general subject of motivation. This general focus may be related to the fact that, by 1943, Hull regarded drive as one of his central intervening variables. As a research professor he was not required to teach any classes. However, in the context of his institute activities, he held weekly seminars. These seminars were designed to facilitate the exchange of scientific ideas among institute members, aid the discussion of fundamental methodological issues, and unify terminology as Hull was convinced that the integration of the social sciences required theoretical and methodological convergence. In this context, he had everybody read Newton's *Principia* as an example of a systematic deductive theory. The meetings were attended by a diverse group of people, including Warren McCulloch, who was going to be instrumental in developing artificial neural networks.[35] As Morawski (1986) points out, Hull's impact on the general orientation of the institute is clearly evidenced by the fact that, at the end of the first decade of the institute's existence, the initial objective of creating an integrated social science had been replaced by that of searching for universal mechanical laws of individual behavior.

It was during these years that Hull started to build his miniature systems, his first attempts to develop a deductive theory (e.g., Hull 1935b). However, he became increasingly frustrated with his own approach (which he referred to as a "geometric method"), and—feeling that axiomatization required a more rigorous grounding in symbolic logic—was excited to have the chance to meet the Oxford biologist J. H. Woodger at the Unity of Science Congress in Paris. Woodger had just completed his axiomatization of biological theory (Woodger 1937). He applied for a Rockefeller fellowship, which enabled him to come to Yale for nine weeks in early 1938 (see Smith 1986), where he was to collaborate with the logician Frederic Fitch (see Smith 1986, 197ff.). However, this project turned out to be more difficult than expected and was not completed by the time Woodger left. The publication that ultimately came out of this (Hull et al. 1940) does not list Woodger as a coauthor.

The 1930s was also the time of the insurgence of European Gestalt psychologists and philosophers of science in the United States. Both these groups provided important impulses. Hull had already heard a lecture by Kurt Koffka at Wisconsin in 1926 that convinced him that Watsonian behaviorism needed to be improved, and Amsel and Rashotte (1984) reprint several excerpts from letters and notebooks showing that he was engaged in lively debates with Gestalt psychologists over the notion of a deductive theory. Herbert Feigl arrived in Iowa in 1930, Kurt Lewin and Gustav Bergmann in the mid-1930s. Kenneth Spence, who had been a student of Hull's since 1932, got a position at Iowa in 1938. Spence remained an important intellectual collaborator, and Hull relied on him to keep him updated with respect to developments in philosophy

---

35. This is interesting because it suggests that, while Hull himself was no longer actively engaged in the project of simulating biological processes, there was still a context in which these kinds of ideas were discussed (see also Abraham 2016).

of science. I suggest that this fact is responsible for some of the inconsistencies in Hull's work in that some philosophical dogmas at the time clearly went against his own intuitions.

1.5. Conclusion

In this chapter, I have provided analyses of the operationist positions of S. S. Stevens, Edward Tolman, and Clark Hull. The aim of these analyses has been twofold. On the one hand, I have attempted to show that, far from adhering to simpleminded semantic or epistemological theses, each of these authors held a set of more or less sophisticated theoretical and methodological—even philosophical—beliefs that I showed to have been interrelated in each of the cases at hand. While in all three cases there was some contact with proponents of positivist and empiricist positions, both their research programs and their methodological approaches have to be placed in the much wider and richer contexts of their academic interests, research projects, and philosophical as well as methodological commitments. On the other hand, I have laid out some key issues in psychological concept formation that I explore throughout the remainder of this book.

Before laying out these issues, let me emphasize that, by highlighting the epistemological aspect of operationism, I do not mean to downplay some of the other factors that may have contributed to the emergence of operationism (in part also contributing to its bad reputation). Among these is the highly contested status of scientific psychology in the early decades of the twentieth century (O'Donnell 1985), a result of its rapid expansion (both within the academy and in applied fields), coupled with a lack of theoretical and methodological unity (Smith 1997, 659). In the years around 1930, a number of books attempting to systematize schools of thought (e.g., Heidbreder 1933; and Woodworth 1931) or the history of the discipline (e.g., Boring 1929) appeared, each with the aim of promoting particular views about the foundations and aims of psychology. In addition to general questions regarding the subject matter of psychology (and, relatedly, a struggle for the limited resources available for psychological research), there was probably a specific context at Harvard, where the fields of philosophy and psychology were still housed together in the Philosophy Department. All three authors covered in this chapter positioned themselves within these debates by pushing for a rigorous and physicalistic approach to psychology. While I have emphasized some academic contexts for this, we also need to take the social climate of progressivism in the 1910s and 1920s into account since it had prepared the ground for the demand that science produce knowledge that can be applied to practical problems (Rogers 1989, 151n).

By the 1930s, this optimism had become as subdued as the economic situation, making it even more important to strengthen the credibility of science by emphasizing rigorous scientific methods. Nonetheless, it bears stressing that, for example, Stevens was explicitly concerned with the practical applicability of his research, as is evidenced in his very active engagement in the American Acoustical Society in the 1930s (Newman, Stevens, and Davis 1937; Stevens 1935c, 1937; Stevens and Davis 1936; Stevens and Jones 1939; Stevens, Volkmann, and Newman 1937).[36] In relation to the real-life contexts of operationism, let me also highlight that the infamous slogan (often attributed to operationism) that "intelligence is what is measured by intelligence tests" was in fact uttered some ten years before anybody used the terms "operationalism" or "operational definition." In 1923, Edwin Boring had published an article in the *New Republic* in which he had argued that, in the practical context of diagnostic testing, it has to be assumed that intelligence is what is measured by intelligence tests.[37] I return to some issues surrounding mental testing in the following chapter. For now, let me highlight the distinction between standardized tests (e.g., IQ tests), on the one hand, and experimental designs in a research context, on the other. In this book, my focus is mainly on the latter, though, as we will see, there was an overlap in the methodological debates about psychometrics and experimentation.

In conclusion, let me return to the main points I want to take from the analyses provided in this chapter. I have offered a contextualized account of the methodological reflections of three early twentieth-century American psychologists, Stanley Smith Stevens, Edward Chace Tolman, and Clark Hull, arguing that we can make out a common thread. All three of these psychologists were deeply concerned with formulating criteria and guidelines for scientific discovery and concept formation for their specific fields of research, and all three shared a basic agreement

---

36. The American Acoustical Society, which had been founded in 1929, had the aim of bringing together research from areas as diverse as psychophysics, psychophysiology, and engineering. This engineering orientation of the society casts yet more light on Stevens's research. It appears that within the field of acoustic engineering there was a demand for a scientific, quantifiable understanding of the experience of tones. Stevens's operational treatment of subjective experience provided such an understanding. In addition, it may be speculated that his active participation in this society, in turn, influenced the shape of his research.

37. This diagnostic practice, in turn, was prompted by a demand that had been created initially by the wave of immigrants between 1905 and 1915 and then by World War I.

with a physicalist methodology, which they each took to be a basic point of behaviorism. This prompted them to think about the ways in which research on experience, demands, cognitions, and molar behavior could be reconciled with the behaviorist doubts about introspection and (mentalist) explanatory terms. Put simply, each of them pursued research objects that are not immediately accessible to the naked eye. In specifying what they took their research objects to be, they faced definitional issues on two levels. First, they needed preliminary specifications of their subject matter, which would enable the design of experiments (see, e.g., Stevens's specification of experience as discriminatory behavior or Tolman's specification of hunger as time since last feeding). I have referred to these specifications as "operational definitions^{wide}." Second, they each attempted to characterize particular research objects in terms of what Tolman called "defining experiments," that is, experiments that produced a quantitative profile for the object in question, thereby legitimating the measurement procedure for the object. The measurement procedure then provided paradigmatic conditions of application for the corresponding concept (e.g., the concepts of *tonal density*, *demand*, and *habit*), that is "operational definitions^{narrow}." Thus, operational definitions^{narrow} simultaneously (*a*) represented empirical results (the outcomes of defining experiments, which, in turn, relied on operational definitions^{wide}) and (*b*) provided the conditions of the possibility of conducting further experiments on the object. It is important to recognize that all three researchers viewed their operational definitions as simultaneously enabling empirical research and reflecting the state of empirical knowledge and, thus, being revisable in principle.

This chapter has highlighted the way in which psychological research objects are often rooted in folk-psychological conceptions, suggesting that researchers explicate and use these conceptions in particular ways for the purposes of experimental research. This raises the question of how the relevant psychological research is, ultimately, related to the folk-psychological objects they are purportedly about. I return to this issue in chapter 6. First, however, I provide a more in-depth account of the research process as such and the methodological and epistemological issues that arise from it.

CHAPTER TWO

Operationism
The Second Generation

2.1. Introduction

In the previous chapter, we delved into the work of three early proponents of operationism in psychology. The point of this historical analysis was to demonstrate that, for an adequate understanding, their writings on the nature and function of operational definitions (and, in some cases, of operationism more generally) have to be situated in the contexts of their research agendas. Early operationists were motivated by a number of concerns, some of which they shared, and some of which were specifically tied to their individual investigative projects. All three were wary of the potential pitfalls of positing phenomena, entities, or processes that went beyond the conceptual resources of a purely observational language. At the same time, their research questions prompted them to use terminology that expressed such concepts. This motivated them to operationally "define" these terms in a way that closely tied them to experimental methods and measurements used to individuate the corresponding objects empirically.

As I have shown, these three authors did not aim to define the meanings of their terms exhaustively. They did, however, engage in conceptual work concerning their research objects. This conceptual work was geared toward the question of how to study these research objects by experimental means, and it was (roughly) situated on two distinct levels.[1] On the first level, the operational definitions advanced by these authors expressed general conceptual presuppositions about the subject matter. These presuppositions, which were typically derived from folk psychology, concerned types of stimuli and responses thought to individuate

---

1. This comes out particularly clearly in Stevens's and Tolman's work.

the research object in a general way. I have referred to these as "operational definitions^{wide}." Stevens's definition of "experience" as discriminatory behavior is a good example of this, as is Tolman's proposal that "hunger" be defined as search behavior as a function of food deprivation. These specifications, I have argued, were intended not to negate the rich meanings that terms like "experience" or "hunger" have in ordinary language but rather to explicate specific aspects of these meanings that were deemed important for the purpose of studying or controlling the referents of the terms.[2] On the second level, I have argued that the psychologists in question explicated specific concepts in terms of specific functional relationships between types of experimental manipulations and their behavioral results, which had been empirically established. I referred to these latter definitions as "operational definitions^{narrow}."

By putting forth an analysis of operationism that departs from standard accounts, I have begun to call into question some of the critiques that attribute to operationism the view that every new experimental operation necessarily introduces a novel concept and that operationism is committed to a strong antirealism regarding the referents of its constructs. However, even if I am right in my contention that operationism in psychology not only arose somewhat independently of the philosophical positions with which it is sometimes equated but was also motivated by a different set of problems, it does not follow that there are no philosophical problems with or raised by operationism as construed thus far. For example, while I have argued that operationists thought of operational definitions as (for the purposes of experimentation) stating sufficient conditions of application for their terms, I have not addressed the normative question of how it can be ensured that they do so successfully. Likewise, while I have argued that operationists thought of their operational definitions as partially capturing the meanings of the corresponding concepts, I have not addressed the question of when (and how) such an assumption is justified.

This chapter addresses systematic questions of this nature by laying out how they were addressed in the two decades immediately following the publication of Stevens's and Tolman's work on operationism. While the previous chapter emphasized that operationism in psychology evolved (somewhat) independently of philosophical concerns, this chapter shows that by the early 1940s there were debates involving both psychologists and philosophers of science that left an impact on both fields. I highlight the ways in which methodological positions articulated

---

2. In this respect, their strategy may be viewed as practicing something like Carnapian explication (e.g., Carnap 1956). I return to this in chap. 5.

by psychologists, while informed by philosophical debates, were concerned with very different issues than those debated by philosophers of science at the time. In this way, I not only provide a thicker articulation of the issues that operationists hoped to address but also emphasize the significant and intriguing philosophical problems—having to do with the epistemologies of experimentation and measurement—that psychologists were grappling with in the 1940s and 1950s. My aim, thus, is twofold. On the one hand, I use historical material in order to further elaborate on and provide evidence for my analysis of operationism as understood by its advocates. On the other hand, I draw out some systematic issues that will continue to be the focus of our attention for the remainder of the book, touching on (*a*) the introduction to and development of concepts within psychological discourse and (*b*) the material and conceptual problems of disentangling relevant and irrelevant factors in an experimental context.

I start in section 2.2 by providing an overview of responses to operationism in the 1930s and 1940s, which reveal that at least some commentators took it to be drawing attention to the complexities of experimental research in the face of relatively poor access to and understanding of the subject matter at hand. The call for operational definitions, thus, was taken (*a*) as voicing the need to explicate the conceptual assumptions going into a given experiment and (*b*) as entirely compatible with the idea that concepts had additional meaning. Beyond that, however, there were some questions about whether operationism could offer any positive recommendations as to how to tackle the epistemic and conceptual uncertainties characteristic of experimental psychology. In section 2.3, I lay out the development of logical positivist philosophy of science in order to bring out more clearly both the points of contact between mid-century American psychology and philosophy of science and the ways in which their research projects were fundamentally at odds with each other. Philosophers took concepts and theories as given and inquired into issues of empirical interpretation and justification. Experimental psychologists, by contrast, tried to develop methodological norms for the actual research process, specifically concerning concept formation and theory construction. Then, in section 2.4, I turn to two (seemingly) competing operationist accounts of how to validate specific concepts and/or empirical methods. I argue that underneath these different accounts was a common assumption, namely, that psychological concepts often are initially derived from ordinary language, which informs the ways in which the concepts are explicated "for the purpose of science" (Stevens 1935b, 521; see also sec. 1.2.1) and, thus, how their purported referents are investigated experimentally. In section 2.5, I introduce the notion of *converging operations* as an answer to this question, coming out of experimental psychology in the 1950s. Finally, section 2.6 concludes this chapter by identifying the most important philosophical questions that my discussion of midcentury methodological debates has brought to the fore and that will be addressed in subsequent chapters.

2.2. Early Debates (1930s/1940s)

A survey of methodological writings on the psychology of the 1930s and 1940s reveals that early commentators interpreted operationism much the same way I presented it in the previous chapter, that is, as closely connected to epistemological questions that arise in experimental research. Underwriting my analysis, these writings display a tension between two seemingly conflicting sentiments. One is the idea that experimental research should drive and be driven by theoretical and conceptual work. The other is the worry that such work might be too speculative, thus requiring a very disciplined method of constraining the ways in which theoretical concepts are introduced into scientific discourse. Accordingly, commentators took operationism to be drawing attention to the epistemic uncertainties connected to the complexity of experimental work, stressing the need to specify how any given concept was being used. At the same time, however, they raised some doubts as to whether—beyond drawing attention to the complexities of experimental research—operationism had the requisite tools to tackle the problems at hand.

2.2.1. Operationism and the Familiar Cry to "Be Careful"

The feature of operationism that was attractive to psychologists was quite similar to the sentiment that had driven Percy Bridgman's (1927) operationism, that is, a worry about the dangers of extending concepts and measurement procedures into novel realms. In Bridgman's case, this worry had been prompted by the downfall of Newtonian physics. Asking what could have gone wrong in the history of physics such that an entire theoretical framework had to be rejected, his answer was that this framework had been built up on too shaky a foundation as scientists had been too quick to abstract from specific observational contexts. Thus, they had illegitimately assumed that a theoretical concept arising from one context could be extended into entirely different contexts and that a theoretically postulated magnitude measured under one set of circumstances could be measured under entirely different circumstances. This warning against unprincipled theoretical speculation and generalization rang true with many psychologists. Indeed, it may almost have appeared trivial from the perspective of any careful researcher. As one commentator wrote: "All good scientists have been operationists in deed, if not in word.... Operationism is the manner in which the present generation utters the familiar cry of science, 'be careful!'" (Pratt 1939, 81). Clearly, however, for many psychologists being careful did not entail a complete eschewal of theoretical work or the introduction of novel scientific concepts. Quite the contrary. As we saw in the previous chapter, Tolman and Hull viewed operationism as part of a methodology for theory construction, and Stevens was quite clear on the need for novel scientific concepts to describe his research objects (tonal volume and tonal density as subjective experiential states). The question, then, was not so much about the admissibility of theoretical posits as such as about how theoretical posits could be both legitimated and constrained by experimental work. For example, two commentators wrote (with respect to the question of the place of operationism in psychology): "Operationism merely re-emphasizes the need of the experimental method to psychological problems" (Waters and Pennington 1938, 422).

But how, specifically, did operationism emphasize "the need of the experimental method"? And what method, precisely, constituted the operationist answer to "the familiar cry of science, 'be careful'"? There are, I suggest, two intertwined answers to these questions. By requiring that crucial terms be operationally defined, operationists demanded that researchers explicate their preliminary understanding not only of the subject matter but also of the material conditions of any given experiment. In this way, rather than simply addressing questions about the (partial) meanings of scientific terms, operationism also demanded a heightened awareness of the complexity of the experimental situation: "The operational scientist may have gained through his operational definitions only the advantage of being in a position of maximal awareness of all the conditions surrounding the observational materials with which he works" (Israel and Goldstein 1944, 179). It becomes clear, then, that there were two interdependent sides to experiments akin to a figure/ground problematic, both of which were addressed by the demand for operational definitions. On the one hand, an adequate operational definition (as reflected and implemented in a particular experiment) was expected to delineate a particular phenomenon or research object. On the other hand, the phenomenon could come into sharp relief only if background and confounding factors were controlled for. Notice that, because of the complexity and epistemic and conceptual uncertainty, this did not simply mean controlling for confounders; it also meant reflecting on the very question of what might *be* potential confounders.[3]

3. I return to this problem in chap. 7.

The general point that operationism drove home, according to these commentators, was the need to be maximally aware of all the factors going into an experiment. In fact, Bridgman himself claimed that this is really all he meant to say in *The Logic of Modern Physics* (1927): "[Operationism] is a technique of analysis which endeavors to attain the greatest possible awareness of everything involved in a situation by bringing out into the light of day all our activity or operations when confronted with the situation, whether the operations are manual in the laboratory or verbal or otherwise mental" (Bridgman 1938, 130).

This statement is telling because it suggests that the notion of an *operational definition* (as focusing exclusively on a particular concept) captures only one part of what operationism was taken to be about. Another major part was that of *operational analysis*, that is, the analysis of all the factors involved in an experiment. Bergmann and Spence gave a succinct answer to why maximal awareness of the conditions surrounding experiments was particularly important for psychology. It was because of the complexity and epistemic opaqueness of the psychological subject matter: "In the less complex and more mature fields of natural science (physics, chemistry) we are reasonably confident that we know and control *practically* all the variables necessary. In the biological and social sciences, on the other hand, this is not the case. Here, complexity of the situation and insufficiency of knowledge tend to preclude successful segregation of all the variables necessary for the complete functional description attempted" (Bergmann and Spence 1941, 5). Thus, they recognized that, whenever a psychological question is articulated and an experiment is designed to investigate it, a number of assumptions have to be made not only as to the nature of the phenomenon but also as to the nature of the variables that need to be controlled for. Given the insufficient understanding both of the phenomena under investigation and of the domain in which they were embedded (both inside and outside the skull), it becomes imperative that scientists explicate the assumptions that go into a particular experimental design. As Bergmann and Spence argued, it can never be ruled out that an individual scientist is working with an inadequate understanding of a phenomenon. However, forcing this scientist to define key concepts operationally would put other scientists in a better position to evaluate his findings critically. An operational definition, Bergmann and Spence argued, tells us how scientists use their terms: "And that is all general methodology can insist upon at this level of the so-called *operational definition of empirical constructs*" (Bergmann and Spence 1941, 3).

2.2.2. Experimental Paradigms and the "Wealth of Meaning"

Apart from the call for caution, however, some authors also recognized a more positive function for operationism, namely, that operational definitions in some sense lay down standardized experimental setups, both for investigative and for didactic purposes. Far from restricting the meaning of a scientific term, those standardized experimental setups serve the positive role of being able to focus scientific attention on specific material-cum-cognitive assemblies assumed to instantiate the object of interest particularly well. Operational definitions, some argued, can perform this function well precisely because there is already so much meaning attached to the concept: "Operational definitions serve their best function when they point out the laboratory conditions and methods by way of which the concept may be illustrated. What instructor does not fall back on this method of explaining the meanings of concepts to his class? . . . But every instructor is keenly aware of the fact that the student's ability to recite or duplicate these procedures does not guarantee his complete understanding of the meaning of the corresponding concepts. These concepts, in other words, embody a wealth of meaning not given in these methods" (Waters and Pennington 1938, 422–23). This statement is intriguingly similar to one of Kuhn's notions of a paradigm in that it suggests that operational definitions were tied not necessarily (or at least not exclusively) to verbal explication but rather to specific ways of doing research. If we unpack this statement, there are (at least) two separate things going on, both of which underwrite the points for which I argued in the previous chapter. On the one hand, we find here a recognition of the fact that definitions are usually intended to provide an explication of a concept's meaning, along with the acknowledgment that operational definitions do not exhaustively capture this meaning. On the other hand, the fact that Waters and Pennington tie such illustrations to situations of scientific instruction suggests that this is not a shortcoming but rather an advantage of operational definitions. They specify experimental conditions that henceforth serve as common reference points for different researchers working on the same topic as well as for training purposes.

These points deserve closer scrutiny. First, with respect to the assertion that concepts in fact have a wealth of meaning that goes beyond experimental operations and their behavioral effects, the question is where this meaning originates from. One suggestion (and this is perhaps the one intended by Waters and Pennington) is that, when students get enculturated in particular ways of doing experiments, their teachers typically possess a richer theoretical understanding of the subject matter than they do. However, I contend that there is a second reading, one that pertains to scientific investigations. In the context of research, scientists often lack detailed knowledge of a term's referent (which is precisely why they study it), and, hence, there is a sense in which they do not fully understand the meaning of the term operationally defined by means of any given experimental setup or paradigm.[4] Yet it is assumed that the concept has meaning—or at the very least has the potential to be meaningful—in ways that go beyond the particularities of any given experiment. Whether such assumptions always pan out is a different matter (I return to this question in subsequent chapters).

One possible answer as to the origin of the "wealth of meaning" intuitively possessed by many concepts in psychology is that it comes from everyday language. After all, many psychological terms are already in use before they are taken up by scientific discourse. S. S. Stevens's operational definition of experience is a case in point. While arguing that "for the purpose of science" experience is nothing but discriminatory reaction, he makes it clear (as we saw in the previous chapter) that there is more to experience than that, citing in particular its phenomenal character. One might speculate that this comes from the established usage of the term "experience." On the other hand, we can speculate that the notion of a *wealth of meaning* being embodied in any given concept points to some dormant qualities of the concept in that it is capable of stimulating research that will ultimately reveal a lot about the domain of research and, thus, in some sense articulate the full meaning of the concept.

Both these readings raise philosophical problems, however. First, the fact that a given term has a particular ordinary language usage does not mean that it will ultimately be viable as a scientific term. (This is a point that has long been emphasized by eliminativists in the philosophy of psychology and neuroscience.) Second, the assumption that a given term may have a meaning of which we are not yet fully aware may commit us to an account of meaning we do not want to endorse (e.g., a causal theory of reference according to which the reference of a term is already fixed). Neither of these points can be developed here, but they are taken up in chapters 3 and 6. Be that as it may, the statement quoted above points to an important set of issues that came to the forefront of methodological debates a decade or so later and that—in close engagement with philosophy of science discourse of

---

4. In chaps. 4 and 5, I spell out my analysis of the relationship between operational definitions and the often ill-understood purported referent of the concepts thus defined.

the 1940s and 1950—inquired into the semantics of scientific terms. I turn to that debate in section 2.3.3 below. For now, let me emphasize what I take to be the descriptive take-home point of this section, that is, that operationism (as understood by some commentators as early as the late 1930s) did not make any strong reductionistic assumptions about the meanings of concepts.

2.2.3. Early Problems and Misgivings

In response to the construal of operationism given thus far, it might be objected that this position is both trivial and vague. It is trivial because the cry to be careful articulated only what would have been self-evident to many experimental researchers anyway. It is vague because it does not give very specific instructions on how to ensure that any given operational definition (or implementation of such a definition) was in fact getting it right.

The triviality charge was made quite explicitly in the early literature on operationism, where it was noted that operationism was no major reform in science. Rather, it told scientists to do what they had always done, that is, to perform experiments (Israel and Goldstein 1944). Worse yet, it was not clear that operationism was going to be able do the job that Stevens had envisioned for it, that is, to "[ensure] us against hazy and contradictory notions and [provide] the rigor of definitions which silences useless controversy" (Stevens 1935a, 323). The reason why operationism was not able to do those things, some writers claimed, was because science always involves theorizing and generalizing (Pennington and Finan 1940; Waters and Pennington 1938). Furthermore, by ostensibly tying the existence of an entity or phenomenon to just one operation, operationism did not offer any insight with respect to the question of how one might confirm claims about the existence of such a posited entity or phenomenon by any other method or how to make more general claims about it. Thus, it seemed to critics of operationism that there were two alternatives. One was to read operationism as not permitting any theoretical posits. The other was to call for a refined operationism. On the former construal of operationism, this was deemed unattractive because (*a*) it seemed unduly restrictive and in conflict with the usual aims of science and (*b*) it was perceived as leading to a "multiplication rather than reduction of the number of concepts" (Waters and Pennington 1938, 418). In addition to those reasons, I would like to argue that both construals of operationism would rule out Stevens, Tolman, and Hull as operationists because, as we have seen, neither Tolman nor Hull were opposed to theorizing (in that they posited the existence of particular entities) and Stevens certainly thought of his results about the attributes of auditory experience as generalizable.

This brings us to the second of the two alternatives just mentioned: If theoretical posits and generalizations were to be allowed by operationism, then this called for not only guidelines as to criteria governing such posits (e.g., that they be introduced by providing criteria of experimental application) but also additional methods regarding the validation of specific posits. As noted by the authors of one paper, if a single operation is used as the inductive basis for positing the existence of a psychological trait, then this operational conceptualization might "camouflage some unspecified factors" (Pennington and Finan 1940, 260). In other words, one might attribute the outcome of a given experimental operation to a particular psychological trait when it was in fact due to (or confounded by) some other factor in the experiment. One plausible way to proceed is to say that we have reason to believe that an operation really indicates an interesting entity or phenomenon if the same entity or phenomenon can also be detected by different operations. In 1945, the *Psychological Review* (vol. 52, no. 5) printed the proceedings of a symposium about operationism. Contributions included Boring (1945), Bridgman (1945), Feigl (1945), Israel (1945), Pratt (1945), and Skinner (1945/1972), and the emerging consensus among most of the contributors was that the idea of multiple operations was in principle compatible with operationism. For example, Edwin Boring remarked: "It is possible to identify two operations in terms of further operations" (Boring 1945, 243). He further stated: "Operationism is not opposed to the validation or extension of a concept" (245). But, of course, this raises the question of how scientists were supposed to know when different operations individuate the same kind of entity or phenomenon. Boring was unclear about how to answer this question.

Interestingly, this weakness of operationism, as formulated thus far, was also recognized by advocates of operationism. For example, Bergmann and Spence (1941) very forcefully warned against overrating the potency of operationism. As they saw it, the "operational principle" is a method of distinguishing between hypotheses that are methodologically sound and those that are not. The criterion of methodological soundness is the operational definability of a concept and, hence, the possibility of formulating hypotheses about its purported referent. However, once a hypothesis has been deemed methodologically sound, operationism has reached its limits. It does not provide any decisive rules as to how to determine whether two experiments individuate the same phenomenon. In other words, the making explicit of conceptual assumptions required for an experiment does not guarantee that the assumptions are correct. As Bergmann and Spence put it: "There is no methodological principle, no 'operational recipe' which guarantees that no relevant factor has been overlooked" (1941, 3).

I believe this to be a very important insight, pinpointing that already in the early 1940s scholars recognized that inferences made from experiments drawing on operational definitions were going to be underdetermined by the empirical data. The task for methodological reflections about operationism was, thus, to figure out strategies for minimizing the risk of overlooking relevant factors in experiments. Such reflections did indeed take place in the 1950s and 1960s. On the one hand, they turned on an engagement with the developments within philosophy of science (specifically, the shift in the cognitive criterion of significance). On the other hand, they were motivated by practical concerns, in particular, concerns with the validation of psychometric tests. In the following two sections, I lay out these two strands of debate in order to bring out commonalities and differences among the psychologists involved more clearly.

2.3. Some Midcentury Developments (Interlude)

Before pressing on with my narrative about methodological debates that took place within psychology, I pause here and provide a brief overview of the (shifts in the) ways in which American philosophy of science talked about the meaning and function of theoretical terms between the 1930s and the 1950s. This serves several purposes. First, it provides a clearer picture of why operationists in psychology are sometimes said to have missed important developments in mid-twentieth-century philosophy of science. Second, it shows that this charge (a) is not entirely true (operationists of the second generation were in fact receptive to some of the changes happening in philosophy of science) and (b) is somewhat misguided (operationists and philosophers were engaged in very different projects, and, hence, it is not entirely clear that psychologists should have taken inspiration from philosophers).

2.3.1. Meanwhile in Philosophy of Science

I focus here mainly on Carl Hempel and Rudolf Carnap, both of whom (among many others) emigrated during World War II from Europe to the United States, where they proved to have a lasting influence on the development of philosophy of science. In particular, there were close intellectual ties between theoretical psychologists and some recent immigrants at the University of Minnesota and the University of Iowa, accounting for some of the writings we see about operationism during those years.[5] To get a better sense of what was happening around that time, let us step back for a moment and recap (*a*) why the verificationist semantics of logical positivism may initially have looked like it was demanding operational definitions of psychological terms and (*b*) why proponents of logical positivism began rejecting this demand.

An important feature of logical positivism (in contrast to the earlier positivisms of the nineteenth century) was the great emphasis it placed on the analysis of language, seeking criteria of meaningfulness as a diagnostic tool to distinguish between "scientific" and "metaphysical" questions. Verificationism famously stated that a sentence is meaningful (i.e., not metaphysical) if and only if it is either a tautology (analytically true by virtue of the meanings of the words contained in it) or empirically verifiable (Carnap 1931b). In turn, the meanings of individual words had to be reducible to simple observational sentences. This raised the question of what types of sentences qualify as belonging to the observational language of verification. Members of the Vienna Circle famously argued that the preferred language of verification was to be a physical language, something Carnap (1931b) deemed to be both intersubjective and universal (i.e., capable of expressing all meaningful sentences). It followed that psychological sentences were meaningful only insofar as they could be expressed in a physical language, which in turn presupposed that individual psychological concepts are coextensive with physical concepts (Carnap 1932). This notion found its application to psychology in "Psychology in a Physical Language," in which Carnap wrote: "Our thesis states that a definition may be constructed for every psychological concept (i.e., expression) which directly or indirectly derives that concept from physical concepts" (1932, 167). He suggested that the relevant physical language was, ultimately, going to be that of neurophysiology but that for the time being the language of stimulus and the resulting behavior would have to suffice for the definition of meaningful psychological constructs (see Feest 2017b). A similar view about the meanings of psychological terms and statements was argued for in Hempel (1935/1980). Given what was just explained, it is not surprising that commentators have assumed that operationism in psychology attempted to define psychological concepts in a physicalist fashion according to a verificationist semantics. The conditions of application

---

5. For example, Bergmann and Spence (1941), discussed above, was coauthored by Gustav Bergmann (a member of the Vienna Circle who had recently emigrated from Vienna) and Kenneth Spence (one of Clark Hull's students), who were at Iowa at the time.

for any given concept would then be provided by stating a specific operation together with the effect it produces. This is indeed how Percy Bridgman's operationism was interpreted by Carl Hempel (1954/1956), who argued that the point of an operationist theory of meaning was to demand the specification of the conditions under which particular observable events would result. Hempel argued that operationism is a special case of this more general idea, with the additional requirement that the conditions in question are brought about by us.

However, if operational definitions have to appeal to counterfactual conditionals in this way, it would seem that the types of terms defined here are disposition terms. But already by the mid-1930s it became clear to logical positivists that disposition terms could not easily be captured by a strict verificationist theory of meaning (thus contributing to the rapid decline of verificationism after the mid-1930s). This shift is most prominently expressed in Carnap (1936–37), in which he remarked that disposition terms could not be provided with necessary and sufficient conditions of application (see also Feest 2005b). Rather, their meanings could be pinned down only by partial interpretations in terms of what he called "bilateral reduction sentences" (Carnap 1936–37). Such reduction sentences described particular conditions that had to be fulfilled in order for the disposition in question to be actualized (thus, in a sense, giving only sufficient conditions of application).[6] The upshot of this was that the very idea of meaning as tied to necessary and sufficient conditions of application began to crumble. Within the logical positivist tradition, this gradually gave rise to the idea that the meanings of individual concepts were implicitly defined by their place in a theoretical network, an idea that found a particularly influential formulation in Wilfrid Sellars's "Concepts as Involving Laws and Inconceivable without Them" (1948). This idea was also articulated by Carl Hempel when he wrote that theoretical concepts are "not introduced by definitions or reduction chains based on observables": "Rather the constructs used in a theory are introduced jointly, as it were, by setting up a theoretical system formulated in terms of them and by giving this system an experiential interpretation, which in turn confers empirical meaning on the theoretical constructs" (Hempel 1952, 32). Likewise, in "The Methodological Character of Theoretical Concepts," Carnap stated that he no longer believed an analysis of disposition terms to be very central to an analysis of the meanings of most theoretical concepts in science. He argued that there were two

---

6. I have, therefore, argued in the past that Carnap's analysis of reduction sentences captures something of how operational definitions were intended by operationists in psychology (Feest 2005b).

ways of introducing scientific concepts into scientific language—as theoretical concepts and as disposition concepts—but he now considered the concepts of the "theoretical language" to be more important: "I think today that, for most of the terms in the theoretical part of science, and especially in physics, it is more adequate and also more in line with the actual usage of scientists, to reconstruct them as theoretical terms in L_T rather than as disposition terms in L_O" (Carnap 1956, 66).[7]

2.3.2. Operationism and the Dynamics of Research

We turn now to a comparison between positivists and operationists with respect to the issues just discussed. As we just saw, positivists of the 1950s no longer believed in the explicit definability of scientific terms, viewing them instead as embedded in and implicitly defined by theories. However, in talking about the meanings of scientific concepts in this way, they did not address the question of the origins of the theories or concepts as such. They took the existence of theories for granted at this point, asking only how they conferred meaning on scientific concepts, and attending to the question of how theories could be empirically tested. By contrast, operationists were interested in the construction of theories and the formation of concepts. In other words, operationists' methodology was focused on the dynamics of research, asking questions about this from the vantage point of the practitioner whose epistemic perspective is marked by uncertainty about the empirical contours and theoretical analysis of his or her very subject matter. Thus, this difference between philosophers and psychologists exemplifies a dichotomy very central to philosophical discourse since the nineteenth century, that is, that between genesis and validity or between discovery and justification.

There is a second notable difference between the two approaches. Philosophy of science in the positivist tradition (even in its postpositivist form, as exemplified by W. V. Quine) continued to concern itself with the analysis of language as a way of doing epistemology, whereas psychologists concerned themselves with issues about operational definitions as deeply entwined with the materiality of experimental practice.[8] Both these features of mid-twentieth-century philosophy (the discovery/justification dichotomy and the focus on linguistic analysis) have since been thoroughly criticized within philosophy of science, and there are by now ample examples of philosophical works that turn their attention to aspects

---

7. Carnap's L_t and L_o are the theoretical language and the observational language, respectively.

8. For a similar point, see Eigner (2010, 61).

of scientific discovery, on the one hand, and scientific practice (especially experimental practice), on the other. This further reinforces my suggestion that, rather than reading operationism in psychology as a symptom of outmoded verificationist and positivist philosophical views, we should read it as having anticipated insights that did not really come to the fore within philosophy until the practical turn several decades later. However, the epistemic uncertainty and conceptual openness characteristic of much scientific research is still not widely appreciated within philosophy (though, as stated in chap. 1, there are notable exceptions to this claim). I therefore believe that operationism, adequately construed, holds some valuable insights for contemporary debates, especially with regard to philosophical analyses of cognitive science. In this section, I continue to argue for this general thesis by looking at some of the ways in which psychologists and philosophers talked about the issues posed by operationism.

First, consider Carnap's notion of scientific theories as being connected to the empirical world via correspondence rules that can be thought of as operational definitions. From the epistemic standpoint of the operationist psychologists at the time, this answer was not really helpful. Surely, if we already know a lot about the referent of a concept and/or have a scientific theory of it, this can help us specify different operations by which it can be detected. But this, of course, begs the question since the issue was how to specify different operations that detect the same entity in the absence of any detailed knowledge about the (presumed) referent of the term. Furthermore, even if we do already have a theory of the phenomenon, this does not necessarily mean that the ways in which we try to individuate it empirically will be successful given the numerous other factors in the experiment that can potentially interfere with data production (as we saw in the previous chapter, both Tolman and Hull remarked on this). Going even further, we can say that it would have been difficult to align operationist research practice with the very notion of *theory* presupposed by positivism. First, while all three of the operationists discussed in the previous chapter took their research as requiring or pertaining to entities or traits that could be only inferred from (not reduced to) empirical data, Stevens regarded his research not as aiming at theory construction but as having taxonomic aims. Second, whereas Tolman and Hull did both regard their work as having theoretical aims, their notion of *theory* was more akin to what more recent philosophy of science might refer to as "mechanistic models."[9] Most importantly, however, I argue that psychology and philos-

---

9. Hull was strongly committed to both a mechanistic and a deductive understanding of *theory*, which complicates his case (see also Feest 2012a).

ophy of science took diametrically opposed positions with respect to the aim of empirical research and the place and shape of theory in such research. We can characterize this difference as one between static and dynamic understandings of theories. While positivists were interested in the ways in which already existing theories confer meanings on concepts or how they can be interpreted and tested empirically, operationists were interested in the question of how to individuate objects of interest empirically and how to construct theories of those objects on the basis of the results of experimental research.

The deep differences between the positivist and the operationist understanding of the relationship between theory and observation did not go unnoticed in the methodological literature at the time (Bergmann and Spence 1941; Koch 1941a, 1941b). For example, Sigmund Koch noted that positivists start out with the idealized assumption of the dichotomy between a formal and an empirical system and then try to give a logical reconstruction of how they can be related. He referred to this as the "interpretive" method of theory development.[10] Most psychologists, on the other hand, insofar as they engage in theorizing at all, simultaneously develop both the formal level and the empirical level. Koch called this simultaneous development of the formal and the empirical side of science the "telescopic" procedure. This procedure does not use an already existing formal system as a tool for ordering data. Rather, "scientists usually proceed by asserting as postulates either certain empirical laws, or prior assumptions as to the functional relationships holding between certain of the empirical constructs (or both)," and then make experimentally testable deductions (Koch 1941a, 20). In light of the case study of Clark Hull's approach to theory construction (see chap. 1), this strikes me as a very perceptive remark.[11] Given the different projects of positivism and operationism, advocates of the latter were probably well-advised

---

10. Koch mentions Kurt Lewin's theorizing as one possible example, though he considers this procedure to be extremely rare in psychology. Koch had only a few years earlier (in 1938) obtained an MA degree from the University of Iowa (Freeman 1996), where Lewin was a professor. In turn, Gustav Bergmann, also a professor at Iowa, having recently arrived from Europe, started out working with Lewin as a research assistant but found his and Lewin's perspectives on theory construction incompatible (Addis 1999).

11. Notice that this way of presenting the difference between philosophers and psychologists is similar to Wimsatt's dichotomy between what, drawing on Feynman, he calls the "Greek" (or Euclidean) approach and the "Babylonian" approach to the structure of physical theorizing. Whereas the Greek approach starts with a few axioms and derives everything from them, the Babylonian approach is more bottom-up and tries to ground a concept in multiple ways (see Wimsatt 1981/2012).

in not taking too many cues from the former. This also underscores the point that operationists had a problem with which philosophy of science did not concern itself. The problem at hand was how to delineate (for investigative purposes) the referents of scientific concepts while at the same time constructing and justifying those very concepts. Within the methodological debates in psychology, there were several attempts to grapple with these issues. Taking the apparent consensus (described in sec. 2.2.3 above) that it was in principle possible that there was more than one operation tied to any given concept (i.e., that the concept's intended referent could be empirically detected in more than one way), the question was how it could be shown that any two (or more) operations did indeed provide epistemic access to one and the same object.[12]

2.4. The Construct Validity of Psychological Tests (1955)

The question of how to determine whether an empirical operation can yield data that are actually indicative of that which they are thought to be indicative was debated in the area of mental testing in the late 1940s and early 1950s, culminating in a report by an American Psychological Association committee entitled "Technical Recommendations for Psychological Tests and Diagnostic Techniques" (American Psychological Association 1954), followed by the more philosophical Cronbach and Meehl (1955). Even though this debate took place in a different context from the (experimental) one we have been looking at so far, there are also similarities. By administering a psychometric test, researchers are essentially performing the operation of providing stimuli along with instructions and treating the responses as indicative of the variable to be measured (e.g., anxiety). The question of how it can be determined that the responses are indeed indicative of the to-be-measured variable was recognized as important as psychometric tests became increasingly popular in various domains.

In psychology, the question of whether a particular test or experimental operation really measures or detects what it is thought to measure or detect is one about the test's *validity*. This notion is commonly contrasted with that of *reliability*. Both these concepts pertain to the goodness of a psychological measurement or detection procedure. A particular testing procedure is reliable if its results can be replicated under similar circumstances, whereas it is valid if it really measures what

---

12. The term "robustness" was not used at the time, but the suggestion was similar in spirit to the more recent notion of *robustness* in philosophy of science. I get to this in a moment.

it is thought to measure (see Goodwin 1995, 100). It is well worth emphasizing the distinction between reliability and validity in psychology by comparison with contemporary reliabilist epistemology, which commonly takes the term "reliable" to refer to the idea that "a belief acquires favorable epistemic status by having some kind of reliable linkage to the truth procedures that reliably produce truth" (Goldman 1992, 433). Within philosophy of neuroscience, we find discussions about whether neuroscientific procedures are reliable in the reliabilist sense (see Sullivan 2007, 2009). While I ultimately believe this latter understanding of reliability to be better suited for an analysis of data quality (see Feest 2022; see also chap. 7 below), the present chapter follows the methodological literature in psychology, which often decouples questions about the reliability of a measurement procedure from questions about the truth of the statements that the measurement procedure can deliver. For example, a test might reliably produce similar data across testing situations, but this does not mean that the assumptions as to what is being measured are correct. The process of establishing that a given test is valid is called "validation." I now turn to two influential early accounts of how such validation works, highlighting the insights and shortcomings of both.

2.4.1. Construct Validation and "Nomological Nets"

In their article "Construct Validity in Psychological Tests," Cronbach and Meehl (1955) begin by distinguishing between three different proposals of how to validate a testing procedure: (*a*) "criterion-oriented validation," (*b*) "content validation," and (*c*) "construct validation." Criterion-oriented validation (which comes in two forms: predictive and concurrent validation) tries to validate a test by showing that the test results are correlated with an external criterion. (For example, a test for academic aptitude might be validated by showing that it either predicts academic performance or is correlated with present academic performance.) Content validation requires that the test items represent the to-be-measured trait (e.g., if the trait of interest is mathematical ability, a test will have content validity if it contains mathematical problems as items).

Both criterion and content validation try to stay true to the antimetaphysical spirit of logical positivism and hence do not wish to commit to the existence of any unobservable mental state or trait and rather evaluate a given test by reference to some external criterion (such as a behavior that the test behavior is expected to be correlated with or to instantiate) (Anastasi 1950). In contrast to criterion and content validity, the notion of *construct validity*, as explicated by Cronbach and Meehl, comes closest to expressing the assumption that there is an independent to-be-measured entity or phenomenon that can be detected by more than one type of operation. Cronbach and Meehl explain their notion of *construct* to mean "some postulated attribute of people, assumed to be reflected in test performance" (1955, 283). I suggest that what they mean is not that they are measuring a construct but that they are appealing to a construct (theoretical concept) as representing "some postulated attribute of people," which they are attempting to measure. In other words, it is not the construct that is reflected in the test performance but the entity that the construct is thought to represent and that the test is thought to measure or detect.[13] Slaney (2017) points out that, in the subsequent development of writings about construct validity, scholars increasingly came to use the term more broadly, for example, to refer to the question of whether a given construct succeeded in serving as a pragmatic sorting device with no commitment to the reality of the phenomenon thus described. It is clear, however, that this was not Cronbach and Meehl's intent, since they wanted to defend the idea that a test measures more than just test performance, that is, something of which the test performance is indicative. This way of interpreting construct validity also captures the ontological commitments I attributed to my historical actors in the previous chapter.

It can be said that what is at stake in construct validation is the simultaneous attempt to justify both a construct (as representing some "attribute of people") and a test (as measuring precisely this attribute as represented by the construct).[14] This seemingly creates a problem of circularity because in order to validate the construct, one must be able to measure its referent; however, in order to measure the referent, one must presuppose the validity of the test. Cronbach and Meehl were well aware of this problem, and their response to it was similar to Chang's (2004) analysis of epistemic iteration. In a section entitled "Specific Criteria Used Temporarily: The 'Bootstrap Effect,'" they suggested that both concepts and tests could contribute to a process of mutual

---

13. I suspect that the continuing insistence of many psychologists that they are only measuring constructs is in part an expression of a legitimate epistemic caution, mixed with midcentury philosophy of science lingo. (For a more recent discussion and critique of the lingering antirealism associated with construct validity, see Borsboom, Mellenbergh, and van Heerden [2004].) This topic cannot be covered in depth in this chapter, but, for a discussion of construct validity, see Feest (2020a).

14. This has prompted Stone (2019) to distinguish the question of the validity of a test from the question of the legitimacy of the construct.

refinement (Cronbach and Meehl 1955, 286). However, in addition, they also considered proposals in philosophy of science at the time (Carnap 1950a; Hempel 1950, 1952; Sellars 1948), which I reviewed in section 2.3.1 above, according to which concepts are embedded in networks of other concepts. This led them to argue that the validity of a construct (and ultimately of the test thought to measure its referent) could be demonstrated by offering a theoretical account of the construct's referent that shows how it is related to other theoretical constructs. They take this to be suggested by midcentury philosophy of science: "We may say that 'operations' which are qualitatively very different 'overlap' or 'measure the same thing' if their positions in the nomological net tie them to the same construct variable" (Cronbach and Meehl 1955, 290). In other words, the assumption was that constructs occur in theories and that, if two distinct measuring operations can be derived from a given theory about one thing, then the two operations in fact operationally define the same concept (or "construct variable," as the authors put it) and thus can be said to measure the same thing. Furthermore, the idea seems to have been that this validates not only the construct itself (as being an adequate representation of the thing) but also the test that is based on the construct. If we abstract away from the slightly antiquated expression "nomological net," there is clearly an important point here that might seem intuitively plausible: at a minimum, two tests purporting to test the same thing by way of different operations should agree on a few central features of what is being measured.

This proposal has several problematic consequences, however. By tying the validation of constructs to the theories in which they occur, this account seems to rule out that a given construct (e.g., the construct of anxiety, or of memory, or of intelligence) can be validated prior to the existence of a theory of the corresponding assumed entity or phenomenon in the world. It also seems to force us to give up the concept altogether should it turn out that the theory is flawed, and it does not give any method for adjudicating between different methods of measuring a trait when these methods are informed by different theories (and hence by different constructs). More generally speaking, this proposal does not do justice to the (operationist) intuition that the construct itself can be, in a certain sense, grounded in experimental operations prior to the existence of a full-fledged theory, and its validity thus does not fully depend on the fate of a theory.[15] There are two possible replies to

---

15. My intention here is not to defend this intuition but merely to remind the reader that it was central to the operationisms of Stevens, Tolman, and Hull, as discussed in the previous chapter.

these problems. One is simply to bite the bullet and grant that, indeed, the validation of constructs is deeply tied to theories and, hence, that constructs cannot be validated prior to having a theory. In turn, this would imply that there is no way of telling whether any two constructs that appear in different theories in fact single out the same entity or phenomenon in the world. The second possible reply is to seek to develop a method of determining whether two or more operations individuate the same thing regardless of the theories that inform our constructs of that thing or perhaps even prior to there being any full-fledged theory.[16]

Something like the second opinion was articulated in a seminal paper by Campbell and Fiske, who wrote: "We believe that before one can test the relationships between a specific trait and other traits, one must have confidence in one's measure of the trait" (1959, 100). In other words, one needs to validate a construct/test before one can start developing a theory of its referent/object (e.g., a particular human trait) and of how it is linked to other constructs/traits. Campbell and Fiske proposed that such confidence could be reached by their "multitrait-multimethod" approach (see below), explicitly contrasting their outlook with that of Cronbach and Meehl (1955), to whom they attributed the view that construct validation requires a theory.

2.4.2. Construct Validation and the Multitrait-Multimethod Approach

The article by Campbell and Fiske (1959) is titled "Convergent and Discriminant Validation by the Multitrait-Multimethod Matrix," and it has been named the most influential psychological paper of the twentieth century (Sternberg 1992), though the authors implied in their retort "Citations Do Not Solve Problems" (Fiske and Campbell 1992) that their proposal—though widely cited—had not really been taken to heart.[17]

The rationale underlying Campbell and Fiske's notion of *construct validation* is the following: (*a*) the trait in question is to be measured by different means; that is, one and the same construct is operationally defined in different ways; (*b*) measurements of this trait are to be distinguished from measurements of different traits; and (*c*) the variance in the data that is due to the method is to be separated from the variance that is due to the trait. Essentially, the procedure is to measure different

---

16. There is a third reply that is, I believe, most appropriate to the actual epistemic situation at hand and that treats validity as a gradual concept. I discuss it in sec. 2.5 below.

17. The second-most-cited paper on Sternberg's (1992) "hit parade" is Cronbach and Meehl (1955), discussed above.

traits by means of different methods and to correlate the results. For example, if there are two measurement procedures (peer rating and an objective test) and two traits (intelligence and effort), we can get the following four results: (i) intelligence as measured by peer rating, (ii) effort as rated by peer rating, (iii) intelligence as measured by an objective test, and (iv) effort as measured by an objective test. Campbell and Fiske (1959), from whom I take this example, argue that a concept is validated by obtaining two measures of validity: convergent validity and discriminant validity.

If the correlation in cell 1 of table 2.1 is high, this would suggest that (i) and (iii) are measuring the same phenomenon (intelligence). Likewise, if the correlation in cell 4 is high, this would suggest that (ii) and (iv) are measuring the same phenomenon (effort). This way of reasoning establishes what Campbell and Fiske call the "convergent validity" of the constructs in question. The notion of *convergent validity* is supplemented by that of *discriminant validity*, which is established by showing that the correlations in cells 2 and 3 are low. The requirement of discriminant validation is supposed to ensure that the (seeming) convergent validity of the construct is not an artifact of the two testing procedures. For example, if it turned out that there is a correlation between effort as rated by peers and intelligence as rated by peers, then we might have to conclude that the results are due to the method (peer rating), not to the entities that are being measured (intelligence, effort).

At first sight, this method bears surprising similarities with more recent accounts in philosophy of science that go by the name "robustness." A canonical formulation of this idea was provided by William Wimsatt (1981/2012), who linked this idea to those of triangulation and multiple determination, summarizing the idea elsewhere as follows: "*Things are robust if they are accessible (detectible, measurable, derivable, definable, producible, or the like) in a variety of independent ways*" (Wimsatt 1994, 211;

Table 2.1: Illustration of the multitrait-multimethod approach

	(III) OBJECTIVE TEST FOR INTELLIGENCE	(IV) OBJECTIVE TEST OF EFFORT
(i) Peer rating of intelligence	1. Correlation between (i) and (iii)	2. Correlation between (i) and (iv)
(ii) Peer rating of effort	3. Correlation between (ii) and (iii)	4. Correlation between (ii) and (iv)

emphasis in original). Specifically, with regard to empirical detectability, Sylvia Culp (1995) has argued that, if two means of gathering data are statistically independent (by virtue of relying on a completely different set of theoretical assumptions), then the likelihood of getting similar patterns of data is so slim that it is justified to invoke the principle of the common cause (Reichenbach 1956; Salmon 1984) in explaining this similarity. She draws on this idea to propose a specific criterion of robustness, according to which a phenomenon can be considered robust if two techniques produce comparable data even though they rely on different theories. Clearly, this is an attempt to answer the question of how a phenomenon can be individuated empirically by multiple methods. The parallel between Wimsatt's proposal and Campbell and Fiske's multitrait-multimethod approach to validation is not entirely coincidental. In fact, Wimsatt's article first appeared in a Festschrift for Donald Campbell, and, while Wimsatt intends for his analysis to have a degree of generality that goes beyond Campbell and Fiske's specific concerns, it is worth noting that he acknowledges Campbell's work on this issue. He even refers to Campbell's (1969/1988) position as "multiple operationalism" (Wimsatt 1981/2012) to distinguish it from the more traditional "definitional operationalism." Quoting Campbell, Wimsatt writes that this type of multiple operationalism introduces "a formal way of expressing the scientist's prepotent expression of the awareness of the imperfection of his measuring instruments and his prototypic activity of improving them" (Wimsatt 1981/2012, 69, citing Campbell 1969, 15).

2.4.3. Construct Validation and Construct Formation

Campbell's statement is instructive in that he talks about "improving" rather than "validating" measuring instruments. Validation has a ring of finality to it and, moreover, suggests that both the tests and the constructs in question no longer undergo any changes. By contrast, Campbell's formulation draws attention to the improvement of tests and constructs as an ongoing scientific activity. By taking seriously this ongoing activity as something worthy of philosophical attention, we can perhaps shed some light on a notable disparity within the literature about robustness. As pointed out by Schickore and Coko (2014), the relevant literature is divided on the issue of how widespread robustness reasoning in fact is in science. While philosophers of science like to present cases that seem to confirm their formal analyses of how robustness can be achieved, historians tend to focus more on the messiness of the process. I suggest that these two perspectives do not need to be mutually exclusive as long as we recognize the following two things. First, construct validity is judged to obtain when a construct explains at least part of the variation of test results. This means that construct validity (as determined by scientists) can come in degrees, because variation is not an all-or-nothing affair. Second, whereas I use the expression "construct validity" here to refer to a scientific result, the term "construct validation" is better understood as referring to a process in which the very construct is still being refined and changed rather than simply being justified (see also Feest 2020a). Some of the messiness encountered by historians who study robustness on the ground might simply be due to the fact that they are looking at ongoing processes of construct development rather than aiming to provide analyses of the validity of a finished product. This may also help articulate my own aims more succinctly as far as it is concerned with a philosophical analysis of dynamic processes of concept formation and validation.

Let me quickly elaborate on this last point to say that, if we look more closely at the two papers just discussed—Campbell and Fiske (1959) and Cronbach and Meehl (1955)—it becomes apparent that their authors were quite aware of the dynamic character of construct validation and of the conceptually indeterminate situation under which it is attempted. As it turns out, in practical terms, the differences between the two approaches are not as big as they might seem at first glance. For example, even though Cronbach and Meehl stressed that construct validation required the relevant constructs to be embedded in nomological nets, they were quite aware of the fact that, at that point in time, there were no full-blown nomological networks to which to turn in search of overlapping measurement operations. As pointed out above, they acknowledged that in such a case one had to start out with some temporary preconceptions as to the nature of the object of interest, thereby addressing the question of how theoretical concepts get introduced and developed in the course of research. Thus, they stated that, at an early stage of concept development, constructs may be no more than inductive summaries of observational laws rather than more abstract theoretical laws. Or, to put it differently, the nomological net may "go beyond the empirical data, only in the sense of making predictions about future observations" (Cronbach and Meehl 1955, 292). Even this modest sense of "theory," they argued, allows for tests, which in turn can lead to revisions of the constructs in question. However, at the (then) present point in time, "psychology works with crude, half-explicit formulations" (294). Cronbach and Meehl went on to say:

> Nevertheless, the sketch of a network is there; if it were not, we would not be saying anything intelligible about our constructs. . . . [T]he

vague, avowedly incomplete network still gives the constructs whatever meaning they do have. When the network is very incomplete, having many strands missing entirely, and some constructs tied in only by tenuous threads, then the "implicit definition" of these constructs is disturbingly loose; one might say that the meaning of the constructs is underdetermined.... We will be able to say "what anxiety is" when we know all of the laws involving it; meanwhile, since we are in the process of discovering these laws, we do not know precisely what anxiety is. (Cronbach and Meehl 1955, 294)

This statement importantly shifts the focus from a narrow concern with validating a fully formed concept to the question of how concepts are formed and improved to begin with. With it, Cronbach and Meehl highlight an observation that is very central to the topic of this book, namely, that research is often marked by deep epistemic uncertainty and conceptual openness. To take their example, research about anxiety is much better characterized as an ongoing attempt to find out "precisely what anxiety is" than as an attempt to validate (once and for all) a specific theoretical concept or test of anxiety. However, by talking about a "sketch of a network," they draw attention to the fact that many psychological terms are already embedded in the linguistic conventions of ordinary usage.[18] While some or all of our folk-psychological concepts might ultimately prove to be untenable as scientific concepts, Cronbach and Meehl thus explicitly acknowledge the ongoing character of theory building and concept formation as starting out from the conceptual presuppositions embedded in our folk-psychological language. These presuppositions may well inform the experimental conditions of application required to get the research off the ground.

In a similar vein, the psychologist Benton Underwood introduced the notions of "literary lead-ins" and "provisional definitions," arguing that literary lead-ins are taken from ordinary discourse and provide the researcher with a provisional understanding of the general nature of the phenomenon in question. However, "we must guard against thinking that our literary statements about a phenomenon which we believe exists, and which we wish to measure, are inevitably identified with the phenomenon we finally measure" (Underwood 1957, 55). I argue that what Underwood's notion of *literary lead-ins* captures the same general idea as my above notion of *conceptual assumptions* about a phenomenon

---

18. Notice that this corresponds with our analysis of the previous chapter, in which we saw that Stevens and Tolman took terms like "experience" and "hunger" from folk-psychological vocabulary.

or object. Those assumptions underlie what Underwood calls "provisional definitions" or what I have been calling "operational definitions^{wide}." Such explications allow for specifying the operations that, when giving rise to particular empirical effects, are treated as indicating a specific research object "operational definitions^{narrow}." As Underwood cautions, such definitions "must be provisional because the investigator may find either that he cannot carry out the operations or if he can, no new phenomenon is discovered" (Underwood 1957, 57). Interestingly, Campbell and Fiske take their proposal to express a type of operationism that pursues and extends Underwood's proposal:

> The requirements of the present paper may be seen as an extension of the kind of operationalism Underwood has expressed. The test constructor is asked to generate from his literary conception or private construct not one operational embodiment, but two or more, each as different in research vehicle as possible. Furthermore, he is asked to make explicit the distinction between his new variable and other variables, distinctions which are almost certainly implied in his literary definition.... His literary definition, his conception, is now best represented in what his independent measures of the trait hold *distinctively* in common. (Campbell and Fiske 1959, 101)

Summing up, then, I argue that, once we abstract away from some of the philosophical language of mid-twentieth-century methodological writings, it becomes clear that the seemingly deep differences between the two approaches to validation (that of Cronbach and Meehl, on the one hand, and that of Campbell and Fiske, on the other) make way for a much more pressing concern that they had in common: how to introduce and develop the terminology pertaining to their subject matter at a point in time at which that subject matter was not yet well understood. And an answer they shared was to start off with a folk-psychological (or "private") conception and derive one or more "provisional" (or operational) definitions from it. This does not sound so dissimilar from the practice I have attributed to Stevens, Tolman, and Hull, though I would like to unpack this some more to argue (in line with my analysis in this and the previous chapter) that we should distinguish between two steps of conceptual explication, that is, (a) one of broadly explicating the folk-psychological concept in terms of environmental features and typical responses and (b) one of formulating specific operational definitions in terms of specific experimental manipulations and measurements.

Now, the fact that these psychologists shared a common concern does not mean that they had a satisfactory solution for it. To begin with, while Cronbach and Meehl (1955) are surely right to point out that any given pretheoretical concept can in principle give rise to more than one operational definition and testing procedure, the existence of several testing procedures does not guarantee that those testing procedures in fact measure the same object. Conversely, Campbell and Fiske's (1959) multitrait-multimethod approach offers a method for determining whether two instruments measure the same thing by tracing the variations in test behavior back to a common source. However, this method does not give any insights as to how further theoretical work can proceed from here. In other words, it does not address the question of how the concept in question can be improved. These two issues are addressed by converging operations (Garner, Hake, and Eriksen 1956), to which I turn next.

2.5. Converging Operations

While the two proposals discussed above come out of methodological concerns with mental testing, the converging operations approach comes out of experimental psychology. Garner, Hake, and Eriksen (1956) has been referred to as "a flagship publication which led to the cognitive revolution" (Logan, Coles, and Kramer 1996, xv). This causal claim is likely too strong (leaving aside the question of whether there really was an event in the history of psychology that can be described as a "cognitive revolution"). However, it is certainly true that Garner, Hake, and Eriksen address head-on the epistemological issues raised by the notion of *intervening variables*, understood to refer to entities or processes that have to be inferred from experimental data. This approach takes seriously an issue that we saw pop up in connection with both Tolman's and Hull's writings about theory construction (and then, again, in this chapter), namely, that for any given operationally defined concept (such as *demand* or *hypothesis*), inferences to a corresponding object are in principle underdetermined by the data. As we will see in a moment, this problem shows up even if no explicit attempt at theory construction is made (e.g., in Stevens's definition of the sensation of tonal density). The reason for this is that, even if the to-be-manipulated intervening variable in fact exists, and even if the experimental operation at hand indeed causally affects the intervening variable (two big ifs), there are bound to be other intervening variables. These other intervening variables can, in turn, selectively influence each other in such a way that the resulting experimental data cannot easily be attributed to the intended intervening variable. This is a problem in a research field where not only the to-be-manipulated intervening variable but also the possible confounders are not well understood.

Garner, Hake, and Eriksen explain this point with respect to sensory experience as an intervening variable. Assuming that we want to conceptualize (as Stevens did) the experience of tonal volume or density in terms of discriminatory behavior in response to certain kinds of stimuli, the question is how we can be sure that the discriminatory behavior is really due to the perception of the tone (as opposed to some other process within the behaving system). The peculiar shape of the behavioral response exhibited by an organism when faced with a particular stimulus might in fact be brought about by a feature of the response system, not the perceptual system (assuming that the distinction between perceptual system and response system is a valid one). Garner, Hake, and Eriksen remark: "We agree with contemporary operationists that the fundamental and prerequisite operation in any experiment on perception is to demonstrate a discrimination between stimuli on the basis of responses. . . . This operation alone, however, assures us only that we have a system which is operating and which is reliably assigning responses. This operation provides us with so little information about perception that indeed we cannot distinguish between perceptual and response processes" (1956, 150).

This statement contains several important points to consider. First, the assumption that the data pointing to a perceptual system might be confounded by the response system presupposes the very concepts of a perceptual system and a responses system. Positing such systems is purely conjectural, however. Garner, Hake, and Eriksen recognize this. However, they argue that such conjectures are needed for experimental work that aims at gradually gaining a better and better theoretical understanding of the domain under investigation. The important point in their proposal is that, in order to move research forward, one needs at least two competing hypotheses to account for a given experimental outcome. For example, if a subject has been presented with sensory material and prompted to respond to it, two competing hypotheses might be either (a) that the response is indicative of the sensation triggered by the stimulus or (b) that it is indicative of some feature of the response system. This calls for an additional experiment specifically to distinguish between these two possibilities. And this is precisely what Garner, Hake, and Eriksen mean by the term "converging operations": "Converging operations may be thought of as any set of two or more experimental operations which allow the selection or elimination of alternative hypotheses or concepts which could explain an experimental result." In other words, the first set of operations (suggested by a particular operational definition) provides data, but these can be explained in several competing ways. The second set of operations is then performed in order to decide between two interpretations of the data: "The two operations taken together provide the convergence" (1956, 150–51).

Garner, Hake, and Eriksen illustrate this with a thought experiment in which subjects are presented with four words—"fire," "save," "shit," and "fuck"—and asked to read them out loud. Two of these words have vulgar meanings. Garner, Hake, and Eriksen invite us to imagine an experiment that reveals the two vulgar words to have a "higher threshold" than the nonvulgar ones, which presumably means that it takes slightly longer to react to them. One possible explanation for this high threshold is that the perceptual system discriminates between the two kinds of words on the basis of offensive content. However, as Garner, Hake, and Eriksen point out, it could also be due to a differential inhibition of responses (by the response system). This means that we are looking at a situation of underdetermination. Garner, Hake, and Eriksen propose to resolve this situation by applying the method of converging operations, that is, an experiment in which the subjects are required to respond to vulgar words by pronouncing nonvulgar words and to respond to nonvulgar words by pronouncing vulgar words.

The rationale is as follows (see table 2.2). It is known from the previous experiment that the threshold is high for condition 1 and low for condition 4. If it turns out that it is high for condition 3, this will be interpreted as showing that the previous results were due to the response system. If it turns out that the threshold is high for condition 2, this will be interpreted as showing that the previous results were due to the perceptual system. If the threshold turns out to be high for both conditions, this will be interpreted as showing that the previous results were due to both systems (in which case one would have to formulate and test a more specific hypothesis about their interaction).

Table 2.2: Illustration of the converging operations approach

	(III) VULGAR RESPONSE	(IV) NONVULGAR RESPONSE
(i) Vulgar stimulus	1. Vulgar stimulus, vulgar response	2. Vulgar stimulus, nonvulgar response
(ii) Nonvulgar stimulus	3. Nonvulgar stimulus, vulgar response	4. Nonvulgar stimulus, nonvulgar response

Now, how does this approach relate to the desideratum (formulated above) for a method of concept formation? I argue that what Garner, Hake, and Eriksen (1956) point to is that experimental research on a specific intervening variable cannot easily be separated from research on how it is causally entangled with other intervening variables. In practical terms, thus, the experimenter needs to find ways to control for other intervening variables. This is not at all trivial because it is not self-evident what other intervening variables there might be, let alone how to control for them. The practical need to control confounding intervening variables thus translates into the need for specific theoretical work: that of modeling the intervening variable in question in relation to its domain.

This way of thinking has several important consequences. First, it suggests that concept formation is closely tied to theory construction in that our growing theoretical understanding of, say, perception is going to be embedded in our growing understanding of the mind/behaving system as a whole. Second, the type of theory in question is going to be (at least in part) mechanistic insofar as it aims to contribute to an understanding of the ways in which various intervening variables causally interact and causally contribute to behavior.[19] Third, the process of theory construction starts out with very specific experimental circumstances, inquiring into the variables and causal processes that explain experimental data. The proposal is, therefore, true to the original operationist spirit of trying to tie intervening variables to particular experiments while putting a specific spin on the recognition that intervening variables are underdetermined by experimental data, that is, turning this underdetermination into a tool of concept formation. Thus, rather than posing deep skeptical problems, the underdetermination in question has a positive role to play in concept formation and theory construction. I elaborate on the relevance of this in chapter 7 below. There, I argue that this approach captures an important aspect of and contains normative recommendations for research practices in experimental psychology, and I offer an account of what its epistemic merits are.

2.6. Conclusion

In this chapter, I have provided an account of the development of operationism from the late 1930s to the late 1950s. In doing so, I have elaborated on my construal of operationism as it was understood by psychologists (and some philosophers) at the time. I have identified

---

19. For a discussion of these two points, see also Feest (2012a).

(*a*) several problems faced by experimental research in psychology as highlighted by debates about operationism as well as (*b*) some of the attempted solutions offered by psychologists in the operationist tradition. The purpose of this chapter has, thus, been twofold. On the one hand, I have continued to show that operationism in psychology is not what it is commonly taken to be (i.e., a simpleminded theory of meaning). On the other hand, I have shown that, as it was debated in the 1940s and 1950s, operationism points to some features of experimental research in psychology (and perhaps in other areas as well) that have not received much attention in mainstream philosophy of science, such as the question of how researchers experimentally individuate and investigate their subject matter when working in conceptually and epistemically relatively uncharted territory. The overview of some of the problems discussed by psychologists contains important insights, but they are only the beginning of a more sustained philosophical analysis, which will take up the rest of this book.

In pursuing this end, I essentially take the analysis offered above of the investigative process in experimental psychology to be descriptively accurate and normatively on the right track given the epistemic situation of the experimenter. In trying to make philosophical sense of the epistemic perspective of scientists (and of the ways in which they deal with it), we will find that we need to rethink some fundamental categories in philosophy of science in general and in philosophy of psychology and cognitive science in particular. Accordingly, and picking up from the analyses in this and the previous chapter, I argue that operational definitions (as I analyze them) are ingrained in specific experimental designs that aim at classifying and describing what I refer to as "objects of research." Chapter 3 develops and defends the thesis that operational definitions are research tools. I introduce two case studies from more recent psychological research: implicit memory and working memory. Chapter 4 provides an analysis of the notion of an *object of research*, arguing that such objects are epistemically blurry. I explain both the notion of *epistemic blurriness* and the notion of *research object*, illustrating them by means of the two case studies. In chapter 5, I contrast my usage of the expression "object of research" with the more common "phenomenon," arguing that objects of research are clusters of phenomena in which specific phenomena (those highlighted by operational definitions) are treated as providing epistemic access to the research objects at hand. Chapter 6 situates my account of objects of research within the literature about natural kinds. Chapter 7 returns to the issues of converging operations and operational analysis.

CHAPTER THREE

Operational Definitions as Tools

3.1. Introduction

In the previous chapters, I presented an account of the usage and reception of operational definitions in early to mid-twentieth-century psychology. In doing so, I highlighted that early advocates of such definitions did not intend for them to provide necessary and sufficient conditions of application. Rather, they should be viewed as preliminary conceptual presuppositions (operational definitions^{wide}) and/or tentative conditions of application (operational definitions^{narrow}), intended for specific investigative contexts (experiments), with the explicit aim of researching a subject matter that went beyond and, hence, had to be inferred from experimental data. As we saw in chapter 2, commentators in the 1940s and 1950s picked up on several aspects of this in that they took operationism to be highlighting the complexity of and the degree of ignorance about the subject matter of psychology. They therefore emphasized genuine uncertainty as to whether any given operational definition really singles out a robust scientific kind (and whether two operations purporting to single out the same kind really did so). This (*a*) made it prudent to relativize a scientific result to the specific ways in which the corresponding concept had been operationally defined and (*b*) gave rise to methodological debates about how the concepts in question (and the instruments thought to detect their referents) might be validated. While this response was one of general epistemic caution, this caution was also seen as having a productive aspect in that it urged scientists to explicate and critically examine the background assumptions inherent in their research designs. Therefore, reactions to early operationism were not only focused on the limits of operational definitions, given what was currently known about the subject matter of psychology. They also emphasized the potential that operationism could play a role in expanding that knowledge.

The current chapter is the first of several in which I explore this productive aspect of operational definitions (and of operationism more generally). I introduce two case studies that will help us appreciate that the issues debated in response to the proposals by early operationists are still relevant with regard to more current psychological research. The main thesis of this chapter is that operational definitions, both narrow and wide, can be thought of as tools of exploration. They can play this role by virtue of defining the objects of research in ways that are deeply ingrained in methods of measurement and experimental designs. More specifically, they are tools that can play a vital and productive role in a dynamic process of conceptual development. A similar idea is articulated by Hasok Chang when he suggests that operationism can provide a "philosophy of extension." Chang argues that "concepts can and do get extended to fresh new domains in which theories are uncertain and experience scant, even if no definite measurement operations have been worked out" (2019, sec. 3.2).

This thesis raises a number of questions, which I address in this chapter. How are operational definitions related to the concepts whose developments they aid? If operational definitions are tools, how do they do the epistemic work I am attributing to them, and what is the notion of *tool* in play here? What is the nature of conceptual development, such that operational definitions are able to play the kind of role in it that I have attributed to them? With regard to the first question, I have emphasized that I do not take the meanings of scientific concepts to be exhausted by the two types of operational definitions I have identified since they are—among other things—infused with folk-psychological content. But the question is how concepts can reach out of the realm of folk psychology into the realm of scientific conceptual development. I claim that both wide conceptual presuppositions and operational definitions more narrowly conceived are vital because they form brackets between folk-psychological presuppositions and experimental actions, which in turn will produce data that can aid in the development of the relevant concepts. With regard to the second question, I argue in this chapter that operational definitions function as tentative and preliminary premises in arguments that allow for specific inferences from the data and that there is no reason why we should not refer to such premises as "tools." In relation to the third question, my primary aim in this chapter is to highlight a few desiderata that a theory of conceptual development in psychology should meet. I develop these by (a) highlighting some shortcomings in traditional accounts of concepts and (b) pointing to approaches that appear to share some of the intuitions informing my approach.

3.2. Operational Definitions and Research Designs in Memory Research

I argue that my analyses of the function of operational definitions (and the ways in which they are derived from folk psychology) can be extended to more recent research practices in experimental psychology. This section begins to lay out two case studies from memory research. These case studies (*a*) illustrate the distinction between operational definitions^{wide} and operational definitions^{narrow} and also (*b*) begin to familiarize the reader with the notion that researchers take their conceptual presuppositions and operational definitions to be directed at objects of research. Such objects, I argue, are invested with more content than is provided by the operational definitions. This content puts normative constraints on the ways in which the concepts can be extended and developed.[1] Operational definitions^{wide} function as tools in the process of experimental research on a given type of research object by providing the schematic outlines that allow for an experiment to be set up in the first place. In turn, operational definitions^{narrow} are also tools in that they encode specific, paradigmatic conditions of application that are used to conduct experimental interventions when investigating specific objects of research. In this section, I begin to explain how this plays out in memory research. I argue that, taken together, the two notions of *operational definition* give us a pretty good first understanding of how conceptual assumptions are ingrained in the experimental designs used by scientists.

3.2.1. Operational Definitions^{wide} as Conceptual Presuppositions

The term "memory" has a fairly straightforward commonsense meaning, being—roughly—connected to the ability to display behaviors that indicate the recollection of something that was previously experienced.[2] While this surely does not exhaust the meaning of the term, it translates easily into a wide operational definition of it, as specified above and in

---

1. My analysis of the tool-like character of definitions is similar to a recent proposal by David Colaço (2022) in that we both hold that definitions play a role in (what Colaço, following Kendig [2016a, 2016b], calls) the "process of kinding" and, thus, do not provide definitive answers as to the exact features of the kinds under investigation. While Colçao argues that the contents transported by definitions can sometimes be regarded as hypotheses, my focus here is the conceptual presuppositions that already need to be in place *in order even to formulate hypotheses* about specific objects. I return to this point in sec. 3.3.1 below.
2. Colaço (2022) rightly remarks that this general understanding is compatible with there being different kinds of memory (e.g., long- and short-term memory, implicit and explicit memory, etc.).

earlier chapters. Just as Stevens defined the term "experience" as discriminatory behavior in response to stimuli (see chap. 1), memory can be (and, for the purpose of research, typically is) defined in terms of behavior that indicates recollection and/or recognition of previously learned stimuli. This broad notion underlies most experimental memory research in psychology (Roediger and Goff 1998), which distinguishes between a study phase, an intermediary phase, and a retrieval phase (Lockhart 2000). This way of setting up memory experiments, in turn, corresponds to a tripartite conception we have of memory phenomena in general involving encoding, storage, and retrieval, encoding taking place during the learning phase, storage taking place during the intermediate phase, and retrieval taking place during the test phase. In this way, then, the rationale of memory experiments tracks a folk-psychological understanding of the object in question. This, I argue, can be seen even in the more advanced taxonomies of memory associated with specific manipulations and measurement procedures.

Before continuing, I give a brief overview of current taxonomic distinctions in memory research, all of which, I argue, are variations of the theme I have just specified (i.e., memory as consisting of three elements: learning, storage and retrieval). Psychologists commonly distinguish between several types of memory. Probably the oldest distinction (within modern theories of memory) is one between short-term and long-term memory, in which the former is thought of as the ability to keep material in mind for the duration of a few seconds and the latter is conceived of in terms of the ability to store and retrieve items over a longer period of time. Something like this distinction is sometimes traced to William James's notion of *primary memory* and was prominently expressed by Atkinson and Shifrin (1968). This taxonomy was subsequently further broken down into different types of both short- and long-term memories. For example, within the category of short-term memory, Baddeley and Hitch (1974) introduced several modality-specific subunits (such as the so-called phonological loop and the visuospatial sketch pad), on the one hand, and an executive control unit with a modality-unspecific storage unit, on the other. Similarly, the past thirty years have seen a proliferation of types of long-term memory: starting with a distinction between declarative and procedural memory (the former storing facts, the latter abilities [e.g., Cohen and Squire 1980]), continuing with distinctions between several kinds of declarative memory (e.g., semantic memory, on the one hand, and "episodic" or autobiographical memory, on the other [e.g., Tulving 1983, 1985, 2000]), and most recently a distinction between explicit and implicit memory, the former thought of as a conscious form of retrieval, the latter as an unconscious form of retrieval and, perhaps, also of storage (e.g., Schacter 1990).

There is no general answer to the question of how these taxonomic distinctions are arrived at, but the introduction of a novel memory concept is often tied to specific methods and empirical findings thought to individuate the corresponding kind of memory. The method in question can be devised in an effort to measure a memory kind posited on theoretical grounds, or the theoretical posit can be the result of novel findings. An example of this latter kind of case is the occurrence of so-called functional dissociations between the results of different methods of measurement. In memory research, for example, this can occur when two different memory tests have diverging results even though the prior learning conditions were identical.[3]

3.2.2. Operational Definitions[narrow] as Paradigmatic Conditions of Application

Regardless of what reasons lead scientists to introduce a novel memory concept, I argue that, once specific effects have been deemed to individuate a novel kind of memory, the conditions that produce the effect are often treated as providing operational definitions[narrow] in the sense outlined above, that is, as providing sufficient conditions of application for the relevant memory concept. Such an operational definition essentially states that, if after being exposed to a learning procedure, a subject is tested with the memory test in question, then an instance of the memory type in question is present if and only if the test registers an effect. I have more to say below about how such paradigmatic conditions of application figure in the larger picture of research designs (in particular, in relation to the notion of a *paradigm*). First, however, let me briefly look at two examples of experimental measurement procedures that illustrate what I have in mind. They concern short-term (or working) memory and implicit memory.

3.2.2.1. *The Case of Short-Term Memory*

The presumed phenomenon of short-term memory was popularized especially by Alan Baddeley and his colleagues (e.g., Baddeley and Hitch 1974),[4] who go back to earlier neurophysiological evidence indicating that certain kinds of brain damage can lead to differential deficits on standard long-term memory tests (Baddeley and Warrington 1970).

---

3. There is also the opposite case, in which there is a difference between performance on a memory test given different populations (e.g., when one population has a specific neurophysiological deficit and the other does not).

4. I am here using the terms "short term memory" and "working memory" interchangeably, but, as we will see below, some researchers today use them to refer to what they take to be different phenomena.

There are two characteristic features that have long been thought to distinguish short-term memory from long-term memory: *duration* and *capacity*. The first characteristic is supposed to answer to the question of what the time span of short-term memory storage is, and the answer is usually taken to be something like twenty seconds. The second characteristic is supposed to answer the question of how many items can be kept in short-term memory. It has long been assumed that the answer to this question is 7 ± 2 (Miller 1956), but more recent research suggests that it is 4 ± 1 (Cowan 2005).

In general terms, short-term memory is conceptualized along the lines described above, that is, as somehow involving learning, storage, and retrieval (operational definition[wide]). It is experimentally investigated by specific measurement procedures that are, I argue, taken to provide operational definitions[narrow].[5] We can roughly distinguish between two classes of tests of short-term memory—*simple span tests* and *complex span tests*—where the former is assumed to tap short-term recollection in the absence of any interfering other tasks and the latter is assumed to tap the way short-term recollection works while one is attending to a different problem. Simple span tests—of which Ebbinghaus's (1885) nonsense syllable method is probably the most well-known example—present subjects with lists of items and then take some measure of their recollection. Complex span tests are basically simple span tests with some additional task. As outlined by Conway and colleagues (2005), three types of complex span tests are especially widespread: the reading span test, the counting span test, and the operation span test. For example, in the original version of the complex reading span test, subjects had to read sentences with the instruction to remember the last words while also judging the logical accuracy of the sentence (Daneman and Carpenter 1980). In recent times, the so-called N-back tests have come to replace complex span tests as a measure of working memory (see Kane et al. 2007). N-back tests present subjects with series of stimuli and instruct them to report whether any given stimulus matches the one that appeared N items ago. This presents subjects with the dual task of remembering a given item while also remembering how many items ago it appeared.

I claim that these measurement procedures for working memory play the function I have attributed to operational definitions[narrow] in previous chapters. They are empirically based and embedded in wider conceptual presuppositions about memory (operational definitions[wide]). They provide paradigmatic conditions of application for the concept in question, thereby enabling the formulation and testing of specific questions about

---

5. In chap. 4, I will return to the difficulties surrounding the issue of how such descriptive features are determined empirically.

the corresponding object of research. For example, once I have defined working memory in terms of the *N*-back task (by stating that the test's results provide sufficient conditions of application for the term), I can ask whether working memory is differentially affected by various treatments, for example, by a concentration training or by specific learning procedures. However, in practical terms, there are two obvious caveats here. First, I have to assume that my treatment (e.g., concentration training) really hits the intended target (short-term memory) and, second, I have to assume that my measurement procedure (e.g., an *N*-back-test) really measures the target. Neither of these points are trivial within the realm of research that is conducted within a domain marked by a high degree of uncertainty.[6] Focusing on the latter issue, an overarching claim of this chapter is that the question of whether a given operational definition adequately individuates a given object of research cannot be answered a priori. Researchers have to rely on such definitions as tools of research prior to knowing whether there even is a stable corresponding research object. That said, I argue that they can still play a valuable epistemic function, even in the course of conceptual change with regard to the very object of investigation. This comes out particularly clearly if we consider the fact that (as mentioned above) there has been a shift in the preferred test for short-term memory from simple and complex span tests to *N*-back tasks. If they do not agree, questions are raised, such as whether short-term memory behaves differently under different measurement conditions, whether one of them simply does not measure short-term memory, or whether the shift signifies a shifted understanding of what typical features of short-term memory are. I argue below that, in this particular case, the shift in measurement procedures does indeed indicate some conceptual developments, debates, and disagreements, thus pointing to the highly dynamic character of this research. For the moment, however, my point has been merely to highlight the tool-like character of the definitions ingrained in paradigmatic conditions of application.

3.2.2.2. *The Case of Implicit Memory*

Let me turn to my second case study: implicit memory. This field, too, can be characterized in terms of specific types of measurement procedures presumed to reveal the object of investigation. Traditional memory tests are either recall or recognition tests. When subjects take such tests, they are explicitly asked to report what they remember—or

---

6. In chap. 7 below, I refer to these two problems as the "manipulation challenge" and the "measurement challenge," respectively.

recognize—from the study phase. We can refer to both kinds of tests as "explicit tests." Such tests can be contrasted with so-called priming tests, which have come into wider use only in the last thirty years or so. In such a test, a subject may be asked to complete a word stem (e.g., "bre__") with two missing letters or to decide whether a given string of letters on the computer screen is a word. In the former case, the subject is judged to have remembered a word if the likelihood of completing a word stem with a word from the learning phase is above chance. In the latter case, a subject is judged to have a memory of an item if the reaction time is shorter for words from the test phase than for others.[7] We can refer to this kind of memory test as an "implicit test" because the subjects are not explicitly asked to report what they recall or recognize.

In the early 1980s, there was increasing evidence for the existence of experimental dissociations between the results of explicit and implicit memory tests. This means that the results on the two kinds of tests differed even though the learning conditions in the study phase had been identical. For example, it was found that normal (i.e., nonclinical) subjects have difficulties explicitly remembering verbal items a week after the learning phase, while their performance on an implicit memory test was as good as it had been right after the learning phase (Tulving, Schacter, and Starck 1982). Another example of such a dissociation can be seen in the fact that the results of explicit tests can be influenced by certain variations of the conditions during the study phase that do not appear to affect the performance on implicit tests. For example, by the 1980s, it was already known that performance on explicit tests for words was better if subjects had been asked to form sentences during the study phase ("elaborative study conditions") than if they were asked to count the vowels in the words ("nonelaborative study conditions") (Jacoby and Craik 1979). But now it was shown that the same was not true for performance on implicit tests (Graf and Mandler 1984). A third example of dissociations between performance on explicit memory tests and performance on implicit memory tests came from cognitive neuropsychology. The experimental dissociations were found in patients who suffered from amnesia (typically patients with lesions in the hippocampus). While it was already known that amnesia is the loss of the ability to form new memory structures, a finding that was taken to support the distinction between short-term memory and long-term memory (Baddeley and Warrington 1970; Warrington and Weiskrantz 1968, 1970), these assumptions had to be revised in light of the finding

---

7. For a classic instance of this kind of research, see Meyer and Schvanefeldt (1971).

that H.M.—probably one of the best-known amnesia patients in the neuropsychological literature—still performed well on implicit memory tests while doing poorly on explicit memory tests (Graf, Squire, and Mandler 1984; Warrington and Weiskrantz 1982). This suggested that at least part of his long-term memory was not damaged, raising the question of whether additional subcategories of long-term memory were called for. This hypothetical possibility was first suggested in Tulving, Schacter, and Starck (1982). It was only three years later that these two posited forms of memory were dubbed "implicit memory" and "explicit memory" (Graf and Schacter 1985; Tulving and Schacter 1990), the former being essentially operationally defined in terms of priming effects.

While the details of this case are quite different from those of short-term memory, I want to draw attention to the main similarity. Implicit memory was early on tied to a specific class of tests, namely, priming tests, which were, for the purpose of research, treated as providing paradigmatic (sufficient) conditions of application for the term. A major difference, of course, is that implicit memory is assumed to be a form of long-term memory. However, like short-term memory, it was closely associated with specific tests that subsequently functioned as operational definitions[narrow] (and, hence, as research tools) for the investigation for the research object in question. There was an additional parallel, namely, that the term "priming tests" in fact denotes a class of tests, not a specific test. The members of this class differ, (e.g., in the modality of the test material), which raise questions about whether all of them in fact measure the same thing.

Summing up, I argue that operational definitions, broad and narrow, function as research tools. By containing rough (and folk psychologically inspired) conceptual assumptions about memory in general, the former lay down the framework in which (types of) memory can be experimentally investigated. By laying down paradigmatic measurement procedures for specific (presumed) types of memory, defined in terms of experimental effects, the latter allow for an empirical investigation of the memory kind in question.

3.2.3. Definitions as Clarifications

Now, the language of "tool" is suggestive, of course, but it needs to be spelled out in more detail. Specifically, what do these tools do exactly, how do they function in research, and is it legitimate to use the term *tool* here? Before addressing these questions in section 3.3 below, let me first say a little more about the question of what these tools are for. As a first approximation, we can say that they are tools of conceptual improvement. This brings to mind Rudolf Carnap's notion of *explication*, as introduced in his *Logical Foundations of Probability* (Carnap 1950b/1962). As has been emphasized by a number of authors in recent years, by "explication" Carnap meant not the descriptive project of explicating implicit aspects of a concept's current usage but rather the activity of explicating what we should mean by a concept, that is, of improving it (Haslanger 2012). Needless to say, this raises questions about the normative standards by which such conceptual improvement can be judged. But it also raises the question of how a current concept can be utilized in its own improvement.

A closer look at *Logical Foundations of Probability* will be instructive here because, even though he proposed using this method for philosophical purposes, Carnap explicitly modeled it on what he took to be a scientific strategy, that is, that of "the transformation of an inexact, prescientific, concept, the explicandum, into an exact concept, the explicatum" (1950b/1962, 1). Moreover, he highlighted that "the explicandum may belong to everyday language or to a previous stage in the development of scientific language" (3). This situation looks a lot like the one I have described above since words like "memory" are obviously taken from our everyday language but the aim of scientific research can be construed as being that of making them more exact. As Carnap recognized, this raises questions about the conditions of adequacy for the new concept. He answers this question by demanding that the new concept be similar to the old one, more precise, more fruitful, and as simple as possible (to the extent that this is compatible with the first three demands). He emphasized, however, that there could also be trade-offs among the criteria. For example, he understood the criteria of precision and fruitfulness to be fulfilled if the explicatum was (*a*) embedded in a network of well-connected concepts and (*b*) useful for the formulation of many universal statements. However, he recognized that these criteria might require a radical shift in the meaning of the concept (a point he illustrates with the example of the conceptual shift between the explicandum *fish* and the explicatum *pisci*, where the former, but not the latter, includes dolphins and whales). In other words, he envisioned the possibility of conceptual change occurring in the course of the explicatory process.[8]

It is easy to see that, historically speaking, Carnap's criteria of precision and fruitfulness are deeply rooted in midcentury developments

---

8. This latter fact has given rise to the charge that explications really amount to a change in subject (e.g., Strawson 1963; see also Justus 2012). I return to the relationship between conceptual change and change of subject in sec. 3.5 below.

in philosophy of science that I covered in the previous chapter, that is, an increasing understanding of concepts as embedded in a network of other concepts, where those connections either constituted rules of application or allowed for the formulation of empirical laws (see sec. 2.3). However, it is not so clear that (both then and now) psychology aims to formulate laws. Hence, I would argue that, if we want to make sense of the explicatum being fruitful, we should keep an open mind as to what this means (including, e.g., allowing for the possibility of formulating regularities, making predictions, or formulating descriptive and mechanistic hypotheses). Now, with respect to the desideratum of the explicatum's being precise, what Carnap envisioned was for the new concept to be embedded in a system of concepts that provide clear rules of application (as opposed to the vague rules of prescientific conceptual contexts). Now, there is a sense in which operational definitions indeed provide clear rules of application. At the same time, they clearly do not do so by virtue of being embedded in a well-worked-out system of concepts. If anything, I argue, operational definitions are constructed precisely because there is no well-worked-out scientific network of concepts (though there is, as we saw in the previous chapter, a "sketch of a network" based on folk psychology). But how could such a sketch conceivably be turned into something more systematic?

I suggest that the question can be addressed within (something like) the Carnapian framework of explication (a) if we understand explication to be an iterative (Chang 2004), dialectical (Carus 2007), or dynamic (Brun 2016, 2020) rather than a linear process and (b) if we adopt a more liberal understanding of what *fruitfulness* means (i.e., as more than just allowing for nomological generalizations). With regard to the former, what I have in mind is that the system of concepts that is supposed to lend precision to mature scientific concepts does not emerge ex nihilo. Hence, if the method of explication is to be illuminating here, it has to refer to a dynamic and piecemeal process of conceptual development and refinement. Georg Brun (2016) suggests a helpful distinction between the *clarification* and the *explication* of a concept where only the latter goes beyond the initial concept while the former brings out existing assumptions in a clear form. With this distinction in mind, we can appreciate that operational definitions (especially operational definitions^{wide}) are clarifications of folk-psychological concepts that are then cast into the shape of a specific measurement operation (operational defintions^{narrow}), which in turn allows for the exploration of the research object, ideally resulting in an explication (improvement) of the initial concepts. With this analysis in place, we can now return to the sense in which operational definitions can be understood as research tools.

3.3. Operational Definitions as Tools: What Do They Do?

My claim is that operational definitions and their implementations provide *premises* relative to which scientists make inferences about their subject matter. So, for example, let us assume that short-term memory is operationally defined[narrow] in terms of the N-back task. This means that the results of the N-back task are treated as paradigmatic (sufficient) conditions of application for the term "working memory." In turn, this means that, when we perform a causal intervention (intervention X) on a sample of human subjects and observe an effect on N-back data, we can infer that intervention X has an impact on working memory. This inference crucially relies on the premise that the N-back task indeed measures working memory. But this is not the only relevant premise here. Say I want to investigate whether short-term memory is affected by a particular learning procedure (e.g., Hebbian learning). In that case, I can administer the learning procedure and register the effects with the N-back test in question. I then make an inference from the test result to a statement about the effect of the treatment on short-term memory, given the assumption that short-term memory exists, that it is adequately defined in terms of the N-back task, and that the experiment is implemented adequately (i.e., without introducing confounders).

Notice that it is far from clear that any of these premises are warranted. There might be no such thing as working memory, and the experimental effect in question might point to another psychological kind altogether. On the other hand, it is also possible that working memory does exist, that the test in question does measure it under some experimental conditions, and that its results are confounded by unidentified variables under other conditions. While a given operational definition might be adequate in principle, it might not be implemented with the proper care in a given experiment. Or there might simply not be an objective fact of the matter as to whether working memory is a natural kind.[9] All these possibilities need to be addressed by a philosophical analysis of the research process. However, none of them negate the basic point of this chapter, which is that, by virtue of supplying premises for inferences from experimental data, operational definitions are essential research tools.

---

9. Javier Gomez-Lavin (2021) has argued that working memory is not a natural kind. By this he means that working memory is not a genuinely explanatory concept but merely redescribes the phenomenon. My focus here is not on the question of kinds (but see chap. 6 below). Rather, the question I ask here is how we might analyze the conceptual and research dynamics with regard to this (presumed) object. Moreover, I argue in subsequent chapters that a lot of research is actually geared not toward explanation but toward describing and delineating the object under investigation and the phenomena we attribute to it.

3.3.1. Tools, Paradigms, and Experimental Designs

As already indicated in the introductory chapter, it will be useful to distinguish between the concepts of *operationalization*, on the one hand, and *operational definition*, on the other. Briefly, operationalizations pertain to questions or hypotheses, and operational definitions pertain to concepts. For example, say I wanted to investigate whether implicit memory is differentially affected by different kinds of learning procedures. In this case, I have to devise an experimental design that operationalizes a specific hypothesis to test with regard to the question. Such a design will consist of an appropriate choice of independent variables (the experimental treatment, which, in this case, is the study condition) and dependent variables (the data resulting from attempts to measure the effect that the treatment has on implicit memory). While there is a sense in which the choice of independent variable also involves an operational definition (e.g., of Hebbian or some other kind of learning), my focus here is on the operational definition of the concept pertaining to the actual object under investigation, that is, implicit memory. As I have argued above, this concept is operationally defined^{narrow} in terms of a specific (type of) experimental procedure, namely, priming, which is assumed to provide epistemic access to the type of memory in question. In other words, when investigating the effect of a treatment on implicit memory, researchers are likely to draw on an operational definition of implicit memory that is specified in terms of a paradigmatic condition of application for this concept, namely, a priming test. A similar logic applies if I am interested in the effects of a treatment on working memory. In both cases, the measurement procedures in question are employed because they are taken to provide paradigmatic conditions of application for the terms "implicit memory" and "working memory," as laid down in their operational definitions^{narrow}. In turn, these paradigmatic conditions of application are embedded within wider presuppositions about memory in general (as consisting of learning, storage, and retrieval), that is, operational definitions^{wide}.

I stated above that the term "operationalization" refers to questions or hypotheses and that the term "operational definition" refers to concepts. Now, there is, of course, a sense in which operationally defined concepts themselves have a hypothetical character in that (as already mentioned) researchers cannot be certain that there is really a genuine kind in the extension of the concept thus defined. This notion of *hypothetical* needs to be disambiguated from the notion that operational definitions are, themselves, hypotheses. The relationship between operational definitions and hypotheses is that, in order even to formulate/

test a hypothesis about a given object, an operational definition of the latter is a prerequisite.[10] This is important for my purposes since (as I show below and argue more fully in chap. 4) psychologists often test descriptive hypotheses about their research objects. Notice that this is entirely compatible with the claim that it is possible to do research on a given object merely by employing an operational definition and without testing a specific hypothesis about it. Friedrich Steinle (1997) has dubbed the latter kind of research "exploratory" experimentation. While his main point is that this type of experimentation does not aim at hypothesis testing, he emphasizes that it nonetheless requires what he calls "conceptual frameworks" (Steinle 1997, 2009; see also Haueis 2017, chap. 3). His point is that these conceptual frameworks can fundamentally shift as a result of exploratory experiments. Here I am arguing that, in psychology, operational definitions can sometimes do the work of Steinle's conceptual frameworks.

The similarity of my operational definitions to Steinle's conceptual frameworks, combined with my talk of such definitions allowing for paradigmatic conditions of application for specific terms, also brings to mind the Kuhnian notion of a *paradigm*. As is well-known, in *The Structure of Scientific Revolutions* (1962), Thomas Kuhn used the concept of a *paradigm* to refer both (*a*) to a larger network of interrelated beliefs, material components, and empirical and theoretical practices that the members of a given discipline share and (*b*) to specific typical instances of a given object or practice. He referred to the former as "disciplinary matrices" and the latter as "exemplars" (as laid out in the postscript to Kuhn [1970]). I argue that my analysis of the origin and function of operational definitions speaks particularly to this latter aspect of paradigms. Operational definitions are exemplars both because they codify paradigmatic measuring procedures for specific objects (N-back tasks for working memory, priming tasks for implicit memory) and because, when these measurements are applied, the resulting effects can be regarded as paradigmatic empirical instances of the research object under investigation. Put differently, operational definitions translate into typical practices employed to measure a given object. This analysis is borne out by the way in which scientists use the term "paradigm," that is, as referring to specific kinds of experimental manipulations or measurement

---

10. Another way of putting this is to say that operational definitions have the function of what Reichenbach (1920/1965) had in mind when he defended the idea that there are constitutive (but historically situated) synthetic a priori principles in play in physics. A similar idea was held by C. I. Lewis (1927), who used the expression "pragmatic a priori."

techniques. The connection between experimental paradigms and operational definitions has also been recognized by Jacqueline Sullivan, who points out: "An operational definition is built directly into the design of an experimental paradigm" (2009, 514). Sullivan intends this insight to contain a word of caution about the potential arbitrariness and context specificity of experimental results. This point is well taken and raises important issues about when it is legitimate to extrapolate from an isolated experimental result to a general claim about a phenomenon. I return to this in chapter 7. However, here, my point is that operational definitions are inevitable and necessary components of experiments.

Summing up, my claim is that operational definitions are built into paradigmatic (if tentative) measurement procedures that are used as part of experimental designs where the latter operationalize specific questions about a given object of research. The basic idea is that, with the operational definitions in place, the data generated by the experiment can be used to make an inference to (some feature of) the object of psychological research. It is only by virtue of operational definitions being employed in experimental designs that it is possible to derive experimental results from experimental data. However, given the preliminary and fallible character of such definitions and their implementation in experiments, the inferences in question can deliver false results. This means that operational definitions are essential but potentially fallible tools of research.

3.3.2. Objections and Replies

It might be objected that my talk of operational definitions as tools is at most an analogy and not a very compelling one at that since (*a*) operational definitions are not physical devices like hammers and screwdrivers, (*b*) hammers and screwdrivers do not perform the kind of epistemic function I have attributed to operational definitions, and (*c*), when we use a physical tool, such as a screwdriver, we typically know the kinds of objects on which it can be used (i.e., screws). But here I am suggesting that there can be tools for something that is ill understood. What does this mean, exactly? Let me briefly respond to all three questions in order to bring out the general nature of my question more pointedly.

A tool is a device that aids us in achieving a particular purpose better than we would be able to without the tool. For example, a hammer is useful when we want to hang a picture. Clearly, tools like hammers are physical objects. Notice, however, that the techniques of mathematics, statistics, or logic can also be considered tools that are used, for example, in the course of scientific inference or data analysis. This latter point makes it clear that scientific tools do not have to be physical devices, though they are often physically implemented. For example, when a scientist uses a statistical rule of inference to test whether a particular empirical effect is significant, it is fair to refer to this method as a "tool" even though it is not a physical machine. This does not necessarily mean that there are no physical machines involved as the rule of inference is often implemented by means of a physical device—for example, a calculator or a computer—that delivers the results of the data analysis. Nonetheless, the example makes it clear that rules of inference can be regarded as tools; hence, tools do not need to be physical.

Now, with respect to the second question, it will be helpful to distinguish between nonepistemic tools and epistemic tools. Hammers and screwdrivers are nonepistemic tools because it is not their purpose to add to our knowledge. Thermometers and telescopes are epistemic tools because they help us learn something about the world. Likewise, psychological tests and questionnaires are epistemic tools. Thus, there is no reason to restrict the word "tool" to things like hammers and screwdrivers. But this brings us to the last objection. Thermometers and psychological tests can function as epistemic tools, it might be argued, because they have been validated, that is, because they have been shown to measure precisely what we take them to measure, for example, temperature or intelligence. The assumption that a thermometer indeed measures temperature is warranted by virtue of the fact that the thermometer has been calibrated, and this is why we can infer from the data delivered by the thermometer to the truth of the inference (e.g., that it is 25 degrees Celcius). But, in that case, it is the thermometer that is the tool, not some operational definition built into the thermometer. There are two points here that we need to consider separately: (*a*) in the case of standardized instruments, it is the physical apparatus, not the operational definition built into it, that we think of as a tool, and (*b*) we are warranted in treating the apparatus as a tool because it has been calibrated/validated. So why am I claiming that operational definitions, qua linguistic entities, are tools, and why am I claiming that they can be tools despite the fact that the very question of what they are measuring is still up for grabs?

My answer is that operational definitions are epistemic tools of a specific type, which we may call "exploratory tools" or "tools of conceptual development."[11] It is their purpose to probe into ill-understood territory. They contain preliminary and revisable assumptions about what is being

---

11. My analysis here is inspired by Heidelberger (1998), which distinguishes between representational tools and tools that add to our knowledge.

explored. And, since the objects of research and exploration are ill understood, the tools themselves and the concepts to which they are tied are still under construction. In this way, they differ markedly from tools that have already been validated. The latter are (to use Latour's term) "blackboxed"—or, as Hans-Jörg Rheinberger puts it, they are "technical things"—precisely because their status as tools is no longer in question. By contrast, operational definitions, as I use the term, play a much more dynamic role because the very boundaries and descriptive features of the objects they individuate are in question.

There is, then, an important sense in which operational definitions as research tools are tentative precisely because they are used as part of an ongoing investigative process. This highlights two things that I want to emphasize. First, by focusing on the ongoing process, I am paying less attention to the question of how a given taxonomic category might eventually be validated and more to the ongoing process of how it is formed and to the epistemic work operational definitions do in the process. This is not to say that questions of validation are uninteresting. However, as already laid down in the previous chapter, coming from the perspective of scientific practice, they are often premature. Putting myself in the epistemic position of experimental researchers, the important question I want to answer concerns the rationality of the investigative process given the epistemic uncertainty and conceptual openness of this process. Second, I take for granted that such investigative processes are typically directed at what are (at least tentatively) assumed to be worthy objects of research that are, however, in important ways ill understood. This means that the process is characterized by conceptual openness and epistemic uncertainty, combined with the assumption that there is a genuine and stable, mind-independent object of investigation.

This raises two (interrelated) follow-up questions. The ongoing research process is characterized by conceptual openness about ill-understood objects. But what do I mean by "ill-understood" objects in general, and how do I conceive of such objects in psychology more specifically? And, given my emphasis on the role of conceptual presuppositions and operational definitions in a process of conceptual development, how does my account connect with existing accounts of concepts and conceptual change? I will have more to say about the former question in the following chapters, where I spell out my notion of an ill-understood object of research as epistemically blurry and corresponding to a subject matter that consists of clusters of phenomena. However, before turning to these questions, I conclude the current chapter by situating my notion of *operational definitions* vis-à-vis existing philosophical accounts of concepts and conceptual change. While my analysis is explicitly not intended as a full-blown theory of conceptual change in psychology (let alone generally), a closer look at some of the issues debated there will serve to highlight some issues my account must address, in particular as they relate to when such change amounts to a growth in knowledge and when it amounts to a change in subject.

3.4. Conceptual Development and Reference: Another Look at the Case Studies

Let me emphasize that, by pushing the thesis of operational definitions as tools of exploration and conceptual development, I do not mean to argue that such change is entirely method driven. On the contrary, there are a variety of factors contributing to the ways in which concepts develop (including, as we saw, folk-psychological assumptions about the subject matter).[12] Operational definitions are one important tool in the research process, not the driving force. They are important because they link the conceptual with the material, semantic content with experimental manipulation.[13] I argue that one central feature of research objects in psychology is that they are not fully understood. This means (a) that operational definitions are, in a sense, educated guesses about the subject matter under investigation and (b) that we are likely to see conceptual developments and shifts occurring as a result of experimental investigations as the educated guesses get revised and adjusted. But this raises the question: What are the mechanisms and implications of such revision and adjustments in practice, in particular with respect to the question of reference? In the remainder of this section, I revisit the two case studies to get a better sense of the nature of the conceptual change that has resulted from experimental research on the two objects introduced above.

3.4.1. From Short-Term Memory to Working Memory

I suggested above that operational definitions can be tools for the change of the very concepts they define. This is, of course, a direct consequence of the idea that they are tools of exploration. The notion of a concept

---

12. In this regard, my proposal differs from a recent proposal about the role of tools in neuroscience (see Bickle 2016).

13. In a series of articles, Eden Smith (2018a, 2019, 2020) has also highlighted the role of concepts as research tools, emphasizing in particular their embeddedness in material practices as well as their historically contingent nature and the "subterranean" meanings of which concept users often are not fully aware and that can both enable and obscure insights about the subject matter. This analysis is obviously highly compatible with the perspective I develop in this chapter.

being defined in such a way that the corresponding object can be explored suggests that there is a stable object. I call this the assumption of *sameness of reference*. I argue that this assumption is crucial but can also be questioned and abandoned in the course of research.

I begin by illustrating conceptual change by means of the shift from short-term memory to working memory (sec. 3.2.1 above). As we saw, researchers initially defined the concept of *short-term memory* operationally in terms of a class of simple span tasks but then moved to include complex span and N-back tasks as well. I have characterized the difference between the two by saying that, whereas the former is intended to measure working memory capacity simpliciter, the latter is intended to measure working memory capacity under conditions of distraction. At least initially, scientists took themselves to be investigating the same object when using these two measures. That is to say, N-back tasks were not treated as measuring something other than what was measured with complex span tests. Rather, they were seen as measuring the same thing under different conditions, thereby contributing to our understanding of how short-term memory behaves under different conditions. This is compatible with my construal of operationism because, as we saw in the previous two chapters, operationism allows for the idea of a stable object that can be measured in more than one way. The notion that simple span tests and complex/N-back tasks measure the same thing began to change, however, when it turned out that the results of complex measures and N-back tasks are highly correlated with those of intelligence tests, whereas the data generated by simple measures are not (Conway, Kane, and Engle 2003). Now, of course this does not speak against the hypothesis that both kinds of tests measure the same object, revealing that some of its features are correlated with intelligence while others are not. Nor does it invalidate the findings generated by means of simple span tests. Yet it draws our attention to the ways in which the very notion of what are central (and interesting) features of a research object can shift, thus contributing to a shift in the way in which the relevant object of research (short-term memory) is being conceptualized. Researchers now take an interest in the ability to perform on complex span tasks in its own right, that is, not merely as a particular variation of a short-term memory task but as revealing a mental process closely related to executive control (e.g., Ricker, AuBuchon, and Cowan 2010). In brief, the idea is that, rather than being a storage space through which information passes en route to long-term-memory, the kind of memory at stake makes information available in the course of complex tasks. For example, we could not understand the end of a spoken sentence if we were not able to remember its beginning. Likewise, we would not be able to solve the simplest math problem if we were not able to keep the individual digits in consciousness. This notion has taken on some cultural and educational significance in recent years as it has been argued that executive control is positively correlated with long-term success in life. As a result, some researchers started distinguishing between short-term memory and working memory (Cowan 2008), the former being measured by simple and the latter by complex span tests. However, some researchers continue to use the term "short-term memory" as the generic term for both types of experimental results (e.g., Jonides et al. 2008), some treat working memory as including short-term memory (Cowan 2008), and yet others refer only to attention-related aspects (as opposed to modality-specific storage) of short-term memory as "working memory" (Conway et al. 2005).

The fact that scientists have responded in different ways to the empirical findings and conceptual developments just outlined is intriguing and does not allow for a clear-cut answer to the question of whether the change in name from "short-term memory" to "working memory" indicates merely a shift in which features of this research object are deemed to be interesting or whether it indicates that a kind splitting has taken place. The fact that, at first glance, both types of tasks require executive control might appear to back those who take the reference to have stayed the same. But this, too, is not so clear as a meta-analysis reveals that the correlations between the data generated by these two paradigms are lower than expected (Redick and Lindsey 2013), once again raising the question of whether this type of memory behaves differently under different manipulations or whether the experimental operations in question in fact manipulate different things.

In addition to these questions about whether we need to subdivide different types of short-term (or working) memory, there is an even larger issue, namely, a debate about the tenability of the very distinction between long-term memory and short-term memory. As described in section 3.2.1 above, this distinction was initially made plausible by neurophysiological data suggesting that certain kinds of brain damage can lead to differential deficits: patients with medial temporal lobe damage fared poorly on tests of long-term declarative memory formation and retrieval but somewhat better on short-term memory tasks (e.g., Baddeley and Warrington 1970). This gave rise to the assumption that the distinction between long-term memory and short-term memory is based on two separate architectures in the brain. However, this assumption has been challenged by some psychological theories of memory (Cowan 2001) and by reevaluations of the neuropsychological evidence (Jonides et al. 2008). According to these accounts, short-term memory is not a special storage space in the brain but rather a specific kind of mental state that occurs when items in long-term memory become activated by virtue of being the focus of attention.

3.4.2. From Semantic Priming to Sensory Priming

In the case of implicit memory, we see that the exact status and boundary conditions of implicit memory were initially unclear as scientists tried to account for the dissociations between explicit and implicit memory test results in terms of an influential theory of semantic memory, the semantic network theory, according to which items get stored in accordance with their semantic "distance" to one another. It was then speculated that, while performance on recognition tests (i.e., on explicit tests) was supported by a recollection of the study phase, that is, by episodic memory (Tulving, Schacter, and Starck 1982), performance on implicit tests was supported by semantic memory and that, therefore, the results on implicit tests were to be regarded as effects of *semantic priming*. However, by 1982, it was no longer clear whether this explanation could hold up as there was now also evidence for *sensory priming*, that is, that the priming effect occurred not by virtue of the meaning of a word but by virtue of some more superficial feature such as the way it was being presented (e.g., Jacoby and Dallas 1981). This made it reasonable to think of implicit memory as a memory system that was responsible for both semantic and sensory priming.

The assumption as to the existence of an overarching implicit system, too, was soon challenged. Importantly, for our purposes, what was called into question was not the existence of semantic and sensory priming effects but the firm conceptual tie of priming per se with implicit memory, which had previously been guaranteed by the operational definition. More specifically, some researchers raised the question as to whether the sensory priming effects warranted a description as *memory effects*. For example, Bowers and Kouider (2003) argued that the effects in question were effects of pattern recognition, that is, of sensory processing, not memory. However, Daniel Schacter, one of the most prominent advocates of the idea of an implicit memory system (e.g., Schacter 1999), argues that pattern recognition effects can still be memory effects. Accordingly, he conceptualizes implicit memory as a *perceptual representation system*, a system that has several domain-specific subsystem each of which is responsible for processing different sensory features of the experimental items (thereby explaining dissociations between and within sense modalities) and all of which work independently of the semantic (explicit) processing of the items (thereby explaining the dissociations between explicit and implicit tests).

In this way, Schacter has significantly shifted his views. While the expression "implicit memory" was initially supposed to refer to a system that was taken to function independently of whatever systems process specific sensory modalities, it is now (at least by Schacter) taken to refer to a system that specifically processes sensory aspects of experimental items. Thus, not only does the scientific conception of this phenomenon appear to have incorporated some of the features with which it was initially contrasted, but the concept is also significantly more narrow. The story told here can also be described in terms of what Hasok Chang (2004, 2016) describes as "epistemic iteration," where the very conceptual assumptions that made the research possible to begin with are subsequently revised so as not to include some of the cases initially thought to fall into the extension of the concept.

However, we also find an entirely different response in the literature. Mirroring Bowers and Kouider's (2003) concern, for example, Roediger (2003) has suggested that the mere fact of behavioral change without conscious recollection of what brought it about does not necessarily indicate that we are looking at a change in memory at all. He illustrates this with the example of an amnesiac bodybuilder who does not recall working out every day yet becomes increasingly bulky. By denying that bodily strength can indicate implicit memory, Roediger is appealing to a commonsense notion of *memory* as involving changes in the cortex. But, as this is not an analytic truth, there is nothing to stop other researchers from referring to mechanisms of pattern recognition as being attributable to implicit memory so long as these mechanisms give rise to priming effects that are dissociated from the effects of other long-term memory tests.

The disagreement just recounted throws some light on the ways in which conceptual and empirical questions are entangled in this research, where debates turn in equal parts on empirical results, existing scientific ideas about the subject matter, and the folk-psychological meanings of central terms, such as "memory." It also illustrates how a scientific finding (the dissociation between semantic and sensory priming) can be interpreted in entirely different ways, that is, either as resulting in a more fine-grained taxonomy (operational definitions^{narrow}) or as challenging the very conception of memory underlying this research (operational definition^{wide}).

3.5. Operational Definitions vis-à-vis Philosophical Analyses of Concepts

The two case studies just presented show clearly how operational definitions are essential in experimental explorations of research objects. They also reveal that the question of when a change in operational definition or even a dissociation between the measurements to which such definitions are tied implies a change in reference does not have a clear-cut answer a priori. On the face of it, this fact seems to make my analysis compatible with the kind of semantic holism associated with the description theory of meaning. I argue, however, that the description theory fails to provide a convincing story of how conceptual change can be grounded in experimental actions and data production. In turn, the causal theory of reference offers too rigid a picture of reference and reference change. I review these approaches with the aim of highlighting their insights and shortcomings (sec. 3.5.1). After that, I (*a*) tie my analysis of the role of operational definitions to a deflationary account of reference while (*b*) arguing that it can account for conceptual change (including reference change) on the ground (sec 3.5.2).

3.5.1. Insights and Shortcomings of Classical and Nonclassical Theories

Let me begin with the classical theory of concepts because (as we saw in the previous chapters) critics of operationism have mistakenly taken operationism to be committed to a version of it. According to the classical theory, concepts have a definitional structure that provides necessary and sufficient conditions of application for the words that correspond to the concepts (Laurence and Margolis 1999). This definitional structure is, furthermore, frequently equated with what Gottlob Frege termed the "sense" (as opposed to the "reference") of the concept (Frege 1892/1980). The sense of a concept contains the mode of presentation of the concept. Possession of a concept therefore enables the subject to identify objects that are in the extension of the concept. The best-known instances of the classical theory of concepts can be found in those forms of empiricism that assume that concepts can be defined in terms of observational features and that the possession of a concept consists in the possession of a list of the observational attributes that a subject uses to classify an object as falling under a particular concept. I have argued that operationists are not committed to the kind of rigid, old-fashioned empiricism just outlined. Nor did operationist psychologists even attempt to formulate a theory of meaning. For these reasons, standard critiques of the classical theory do not challenge operationism as I construe it.

Nonetheless, those standard critiques and alternative proposals are helpful in highlighting some of the pressures under which concepts can come in the course of the scientific developments outlined above.

There are two groups of standard reasons why the classical view must be regarded as having failed. The first type of argument against the classical view points to counterexamples that show how almost any definition in terms of necessary and sufficient conditions runs into cases in which either the definition includes instances that we intuitively do not take to be in the extension of the concept or cases in which the definition does not capture instances that we intuitively do take to be in the extension of the concept. The second type of argument points out that there is not much psychological evidence to suggest that in our ordinary lives we do in fact draw on the definitional structure posited by the classical theory when applying a concept. Not only that, but we seem to be able to apply concepts correctly when we are ignorant of their defining features or when we have positively false beliefs about the objects in their extensions. This latter point has led to theoretical and empirical work about the question of how we in fact manage to apply concepts correctly even though we do not seem to draw on a list of necessary and sufficient conditions in order to do so (e.g., Rosch 1978).

Two well-known theoretical responses to the problems with the classical theory of concepts outlined above are the description theory of meaning and the causal theory of reference, which are associated with the names of W. V. Quine and Hilary Putnam, respectively. The former holds on to the idea that meanings are provided by descriptions while relaxing the requirement that such descriptions be given in observational language only or that they provide necessary and sufficient conditions of application. The latter suggests that the meaning of a concept is only in part provided by the criteria by which we identify instances of it (the "perceptual stereotype") and in addition provided by the referent itself, regardless of how well we are currently able to describe it theoretically. Both these theories of meaning, I argue, explicate some intuitions we have about scientific concepts. Thus, for example, I claim that, when scientists choose one or several operational definitions for the purposes of research, they do so on the assumption that the definitions are associated with a concept that has a stable referent that can be investigated by means of the operational definitions (we can call this an assumption of *sameness of reference* across procedures). In some cases, this intuition might pan out in that a given research object is operationally individuated and subsequent research merely fills in the descriptive details. But, in many other cases, things turn out to be more complicated. For example, it is not entirely clear whether the shift from short-term memory to working memory is merely a development whereby different features of the research object are better understood now or whether we are looking at a genuine shift in reference in that the expression "working memory" refers to a different object than the expression "short-term memory." Likewise, with implicit memory, it is not so clear whether we should say that implicit memory was really a feature of sensory processing all along or whether we should say that Schacter is simply no longer talking about the same thing as before. The causal theory of meaning would maintain something like the former (i.e., there is a fact of the matter as to the referent of a given expression even if scientists do not know what it is), whereas the descriptive theory of meaning would say that, every time we change a description, we change the referent. In what follows, I argue that, while each of these points contains insights about scientific practice in experimental psychology, both fall short of providing satisfactory analyses of our case studies.

I begin with the descriptive theory, which (following the classical theory) maintains that intension determines extension, that is, that the reference of a concept is determined by a set of descriptions but (departing from the classical theory) that those descriptions are embedded in a network of empirical beliefs we have about their referents. It found an influential expression in Quine's "Two Dogmas of Empiricism" (1951), which famously rejected the notion that it is possible to draw a clear line between factual (synthetic) and conceptual (analytic or definitional) truths. While acknowledging that scientific theories contain statements that may look like definitions, they are holistically connected to factual elements of theories. But, since our factual knowledge of the world changes, so will our definitions, and, hence, there is nothing that can robustly fix the reference of a concept. Now, as we saw, the notion that definitions make factual assumptions and, hence, can change is clearly compatible with this picture. On the other hand, it is not self-evident that such shifts in operational definitions (and shifts in other descriptive features of the research object) necessarily imply shifts in reference. They may simply reflect an improved understanding of what scientists were trying to talk about all along. Moreover, the "hazy holism" (Wilson 2006) associated with descriptive theories of meaning fails to do justice to the very tangible experimental interventions that my account of operational definitions as tools has highlighted. Thus, I argue that my account of the role of operational definitions as enabling conceptually informed yet physically grounded data production offers a welcome middle ground between what Joseph Rouse (2015, chap. 7) has called the Scylla of "the Given" and the Charybdis of "freewheeling holism."

Nonetheless, it seems plausible that researchers often operate with an intuitive assumption of sameness of reference across changes in description. The question of what might justify this intuition was addressed by Hilary Putnam. In "The Analytic and the Synthetic" (1962/1975a), Putnam essentially agreed with Quine that the analytic/synthetic distinction is untenable for statements pertaining to objects of scientific study and that natural kind terms (the theoretical concepts of the sciences) therefore cannot be provided with definitions in the traditional sense. However, he pointed out that there are historically documented cases where people changed core definitions of their concepts yet continued to talk as if they were still talking about the same referent. As an example, Putnam cites the fact that, even though non-Euclidean geometry fundamentally overthrew the very definition of physical space, scientists continued to think of geometry (both before and after the shift) as being *about* physical space. Putnam's famous "The Meaning of Meaning" (1975b) was intended to provide a justification for this assumption. As is well-known, Putnam argued that the referents of natural kind terms get fixed independently of the beliefs people have about those referents. Briefly, (*a*) meanings cannot be identified with intension, (*b*) intension does not determine extension, and (c) we do not always know what we mean by the words we utter. Instead, the meaning of a word consists of a perceptual stereotype and an extension where competent users of the word do not necessarily know the extension even though the perceptual stereotype enables them to apply the concept correctly. Our words, then, have referents by virtue of the existence of a causal link between our current language use and the substances that were present when the words were introduced. And we are, thereby, justified in the belief that we are still referring to the same phenomenon—despite radical changes in beliefs about the phenomenon.

3.5.2. A Deflationary Account of Sameness of Denotation

As in the case with the description theory of meaning, the causal theory gets something right while ultimately not being satisfactory. What it gets right is that scientists typically assume that their experimental operations indeed individuate an object that exists independently of their conceptualizations, that is, that they do not take the reference of their terms to be determined solely by descriptions. Hence, they think that it is possible for different people to refer to the same object while having conflicting accounts of it. Where the causal theory fails is in its attempt to give an a priori account of how the meanings of terms are settled. Standard accounts of causal theories of reference capture the intuition of sameness of reference but fail to do justice to the genuine conceptual indeterminacy that can at times exist. This point is made succinctly in Mark Wilson's "Predicate Meets Property" (1982) as well his *Wandering Significance* (2006). In the former, Wilson invites us to imagine a group of Druids who settled on a deserted island a few centuries ago. Upon first seeing an airplane in the sky, they might extend their concept of *bird* to describe this new object. If, however, they encounter the same object on the ground, they might extend their existing concept of *house*. Wilson intends for this thought experiment to drive home the point that it seems far-fetched to assume that there was previously a fact of the matter about which of these two concepts already had planes in their extension, as the causal theory would have to posit. A similar point is made by Gary Ebbs (2000, 2003), who emphasizes: "What we actually say when we find ourselves in previously unimagined situations almost always trumps our earlier speculations about what we would say if we were to find ourselves in that situation" (Ebbs 2003, 252). Ebbs concludes from this that we should take seriously the "practical judgment of sameness of extension across time" (Ebbs 2000, 259) without appealing to the justificatory strategies offered by the causal theory of reference or the more recent writings about primary and secondary intensions, arguing that this type of judgment is much more robust than are any of the strategies to which people have appealed in order to justify it. He therefore suggests a deflationary model of such judgments, according to which: "To *describe* our practice of disquoting our own words and taking other speakers' words at face value, it is enough to trust our practice of taking other speakers' words at face value.... We can describe our practical judgments of sameness of denotation in this way without justifying them" (Ebbs 2003, 259; emphasis in original).[14]

This deflationary account of the sameness of denotation fits my account of operational definitions as tools very well, except that, in an experimental context, the judgment of sameness of denotation is not just one across time but also one across procedures and contexts. In this way, one commitment that is fairly basic to scientific practice is the intuition that concepts can be extended beyond the circumstances of

---

14. It might be tempting to argue that reference gets fixed along the lines suggested by causal theories of reference by experimental operations associated with a concept. Arabatzis (2012) attempts to do something like this with regard to typical experimental operations used to individuate electrons, even across theoretical change. However, I argue that this does not work in this case, precisely because scientists use more than one operational definition to investigate a given object. Thus, they might be judged to be referring to the same things, but this judgment does not constitute a justification.

their initial use and still refer to or instantiate the same kind of thing, both in a synchronic sense and in a diachronic sense. That is to say that scientists tend to think (*a*) that whatever phenomenon they have identified under very specific circumstances within the lab also exists outside these circumstances and (*b*) that we will continue to be able to talk about a thing even if our beliefs change. The former assumption is perhaps what Joseph Rouse has in mind when he remarks that "what makes experimental phenomena significant is that the pattern they embody is informative beyond its own occurrence" (Rouse 2015, 233), whereas the latter is what both Hasok Chang (2004) and Mark Wilson (2006) point to when remarking on the often unexpected ways in which concepts can be extended in the course of their development.

The deflationary account of sameness of reference can accommodate both the idea of the stability of a research object and the dynamic character of our knowledge about it. Given the notion of *stable research objects*, researchers need methods to pin down precise conditions under which the object can be observed. Given the facts of conceptual extension and even reference change, researchers need methods that allow them to make observations and gather novel information. I claim that these are precisely the two functions operational definitions can serve. They (temporarily) fix the reference of a concept, and they allow for experimental operations that explore the (purported) object in the extension of the concept. They therefore presuppose the judgment of sameness of denotation across time and space, but, as I emphasized, this does not mean that this judgment is not itself revisable. As Ebbs writes: "We trust these judgments unless we have some special reason to doubt or revise them" (Ebbs 2003, 259). This means that, even though the judgment of sameness of denotation is fairly robust, it is not unshakable. Indeed, I argue that, in the case of experimental research, such judgments—and the background assumptions that stabilize them—are constantly being questioned. Accordingly, I adopt the idea that the judgment of sameness of denotation is a prerequisite for research on a given purported object, but, in addition, I argue that, when there are reasons for doubt, this is in fact a vital and productive part of the process of experimental research. The explication of such doubts is, I argue, the central ingredient of what I refer to as "operational analysis" (see Feest 2016; see also chap. 7 below).

3.6. Scientific Concepts and Investigative Practice

In the previous section, I probed two well-known twentieth-century accounts of concepts with respect to the issue of whether they can be turned to for an analysis of scientific practice. Far from being a detailed or exhaustive account of philosophical treatments of concepts, the point of this passage was to bring into sharper relief the nature of my own project, that is, to provide an analysis of the crucial role of preliminary conceptual presuppositions in the processes of experimental investigation. In particular, I highlighted that descriptive theories of meaning neglect the specific materiality of experimental operations as grounding scientific concepts in experimental research and that causal theories are too rigid in their assumptions about reference fixing.

There are also other recent approaches to concepts that are more explicitly geared toward scientific practice and, thus, more in line with my emphasis on the role of scientific concepts in the investigative process. In particular, Nancy Nersessian has since the 1980s pioneered investigations of the cognitive role of concepts in scientific reasoning (e.g., Nersessian 1984, 1992, 2008). Andersen, Barker, and Chen (2006) give a reconstruction of well-known historical cases of conceptual change in terms of Barsalou's dynamic frames theory of concepts (e.g., Barsalou 1987). In turn, Mark Wilson (2006) has importantly spearheaded what is sometimes referred to as a "patchwork" account of concepts (e.g., Bursten 2018; Haueis 2024), which emphasizes, in a spirit not so different from Bridgman's, that there is no reason to suppose that scientific concepts are glued to the same kinds of objects under different circumstances. What each of these approaches emphasizes, albeit in different ways, is that scientific concepts often carry within them the potential to be extended in surprising ways that cannot be historically anticipated.[15]

In the remainder of this section, I articulate my methodological approach to conceptual development more clearly by juxtaposing it with Nancy Nersessian's cognitive-historical approach, which goes back to her *Faraday to Einstein* (Nersessian 1984). Nersessian's approach shares with mine the contention that, rather than trying to develop an account of the meanings of concepts and then worry about conceptual change, scholars of scientific concepts ought to begin with the fact that "scientific concept formation is a process, involving the struggle to articulate what we mean when we say, e.g., that something is a 'field.'" Nersessian continues: "The key to understanding the concepts

---

15. Haueis (2024) formulates a normative theory of when the introduction of new patches (and, thus, the extension of a concept into novel territories) is legitimate. His account is compatible with mine in that he, too, recognizes the importance of operations as individuating objects. According to him, two operations are relevantly similar if they instantiate the same "reasoning strategy." (For a recent analysis of the relationship between operationism and patchwork accounts of concepts, see Novick and Haueis [2023].)

of science lies in examining the various places of that process as it takes place in scientific practice" (1984, xi). Her early work was motivated by the aim of providing an answer to the problem of incommensurability between the conceptual structures before and after a conceptual shift. She and others at the time (e.g., Shapere 1982, 1984) attempted to disentangle the fact that concepts before and after a shift are radically discontinuous from the claim that such shifts can take place in a radically discontinuous (and, thus, possibly nonrational) manner. With regard to this latter claim, these authors argued that it is often possible to reconstruct the chain of reasoning connecting two successive versions of a concept. Nersessian (1984, 144) spells this out by way of a detailed account of the development of the concept of the *electromagnetic field*, suggesting that we see several stages: a "heuristic guide" stage, an "elaborational" stage, and a "philosophical" stage. She concludes *Faraday to Einstein* by proposing—albeit in a programmatic fashion—a notion of *concept* (or meaning schema) that is similar to a prototype theory of concepts and that construes meaning as "a two-dimensional array which is constructed on the basis of its descriptive/explanatory function as it develops over time" (156). The two dimensions she has in mind concern synchronic and diachronic variation, and as the most important descriptive/explanatory functions she makes out "stuff," "function," "structure," and "causal power" (Nersessian 1984, 157; see also Arabatzis and Kindi 2008).

In some of her subsequent work, Nersessian spells out the implications of her analysis for questions about the creation of novel scientific concepts (e.g., Nersessian 2008). Drawing on the cognitive science literature about model-based reasoning, which she links to analogical reasoning in complex material contexts (see also Nersessian 2012), she argues that "concepts specify constraints for generating members of a class of models," thereby enabling a type of reasoning that "promotes conceptual change" (Nersessian 2008, 187, 201). In a similar spirit, Hanne Andersen, Peter Barker, and Xian Chen (2006) also emphasize the kinds of inferential possibilities and constraints contained in existing concepts. They hold that this allows for novel (yet continuous) conceptual structures to emerge, building their analysis on Barsalou's theory of concepts as dynamic frames (Barsalou 1987; see also Barker 2011). Bloch-Mullins (2019, 2020), also drawing on psychological research, captures a similar insight by referring to scientific concepts as "forward-looking." Likewise, Jürgen Renn and his colleagues have also drawn on (among other things) the cognitive science literature (in their case, that involving mental models) to model scientific change (e.g., Renn 2008; see also Feest and Sturm 2011).

I share many of the intuitions underlying the approaches just outlined and especially the broadly naturalistic outlook that informs them. Like theirs, my focus is on reasoning practices in the sciences, and, like them, I wish to account for the dynamics of knowledge generation. Unlike them, however, I do not turn on cognitive theories of reasoning in general but rather focus on methodological considerations concerning *scientific* reasoning, and I do not turn to concepts as basic components of human cognition; rather I focus on *scientific* concepts. In making these choices, I hope to remain neutral with respect to complicated issues regarding (a) the continuity between ordinary and scientific reasoning (see Cheon and Machery 2010), (b) the status of the notion of a *concept* as a theoretical entity in cognitive science (Machery 2009), and (c) the relationship between the role of concepts in scientific practice and the meanings of concepts (in a more substantial sense of a theory of meaning).[16] Moreover, my approach to concept formation aims specifically to elucidate questions about the ways in which experimental and conceptual practices are intertwined in experimental psychology.

3.7. Conclusion

In this chapter, I have argued that my analysis of the role of operational definitions presented in earlier chapters carries over to contemporary research in cognitive psychology. Accordingly, I hold that the enduring importance of operationism in psychology points to the close relationship between exploration and concept formation in a domain of study that does not typically start with full-fledged theories or fully formed concepts. This, however, does not mean that there are no prior assumptions and empirical findings in place with which scientists work in their attempts to gain a better understanding of their subject matter. I argued that such assumptions/findings are often codified in operational definitions^{wide}, which in turn are exemplified in operational definitions^{narrow}. The latter are deeply ingrained in specific (paradigmatic) experimental procedures of manipulation and detection. At the same time, the concepts thus defined retain some of the surplus meaning provided by theoretical and folk-psychological contexts (Smith 2018b). I have referred to both kinds of operational definitions as "tools" (a) because I want to indicate their pragmatic and preliminary nature, (b) because they supply the concepts in question with fairly specific instructions about how the

---

16. This last point also distinguishes my approach from that of Ingo Brigandt (e.g., Brigandt 2010, 2012), who argues for an account of a concept's content that includes epistemic aims and values.

corresponding research object in question can be investigated, and (*c*) because they function as premises relative to which the resulting data can be used for inferences about the objects under investigation.

As such, operational definitions can be regarded as epistemic tools. The kind of research they enable is simultaneously anchored by experimental interventions and marked by a high degree of conceptual indeterminacy. The conceptual indeterminacy concerns two dimensions, namely, (*a*) the question of stability of reference across time and (*b*) the question of how well and/or far concepts are extendable beyond the specific experimental circumstances of their operational definitions. As argued in section 3.5.2 above, researchers in psychology (rightly or wrongly) often assume a certain stability of reference across time and contexts to be guaranteed by experimental operations while at the same time treating the scope of their concepts as fairly elastic.

The analysis offered above elucidates both how research about specific objects is held together and how it can come apart. Memory researchers typically agree on what constitutes an experimental design for the study of memory (it involves having subjects memorize and retrieve information), and they also agree on what constitutes a paradigmatic procedure for the detection of specific types of memory (e.g., a priming test for implicit memory or an N-back test for the study of working memory). While this holds the research together, research objects can also come apart if the conceptual connections between them and paradigmatic methods (operational definitions) are questioned by either conceptual or empirical considerations. The conceptual connection between priming and implicit memory, for example, can be severed either altogether or merely for some priming tests. Experimental designs can persist while dropping or adding some exemplar paradigmatic methods. In turn, paradigmatic methods and the effects they produce can be dropped as irrelevant, they can be investigated on their own terms, or they can be conceptually tied to a different research object.

After introducing the general idea of concepts as tools (secs. 3.1–3.3), section 3.4 illustrated my thesis by means of two case studies. Section 3.5 highlighted that traditional theories of concepts cannot do justice to the dynamic character of conceptual change and the material practices that often drive such change, adopting instead Gary Ebbs's deflationary account of reference. Section 3.6 situated my account within the context of other accounts of conceptual change and emphasized, as a distinguishing feature of my analysis, that it is specifically geared toward the role of concepts in experimental practice. While this chapter has made the case that operational definitions should be viewed as tools of conceptual development and exploration, little has yet been said about the status of the objects that are being thus explored. This question breaks down into two parts. First, how can we make sense of the notion of an *object of research* if such objects are, as I put it, "ill understood"? Second, assuming that such objects are the targets of research, what is their ontological status, and what are the criteria of success for their exploration? The following chapter considers the first of these questions. I return to the second in chapter 6.

CHAPTER FOUR

Objects of Research as Targets of Exploration

4.1. Introduction

In the previous chapter, I argued that operational definitions in psychology are tools for the experimental investigation of objects of research and that they perform this function within the frameworks of existing disciplinary practices and folk-psychological assumptions about the subject matter at hand. In this chapter, I elaborate on my notion of *objects of research* and say more about what the investigation of such objects amounts to. Briefly, (*a*) objects of research are the (presumed but not yet well understood) objects at which research is directed, and (*b*) the aim of much experimental research in psychology and cognitive science is the delineation and empirical description of such objects. I refer to these two activities taken together as "exploration."

Let me begin by reminding the reader of my argument (in the previous chapter) that, on the one hand, scientific investigations are typically characterized by the assumption of a stable referent across time and research contexts (we can refer to this as a *realist presupposition*) and that, on the other hand, this very assumption can also be revised in the course of the research process since—from the epistemic perspective of researchers—the question of what their research is about is very much at stake. Attempts to secure reference by way of a particular (i.e., causal) theory of meaning thus seem heavy-handed and top-down. In this respect, I have more sympathies with the basic spirit of traditional descriptivist accounts, which have emphasized that reference can shift in sometimes unforeseeable ways. However, owing to their linguistic holism, traditional descriptivist accounts mostly do not capture the very real and material operations and constraints characteristic of experimental research. Experimental designs and the operational definitions ingrained in them are the sites where conceptual and material aspects of scientific research interlock. It is here that we see the labor of trying to get a taxonomic and descriptive grip on the subject matter of cognitive psychology.

What, then, are objects of research? They are the things assumed to correspond to our preliminary taxonomic categories insofar as they become targets of our investigative efforts. This sentence may appear dangerously slippery as it seems to conflate talk of "real" (mind-independent) objects with the notion of objects as posited and conceptualized *by us*. Of course, the fact that we posit and conceptualize objects in certain ways does not guarantee that there really are corresponding (natural) kinds of objects in the extension of these terms. Nor does it guarantee that the descriptions we give of them are correct. This highlights the need to distinguish between two separate issues. The first concerns the role of concepts as manifesting a particular understanding of the taxonomic structure of a domain. The second concerns the question of what the ontological status of the corresponding (presumed) objects in fact is. Ultimately, these two questions cannot be treated entirely separately. However, in the current chapter, I focus on the first. I take it as given that researchers think of their research as being directed at specific objects (such as memory, attention, emotion, perception, etc.) that they attempt to capture and investigate by means of operational definitions and corresponding experimental operations.[1] The two main theses of this chapter are (*a*) that from the epistemic perspective of practicing researchers these objects are in a very fundamental way ill understood (or, as I call it, "epistemically blurry") and (*b*), consequently, that much experimental research in cognitive psychology aims at the exploration (delineation and description) of objects of research and, thus, the development of taxonomic categories.

My account owes much to that of Hans-Jörg Rheinberger (1997), who has also emphasized the centrality of objects (as that toward which research is directed) and the fact that such objects are typically not well understood. In relation to this, he has coined the expression "epistemic things," describing them as "blurry" (*verschwommen*) (Rheinberger 2006), "vague," or characterized by what we do not know about them (Rheinberger 1997, 28). With regard to the terms "vague" and "blurry," it is useful to distinguish between semantic and ontological questions (pertaining to the vagueness of concepts and the blurriness of objects),

---

1. I return to the second question—concerning the ontology of psychological objects—in chap. 6 below. There I also challenge the assumption that the reality of psychological objects implies mind-independence.

on the one hand, and epistemological questions (pertaining to our limited knowledge about objects), on the other (see also Feest 2011a). Looking at the latter (epistemological) question, I find the notion of *blurriness* (or perhaps, more specifically, of *blurry vision*) to be a very apt metaphor, one that describes the epistemic situation of scientists as trying to get their objects "into clear focus." On my construal, then, our understanding of a given (presumed) object can be blurry in the sense of being clouded in uncertainty, where this blurriness extends to the very question of whether we will ultimately find that there even is a unified, coherent object. I refer to this predicament as "epistemic blurriness," and I argue that it is constitutive of objects of research that they are epistemically blurry.

It is a basic contention of this book that the epistemic blurriness of the objects of psychology has been vastly underappreciated within mainstream philosophy of psychology and neuroscience. As a consequence, much philosophical work has focused on issues of explanation and reduction, relying on the idealized assumption that it is already clear what the relevant phenomena are or (for that matter) what the conditions of identity for phenomena in cognitive science are in the first place. While I do not mean to dismiss the value of such work, my own approach is different in that I analyze the ways in which scientists cope with the predicament of uncertainty about the nature, boundaries, and descriptive characteristics of their objects of investigation. The process I wish to capture in this book is one whereby scientists try to gain an understanding of what their objects of research really are. Or, to use a formulation cited in the previous chapter, I am interested in "the [scientific] struggle to articulate what we mean" (Nersessian 1984, xi) when, for example, we use expressions like "explicit memory" or "working memory." More specifically, I inquire into "the systematic character of experimental operations" (Rouse 2011, 247) at work in this struggle as it plays out in psychology.

In what follows, I return to the case studies of the previous chapter to illustrate and spell out in more detail my contention that objects of psychological research are epistemically blurry and that much empirical research is aimed at exploring these objects with the aim of reducing their blurriness. Section 4.2 begins by briefly outlining what is unique about my focus on (and analysis of) research objects. I argue that the blurriness of such objects can be cashed out neither in terms of imprecise empirical data nor in terms of a lack of explanatory knowledge. Section 4.3 provides detailed examples from implicit and working memory designed to illustrate the extent to which and the ways in which these objects of research are epistemically blurry. I show that, when scientists try to provide empirical descriptions of implicit or working memory, they are neither reading these descriptions straight off the empirical data nor deriving them from some general theory of the relevant phenomena. They are, however, importantly doing work that should be construed as theoretical. In section 4.4, I sharpen my analysis of psychological investigations as being directed at the exploration of objects of research by situating it vis-à-vis existing work on exploratory experimentation. I argue that my analysis shares important features with accounts of exploratory experimentation (in particular the emphasis on conceptual openness) but draws attention to the ways in which exploratory research in psychology is conceptually structured and directed toward objects.

4.2. Delineating and Describing Objects of Research: A First Approximation

It might be objected that my emphasis on epistemic uncertainty does not convey anything new. Surely it is well-known that there is a lot of epistemic uncertainty in science insofar as science relies, part and parcel, on inductive inferences, which are, by their very nature, underdetermined by empirical observations. While I agree that all inductive scientific inferences carry a certain degree of epistemic uncertainty (by virtue of being inductive), I contend that typical accounts of underdetermination do not do justice to the issue I am getting at here since they tend to assume as unproblematic the very notions of *theory* and *observations* as well as their presumed roles in scientific investigations. According to the received view, descriptions of the observable world are (*a*) taken as the inferential basis for theoretical claims about the unobservable world where theories are (*b*) taken to explain observable phenomena and (*c*) already fully formed prior to their empirical test. My account challenges all three assumptions in the case of experimental psychology. First, the data that form the inductive base of experimental inferences are carefully manufactured in accordance with (among other things) operational definitions and, thus, are not unproblematic observational primitives. Thus, the question of what constitutes good data and how high-quality data can be ensured in any given epistemic context is part of the investigative process (see also Feest 2022).[2] Second, the point of experiments is frequently not to test an explanatory theory but rather to explore descriptive features of specific objects of research. Third, psychologists typically start out not with a fully formed (descriptive or explanatory) theory of their objects but rather with preliminary concepts

---

2. I return to the issue of data quality in chap. 7 below.

that are operationally defined for investigative purposes. By this I do not mean to deny that there are scenarios in which researchers face issues of underdetermination of theoretical claims by empirical data. But I claim that the focus on objects of research, as opposed to theories, highlights that it is not just the unobservable or "theoretical" features of such objects that are epistemically blurry; the blurriness, rather, extends to observational/descriptive features as well.[3]

Let me be clear that I do not restrict the notion of an *object of research* to middle-sized physical things; rather, I take it to include whatever entities, phenomena, effects, or mechanisms scientists happen to be interested in. In this book, I devote special attention to macro-level psychological objects like memory, but I am fully aware that such objects are partially characterized by phenomena—such as the primacy effect, the priming effect, and the chunking effect—that can also become objects of research in their own right.[4] Notice, however, that scientists often treat the effects just mentioned as specific *kinds* of effects (e.g., memory effects), thereby tacitly acknowledging broader taxonomic assumptions about their subject matter. Thus, while my notion of an *object of research* extends to anything to which scientists turn their epistemic attention, I am particularly interested in the ways in which preexisting taxonomic categories (which are often taken from folk psychology) such as memory, perception, and attention impose a conceptual structure on their research fields. Needless to say, it does not follow that the objects that populate these fields are clearly understood.

What the discussion offered above highlights is that the empirical delineation of objects of psychological research is by no means trivial. Moreover, the case studies in the previous chapter have already provided us with a first taste of the ways in which such objects are epistemically blurry. Rather than simply describing their research objects empirically and then searching for explanations, researchers have continued to investigate questions about the precise extension of these concepts, thereby probing the issue of what exactly their initial delineations delineated. As we saw, this process can—but need not necessarily—result in a shift in a concept's reference, though it is equally conceivable that a concept might be abandoned altogether. In this chapter, I argue that the notion of a *blurry object of research* is a helpful way to accommodate

---

3. I spell this out in more detail below by arguing (among other things) against the common conflation of *observational* with *descriptive* and *theoretical* with *explanatory*.

4. Chap. 5 argues that objects of psychological research are clusters of phenomena. There I provide a more detailed discussion of the relationship between phenomena and objects of psychological research (see also Feest 2017a).

these possibilities in that it highlights conceptual openness on three levels. First, it is not obvious that a given concept will prove to have a sustainable referent at all. Second, it is not obvious that a given operational definition delineates this referent adequately. Third, even once a specific operational definition has been chosen, it is not obvious *what are additional descriptive features of the object* thus delineated. I argued in the previous chapter that operational definitions are tools of exploration. Here, I specify that exploration often amounts to providing descriptive characterizations of objects of research with the aim of making them less epistemically blurry.

With this thesis I challenge a well-ingrained assumption in philosophy of science according to which science mostly aims at explanations, though it must be emphasized that the importance of taxonomies and descriptions is well-known in some areas of the history and philosophy of biology. In general terms, Dudley Shapere (1976) addressed a similar question some decades ago. Shapere distinguishes between individual things ("items") and classes of things ("domains"), pointing to the question of how scientists delineate one domain from another and how items within a domain are delineated from one another. Given the kinds of research objects on which I focus in this book, we might say, for example, that the relevant domain is the cognitive system and the items within the domain are working memory, implicit memory, long-term memory, episodic memory, etc. In response to the question of how items and domains are individuated, Shapere argues that, whereas early taxonomies can be guided by superficial sensory features (color, shape), this is not the case in more "sophisticated science" (1976, 284), hence raising questions about the ways in which additional descriptive features of the items in question can be determined.[5] He concedes that, even in such (sophisticated) sciences, objects are identified in terms of observable features but emphasizes: "That those features should be considered as identifying or indicating the sorts of entities to be studied is not a matter of anything that could be called immediate or obvious sensory characteristics" (284). Shapere highlights the topic of description as worthy of philosophical analysis, emphasizing that description in science should not be taken to rely on what is observationally obvious. Coming up with good descriptions is, thus, in an important sense, theoretical work precisely because it is not observationally obvious what the right descriptions are. I emphasize this here because the term "theory" (and the kind of work it does) is in philosophy of science often taken to be

---

5. This claim is also familiar from Quine's "Natural Kinds" (1969), to which I return in more detail in chap. 6.

synonymous with *explanatory theory*. By contrast, I argue that much empirical work in psychology is closely intertwined with theoretical work even though it is not aimed at formulating explanatory hypotheses.[6]

With respect to the cognitive and behavioral sciences, the statement by Shapere quoted above needs to be qualified and sharpened in two ways. First, the descriptive characterization of memory (or other mental entities), even in the less "sophisticated" realm of folk psychology, is tied not so much to sensory modalities as to specific ways of classifying observable types of behaviors that organisms exhibit under specific conditions, that is, to conceptual assumptions about what the behavior signifies. Second (as argued in the previous chapter), this fact informs the way memory is studied scientifically, namely, by creating and varying the conditions under which memory is thought to reveal itself behaviorally. This is to say (*a*) that, already in the realm of folk psychology, descriptive features of objects cannot be treated as being somehow naturally provided by our senses and (*b*) that, even in the realm of scientific psychology, the way in which data pertaining to research objects are produced is rooted in folk psychology.

4.3. Describing Empirical Features of Objects of Research

Having explained my contention that empirical research in psychology is often geared toward the exploration (i.e., delineation and description) of specific objects of research and that those objects are in an important way epistemically blurry, let me now develop this thesis by looking at two such objects already discussed in the previous chapter: implicit memory and working memory. Specifically, I analyze one salient example of a feature thought to describe implicit memory and a feature thought to describe working memory, namely, that implicit memory is an *unconscious* form of retrieval and that working memory has a definite *capacity limit* of 4 ± 1.

4.3.1. Implicit Memory: The Case of Nonconscious Recollection

When we think of descriptive features of implicit memory, one such feature that immediately comes to mind is that it is a type of memory that is implicit. But what does it mean to describe a process as "implicit"? In pursuit of this question, we might hope to gain some insights from Graf and Schacter (1985). Not only did they coin the expression "implicit

---

[6]. By this I do not mean to deny that theoretical work explanatory hypotheses can often provide explanations.

memory," but they also stated that they were not (at least at that point in time) committed to the idea of an independent memory system in the brain. Rather, they claimed to be using the expression "implicit memory" descriptively. But, if they did not posit an implicit memory system, what did they take themselves to be describing? While it is tempting to think that they were merely describing empirical data or test performance, I argue that they wanted to describe an object of research (implicit memory), albeit without much of a commitment as to what kind of thing it was. And they wanted to derive the description from salient features of the circumstances under which the data were generated.

Let me begin to explain this by distinguishing between two possible readings of the dissociation between the results on priming tests and the results on explicit memory tests. According to the first reading, what is dissociated, strictly, are only *behavioral performances* on two different types of memory tasks. According to the second, what is dissociated are two different types of *memory processes*. Researchers investigating implicit memory early on warned against conflating the two. One pair of authors pointed out: "The terms implicit and explicit memory are used in two logically different ways; both to classify different memory tasks and to describe the memory processes underlying these tasks" (Dunn and Kirsner 1989, 17). As we just saw, early implicit memory researchers wanted to remain neutral with regard to the existence of a specific system or type of neural process. However, they also wanted to do more than simply point to the dissociation between performances on two types of memory tasks. Rather, they aimed at the description of implicit memory in terms of a specific feature that distinguished task performance on explicit and implicit memory tests. There emerged early on a consensus that performance on implicit tests did not require consciousness and that performance on explicit tests did and, therefore, that the *absence of consciousness* could be regarded as the distinguishing descriptive feature of implicit memory. For example, two prominent researchers argued: "Implicit memory is revealed when performance on a task is facilitated in the absence of conscious recollection; explicit memory is revealed when performance on a task requires conscious recollection of experiences" (Graf and Schacter 1985, 501). The rationale for this assertion was that, in traditional (explicit) memory tasks, subjects are provided with tests that require them either to state whether they have seen a given item during the test phase (this is also referred to as a "recognition paradigm") or simply to recollect what was presented to them during the study phase (this is also known as a "recall paradigm"). By contrast, implicit memory tasks work by way of priming. Priming is a method where (*a*) subjects are not instructed to memorize test items and (*b*) subjects are presented with additional stimuli, so-called primes. If specific primes result in better-than-chance of memory for test items, this is called a "priming effect." In implicit memory research, priming effects are counted as indicating implicit memory.

Now, given the dissociation between performances on the two types of tasks, it is not unreasonable to look to the tasks to derive descriptive characteristics that distinguish the two types of performance from one another. At first sight, the descriptive distinction between conscious and unconscious processes seems reasonable. But, of course, the fact that subjects in implicit tests are not instructed to use consciousness does not guarantee that they do not use it. Moreover, there are significant difficulties in establishing whether they do or do not. One reason is that the notion of *conscious recollection* is ambiguous (Schacter, Bowers, and Booker 1989). It can mean either that the subject intentionally retrieves previously studied material (i.e., makes a conscious effort to recall the study period) or that the subject is phenomenally aware of having previously studied the test items (regardless of whether a conscious effort was made). Absence of conscious recollection, therefore, might mean either that the subject does not make an intentional effort to recall the previously studied material or that the subject does not—during the priming experiment—have a phenomenal recollection of having seen the test items or primes before. If we want to know in which of these two senses implicit memory is an unconscious form of retrieval, we need to find a way to distinguish between them empirically. And, indeed, this has been a subject of ongoing debate for some time.

When they first pointed to this issue, Schacter, Bowers, and Booker acknowledged that they could not think of an adequate testing procedure (which they explicitly referred to as an "operational definition") for phenomenal awareness or the lack thereof. Therefore they suggested that implicit memory should be defined in terms of the absence of intentional retrieval rather than the absence of phenomenal awareness. Accordingly, they endorsed what they called the "retrieval intentionality criterion" over the "phenomenal awareness criterion," primarily because they could "develop rigorous criteria for making the former, but not the latter, distinction" (Schacter, Bowers, and Booker 1989, 49). In practical terms, they proceeded by asking the subjects after the priming experiment whether they had intentionally tried to retrieve the items from the study period, not controlling for phenomenal awareness. By leaving out any reference to the presence or absence of phenomenal awareness, this characterization of implicit memory fails to distinguish between the case in which someone exhibits a priming effect without any phenomenal recollection of the study period (the paradigmatic example of this is supposed to be the amnesiac patient) and the case in which someone exhibits a priming effect but also unintentionally retrieves a phenomenal memory of the study period. The latter possibility has been referred to as "unintentional explicit memory" (Schacter 1987), "involuntary explicit memory" (Hirst 1989), and "involuntary aware memory" (Mace 2003). Moreover, the fact that Schacter, Bowers, and Booker did not know how to define phenomenal awareness operationally means that they were also not able to control for phenomenal awareness and, hence, by their own admission could not distinguish empirically between unintentional retrieval that involves awareness and unintentional retrieval that does not. Relating to these questions, one commentator stated: "Researchers agree that implicit memory is memory that does not involve conscious recollection: The main issue of contention is whether conscious recollection should refer to intentional retrieval or phenomenological awareness of the study episode" (Kinoshita 2001, 59).

Now, if it were determined that the relevant sense of "conscious recollection" was phenomenal awareness, then the presence of phenomenal awareness in priming tasks could have one of two consequences: (a) either it would throw doubt on the adequacy of the most central supposed descriptive feature of implicit memory, that is, that it does not involve consciousness, or (b) it would raise the possibility that many instances of implicit memory effects (priming effects) are not, in fact, due to implicit memory. This point is relevant for our purposes because there is, indeed, some evidence to the effect that unintentional phenomenal recollection is often present both in perceptual and in conceptual implicit memory tasks (e.g., Mace 2003). This thereby gives rise to the worry that the priming results might in fact not be due to implicit memory (see also Richardson-Klavehn, Gardiner, and Java 1994). In turn, this has prompted other authors to "suggest that there is . . . a lack of strong evidence for implicit [i.e., unconscious] memory in normal subjects" (Butler and Berry 2001, 192). They conclude that, if the term "implicit memory" is to be retained, much more care needs to be taken to ensure that performance on implicit memory tasks is not contaminated by consciousness.

Notice that a revealing shift has occurred here. Whereas the descriptive characterization of implicit memory as *unconscious* was initially derived from a feature of the relevant tests (i.e., that they did not require conscious effort), subsequent debates turned to the question of whether those tests really measure implicit (i.e., unconscious) retrieval. This means that, while scientists started out by operationally defining their object of research (implicit memory) in terms of priming tests, the issue subsequently became how to devise more precise tests for what was now taken to be a central descriptive feature of the object of research: unconscious retrieval. Like the case discussed in the previous chapter (sec. 3.4.2), this situation is very similar to one that Hasok Chang (2016) has described as typical of the iterative nature of research. A given test can serve as a classificatory procedure and, thereby, kickstart the empirical investigation of a given object of research, but this does not stop subsequent researchers from questioning the adequacy or the precision of the very classifications they started out with, all the while operating with a descriptive characterization derived from the initial tests. In the case at hand, the issue is whether the very notion of *implicit memory* needs to be rejected—or at least fine-tuned—in light of the two readings of *unconscious retrieval*. In practical terms, one relevant question is whether it is possible to develop methods of empirically distinguishing between involuntary awareness and automaticity.

Dawn McBride (2007) discusses two such methods: (*a*) on-line recognition measures and (*b*) process dissociation procedures. The former class of procedures, devised by Alan Richardson-Klavehn and his colleagues, tries to distinguish between involuntary aware memory and automatic memory by asking subjects to identify items on the priming test that they recognized from the study phase.[7] As McBride (2007, 205) points out, however, this procedure relies on a questionable assumption, namely, that processes can be isolated and measured directly by means of specific task performances. By contrast, other researchers argue that any given task performance will most likely involve more than one type of process (e.g., intentional and automatic). This calls for a method for separating them out. Jacoby's (1991) process dissociation procedure is an attempt to address this desideratum. Subjects are subjected to two tests, both of which require intentional and automatic processes, with the first test (the *inclusion* condition) requiring them to work together and the second (then *exclusion* task) requiring them to work at cross purposes.[8] The rationale of this method is that one gets the pure automatic process by subtracting the one from the other. As McBride points out, however, while this method appears to capture the distinction between involuntary awareness and

---

7. Using a similar method, Mace (2003, 2005) presents evidence that phenomenal awareness of the study period does enhance priming effects.

8. More specifically, in the inclusion condition, subjects are asked to retrieve an item on the study list (thought to trigger intentional processes) and to revert to guessing if they cannot remember (thought to elicit a priming effect). In the exclusion condition, on the other hand, subjects are instructed to produce items that were not on the study list.

automaticity, it cannot distinguish between intentional and unintentional forms of conscious recollection.

I find this case to be instructive in several ways. Researchers are clearly aware that absence of conscious recollection as a presumed descriptive property of their object of research (implicit memory) is underspecified by their measurement procedures (operational definition) in terms of implicit tests. This leads them to inquire into more fine-grained procedures to figure out the sense in which implicit memory might be unconscious. The case study also reveals that it is a mistake to assume that a given experimental effect (in this case a priming effect and its dissociation from the results of other memory tests) figures as a straightforward explanandum phenomenon in memory research. Instead, researchers treat the effect as pointing to an object of research, on which they try to get a descriptive handle. Another point to highlight here is that, in trying to get a conceptual and empirical grasp on implicit memory, they inevitably contrast its descriptive features with those of explicit memory (i.e., implicit memory is understood as the kind of retrieval that is not intentional or that happens without phenomenal awareness). However, as Robins (2022) points out, the question of what we mean by "explicit memory" also has no simple and straightforward answer. This means that the epistemic blurriness I have identified here with regard to implicit memory quite possibly can be addressed only by also investigating implicit memory's supposed contrast classes.

To be completely clear, my point here is not that these issues are in principle irresolvable. However, the case drives home the reality of the research situation at a given point in time as characterized by an acute awareness of the discrepancy between conceptual possibilities and available techniques and measurement tools. Thus, scientists are painfully aware of the extent to which their objects are (to use my expression) epistemically blurry, and this gives rise to sophisticated methodological puzzles, conceptual debates, and research designs, such as the ones covered above.

4.3.2. The Case of Short-Term Memory Capacity

We turn now to a second example of the methodological and conceptual difficulties that arise when trying to give descriptive characterizations of research objects in terms of features of the empirical data that—at first glance—seem perfectly straightforward: short-term (or working) memory. Here, as in the implicit memory case, we find an intriguing dynamic relationship between testing procedures and descriptive characterizations as well as theoretical conceptions. However, the descriptive characterizations in question are derived not from task instructions but from seemingly direct quantitative measures of task performances, namely, performances that point to the *duration* and *capacity* of short-term memory. The first characteristic is supposed to answer the question of what the time span of short-term memory is, that is, how long an item can be kept in short-term memory. The answer is usually taken to be something like twenty seconds. The second characteristic is supposed to answer the question of how many items can be kept in short-term memory. It was long assumed that the answer to this question is 7 ± 2 (Miller 1956). In what follows, I focus on this latter characteristic.

On the face of it, one might think that the issue of how to measure capacity can be attacked by simply presenting subjects with items and measuring how many they can reliably reproduce within a given time span. The difference between long-term capacity and short-term capacity would, then, be established by comparing performance on long-term memory tests with that on short-term memory tests where the latter tap into processes within twenty seconds of subjects' being presented the stimulus material. While this is, of course, the basic rationale that underlies efforts to measure capacity, the situation is complicated by a number of things. In this regard, questions concern the issue of how to differentiate between capacity and performance limits (e.g., Meyer and Kieras 1997) as well as between storage and processing (e.g., Daneman and Carpenter 1980). But, even if we assume for the moment that retrieval can unproblematically be treated as a measure of storage capacity, the question of how many memory items a given subject's experimental response corresponds to arises. This is a question that arises for both long-term memory and short-term memory, but with regard to short-term memory there is a special problem. The problem can be clarified if we keep in mind that one of the significant empirical findings in relation to short-term memory is a phenomenon called "chunking," that is, a presumed process whereby individual items are grouped together so that they form one larger item (Miller 1956). According to the chunking hypothesis, the number of items that can be stored in short-term memory at any given time is fixed, but, once chunking takes place, the amount of information that can be kept in short-term memory radically increases because the items are now composed of chunks of smaller items. As has been observed in the literature, Miller's account of chunks pulls in two different directions in that it posits the existence of a static capacity of the human cognitive system (storage capacity in terms of number of chunks) while at the same time accounting for our "expansive possibilities," that is, our ability to increase the size of chunks (Pramling 2011, 280). If we focus our research efforts on establishing a measure of the former (short-term memory capacity), then the latter will inevitably pose potential obstacles to attempts to describe this capacity accurately. At the very least, it alerts us to the necessity of ensuring that the capacity limit we come up with is not inflated by the effects of chunking. In other words, in order to produce an adequate measure of capacity, we need to ensure that we have a way of empirically identifying the chunks (as opposed to their parts) that subjects keep in short-term memory. This, then, raises the question of how to control for factors that could conceivably give rise to a distorted measure of short-term memory capacity.

In an influential article in *Behavioral and Brain Science*, the psychologist Nelson Cowan (2001) has pointed out that, even though Miller's famous 7 ± 2 estimate of working memory capacity has become firmly established as a popular scientific fact, it was speculative at the time it was proposed and has remained controversial over the years. Cowan argues that there is some evidence to suggest that there is indeed "a relatively constant limit in the number of items that can be stored in a wide variety of tasks" (2001, 88). However, that limit is between 3 and 5 (i.e., 4 ± 1), not 7 ± 2. If this is true, it calls for a modification of a central descriptive characteristic of short-term memory. Cowan's methodology is well worth reviewing as it gives us some glimpses into the considerations that enter into the kind of descriptive work discussed here. Specifically, Cowan strives for an account of short-term memory capacity that is descriptive in the sense that advocates of different theoretical approaches can agree with while also recognizing that the methodological task at hand calls for theoretical considerations. He approaches the question in three ways, that is, (*a*) by providing a theoretical definition of a chunk, (*b*) by suggesting four conditions under which a pure capacity limit might be observed, and (*c*) by suggesting a theoretical framework for interpreting the results of his studies. This broader theoretical framework involves the hypothesis that the capacity limit in question is one of *attentional focus*, which is underwritten by a common mechanism. Notice that theoretical assumptions enter into Cowan's methodology of description in two ways. In the following discussion, I focus on the former, which has to do with the nature of chunks (as opposed to the nature of short-term memory).

According to Cowan's theoretical definition, a chunk is "a collection of concepts that have strong associations to one another and much weaker associations to other concepts currently in use" (2001, 89). The notion of *association* invoked here points to a presumed feature of long-term memory. Cowan and others assume that chunking processes in short-term memory can draw on associations stored in long-term memory. The practical consequence of this is that, if one aims to establish pure estimates of the storage capacity of short-term memory, one has to take care not to use items that are likely to be strongly associated in long-term memory already. For example, if I present subjects with three separate items, F, B, and I, those are very likely to be chunked together because they are already strongly associated within long-term memory. In other words, the majority of experimental subjects will store FBI as one chunk, not three. Failure to control for such cases might lead researchers to overestimate the number of chunks that can actually be stored in short-term memory. As Cowan puts it: "*The purest capacity estimates occur when long-term memory associations are as strong as possible within identified chunks and absent between those chunks*" (Cowan 2001, 90). These considerations give rise to four strategies designed to rule out the occurrence of inadvertent chunking in the course of an experiment. One such strategy is to overload subjects with stimuli. A second is to block the rehearsal or recoding of items into larger chunks, for example, by requiring subjects to repeat a single word over and over again. A third way analyzes subjects' behavior to identify abrupt changes and discontinuities as indicating a shift in strategy, and a fourth takes into account priming effects that might give rise to inadvertent chunking. Cowan argues that these measures yield fairly similar measures of chunk capacity (between three and five chunks) and that this supports their validity as well as the accuracy of the chunk capacity measures thus obtained.

Of the many responses to and critiques of Cowan's target article, two are of particular interest for our purposes. The first objection points to an apparent circularity that results if we identify chunks by reference to particular strategies for ruling out artifacts (i.e., by ruling out chunking). The second objection worries about the legitimacy of extrapolating from particular experimental protocols to more general statements about the capacity of working memory. Regarding the first point, the question is, How are we to know whether a strategy of preventing higher-order coding has in fact succeeded? As put by one critic: "When steps have been taken to prevent coding of an item to a higher level, then the item will be the chunk. How do we know whether the experimental conditions have been successful in preventing recoding? Because the item will act like a chunk" (Beaman 2001, 118). Beaman himself remarks, however, that Cowan has an answer to this objection in terms of the correlation of the results obtained by his four strategies, all of which converge on 4 ± 1. In fact, Cowan's strategy here is quite similar to a version of the multiple-determination method discussed in chapter 2. But, as I argued there, the fact that different methods of measuring a particular presumed phenomenon produce similar effects does not ensure that they measure the same entity (i.e., they might be measuring two distinct entities whose phenomena happen to be correlated). Moreover, what if we vary not only the measurement technique but also the experimental stimuli used (e.g., pictures, dots, numbers)? There is, in fact, some evidence of dissociations between task performance with different types of stimuli, that is, of different capacity limits for different modalities of stimuli. This raises the question of how such modality-specific effects are to be reconciled with the presumably modality-unspecific nature of working memory. Other critics of Cowan's have therefore cautioned not to overgeneralize from one type of experimental setting. For example: "How do we know that 4 dots, screen locations, or digits correspond to 4 chunks as for example four words might? The main question is: How can we measure or define chunks independently of the experimental context we are dealing with in a concrete experimental situation?" (Schubert and Frensch 2000, 146).

Notice that there is a sense in which the two objections make similar points. Both agree that Cowan's theoretical definition of a chunk (in terms of associations in long-term memory) is too broad to narrow down sufficiently the conditions under which an experimental response can be regarded as indicating a chunk in a non-question-begging way. This complaint is especially instructive in that it points to the tension between the particularities of specific experiments (with their varying stimulus conditions) and the demand for a notion of *chunk* that transcends those specifics. The two objections thus come down to the demand for a measurement procedure (a) that differentiates between pure chunks and artifacts (possible contaminations in terms of inadvertent chunking) and (b) that is applicable across domains. For example, two critics of Cowan's mentioned above argue: "Cowan . . . does not formulate a convincing operational definition of what exactly a chunk is" (Schubert and French, 146).

What Schubert and French put their finger on is that there is a tension between providing precise descriptions of the experimental conditions under which an experimental finding will be regarded as a chunk (i.e., precise operational definitions) and providing an analysis that explains when different experimental conditions can be treated as producing evidence for the same kind of item (e.g., a chunk).[9] One possible response to this tension is to refrain from the assumption that different experimental effects in various domains are related to each other by some

---

9. Notice that this is quite similar to a problem discussed above in relation to priming as an effect of implicit memory. On the one hand, this effect is formulated at a level of abstraction that transcends the specifics of any given priming effect, but, on the other hand, it glosses over the possibility that different priming effects might be due to different underlying processes. In chap. 7, I argue that addressing these kinds of worries in an experimental context is one of the central ingredients of a principled approach to concept formation in psychology.

common, underlying process or by some underlying feature of working memory. This amounts to saying that, for the time being, we should not assume that different experimental stimuli measure the same thing. A second possible response might be to go the opposite way and argue that a theoretical account is needed in order to underwrite the assumption that different experimental findings (e.g., findings of a retention of 4 ± 1 items under different stimulus conditions) are, indeed, instances of the same kind of fact, namely, a fact about a particular descriptive feature of working memory. Once such an analysis is available, one might further argue that this would also give a more solid foundation to attempts to distinguish between facts and artifacts with regard to measures of working memory capacity. In line with this latter response, some commentators suggest that Cowan should have provided a more careful theoretical analysis of the nature of chunks and storage requirements (e.g., Halford, Phillips, and Wilson 2000; Milner 2000; Pascqual-Leone 2000), and one of the authors cited above argues: "Cowan's argument would benefit from a stricter definition of how to prevent higher-level recoding than he currently provides" (Beaman 2000, 118).

To sum up, critics of Cowan are divided between those who feel that his definition of a chunk is too theoretical (insofar as he draws on models of long-term memory, thereby illegitimately subsuming many different findings under one label) and those who worry that it is not theoretical enough (insofar as a more thorough understanding of memory mechanisms might provide a more systematic picture of the kind of phenomenon chunking is). In his reply to his critics, Cowan argues that, while it is true that his definition of a chunk does not give rise to a precise operational definition, it does draw attention to the necessity of very careful task analyses, which in turn does some useful work. Moreover, he maintains that, given the wide variety of contradictory theoretical accounts, "I staunchly continue to view the theoretical indecision of the target article as a strength, not a weakness" (Cowan 2001, 156). Replying specifically to Schubert and French, moreover, he argues (*a*) that, in contrast to their assertion, the similar outcomes of very different methods do, in fact, provide credibility to his hypothesis concerning a common underlying mechanism and (*b*) that, in the absence of detailed knowledge of the explanatory mechanisms that hold together different instances of chunks, one must rely on one's good judgment (Cowan 2001, 157).

This response suggests a particular picture of how the empirical exploration of descriptive features of the objects of psychological research proceeds, that is, neither by collecting brute empirical regularities nor by testing a theory of the phenomenon in question. In the case at hand,

Cowan attempts to describe a particular feature of working memory (its capacity) in a way that remains neutral with respect to theoretical disagreement about the nature of working memory while drawing on theoretical assumptions about the nature of chunking and relying on his "good judgment." I argue that this approach is typical of other research as well, highlighting the extent to which the descriptive work in question relies on background assumptions, some of which are explicit (as in the case of theories of chunking) and some of which are encapsulated in one's judgment.[10]

As revealed by the two case studies described above, some of the central descriptive features associated with objects of research (such as the fact that implicit memory functions in the absence of consciousness and that short-term memory has a capacity of 4 ± 1) are not straightforwardly observable insofar as they involve notions like *consciousness* and *chunk* that also do not have obvious or uncontested empirical descriptions or indicators. Experimental work aimed at describing objects of research is, thus, fraught with conceptual and empirical difficulties. The cases thereby illustrate some of the ways in which research objects can be epistemically blurry.

4.4. Exploratory Research

Given my contention that the aim of much psychological experimentation is to explore objects of research in order to delineate and describe them, the question might be raised whether this is exploratory experimentation in a technical sense. In this section, therefore, I develop and juxtapose my analysis of exploration with existing accounts of exploratory experimentation. I argue that, while the kinds of experiments enabled by operational definitions are not exploratory experiments in a technical sense, they do play an important role in exploratory *research*, that is, research geared toward the descriptive characterization of objects of research. I begin (in secs 4.4.1 and 4.4.2) by distinguishing between two features that extant discussions of exploratory experimentation have emphasized (conceptual openness and non-theory-guided experimentation) to bring my account into sharper relief. After that (in sec. 4.4.3) I explain my notion of *exploratory research* in more positive terms as emphasizing the epistemic *process* of scientific research as opposed to scientific *results*.

---

10. This latter point raises intriguing issues about the extent to which experimental research relies on tacit knowledge (see also Feest 2016).

4.4.1. Exploratory Experimentation and Conceptual Openness

The expression "exploratory experimentation" was introduced into the literature independently by Richard Burian (1997) and Friedrich Steinle (1997), though, as Schickore (2016) has pointed out, there were also significant differences between their analyses.[11] Steinle, in particular, situates himself within the context of experimentalist approaches to philosophy of science and uses the expression "exploratory experimentation" in contrast to "theory-driven experimentation." One central component of his account is the notion that exploratory experiments are characterized by the systematic variation of experimental parameters, typically with no theoretically guided expectations about the outcome, whereas "theory-driven experiments are typically done with quite specific expectations of the various possible outcomes in mind" (Steinle 1997, S70). The theoretical/conceptual openness of exploratory experiments is also emphasized by Richard Burian, who states: "Experiments count as exploratory when the concepts or categories in terms of which results should be understood are not obvious, the experimental methods and instruments for answering the questions are uncertain, or it is necessary first to establish relevant factual correlations in order to characterize the phenomena of a domain and the regularities that require (perhaps causal) explanation" (Burian 2013, 720).

On the basis of what was just said, we can—to a first approximation—distinguish two aspects of exploratory experimentation as articulated by Steinle and Burian, namely, (*a*) the notion that it is not geared at testing theoretical hypotheses (and perhaps does not even rely on theories at all) and (*b*) the notion that such experiments operate within the context of a certain degree of conceptual openness. Of those two aspects, the former has been subjected to critical scrutiny in the literature (I get to this in sec. 4.4.2), whereas the latter has not been remarked on as much (with a few exceptions, e.g., Haueis 2016, 2023), perhaps because the two are often viewed as amounting to the same thing. (If a domain has an undeveloped theory, then the concepts will be undeveloped as well.) However, this misses two important insights, namely, that undeveloped concepts can do epistemic work even in the absence of developed theories (Feest 2005b, 2010, 2011a; Haueis 2016) and that it is possible to do descriptive conceptual work in the absence of an explanatory theory

---

11. Steinle's aim, she argues, was to provide empirical evidence of a mode of research that does not fit the mold of hypothesis testing, whereas Burian was interested in the question of how varied experimental systems (in Rheinberger's sense) can be used for experiments that ultimately allow them to converge on the same subject matter.

(Feest 2017a). This highlights two separate points I wish to make here. First, concepts are often needed to delineate the very objects that researchers want to investigate (as I argued in chap. 3, this is where operational definitions as tools come in). Second, investigating and exploring an object is not the same as explaining it.

As I showed in section 3.3.1 above, Steinle (1997) acknowledges the distinction between theories and (what he calls) "conceptual frameworks." Indeed, the distinction is quite central to his approach. While rejecting the notion that all experiments aim at testing theories or theoretical hypotheses, he argues that existing conceptual frameworks enable experimental research but can prove to be inadequate for understanding the experimental data. In such cases, exploratory experimentation can give rise to the emergence of new concepts. With this analysis, Steinle draws attention to the important fact that concepts can fail, thereby necessitating the introduction of novel concepts. While I agree with this, I operate with a broader understanding of *exploration*, one according to which exploratory work does not necessarily give rise to novel concepts but can also play a role in the elaboration and extension of existing concepts. Thus, conceptual openness occurs not only when there is epistemic blurriness with regard to the question of what are adequate concepts for a given domain but also with regard to the question of how existing (and prima facie adequate) concepts should be developed and extended.[12]

Summing up, I share the contention of authors like Steinle and Burian that exploratory experimental research occurs in domains that are characterized by conceptual openness, though I emphasize the fact that in psychological research such openness can exist within the confines of preliminary concepts that individuate objects of research (i.e., objects of exploration). As I have argued above, such exploration often amounts to providing descriptions of objects of research. However, if we include as descriptive features of psychological objects attributes like *unconscious* and 4 ± 1 that are not straightforwardly observable, this kind of exploration has features of theoretical work. How does this relate to Steinle's contention that exploratory experimentation is not geared at testing theoretical hypotheses? In response to this question, we can remark that research can be theoretical without necessarily testing a fully formed theory. While I am on board with this point, I want to go further

---

12. The distinction between the possibility of a concept's failure and the possibility of a concept's extendibility gives rise to the question of whether there is a principled account of the difference (see Haueis's ongoing efforts in that direction [e.g., Haueis 2023, 2024]).

here and say that, when experimental researchers explore descriptive features of their objects of research (such as implicit memory or working memory), they frequently ask specific questions about those objects and design experiments that are specifically geared toward those questions. This can involve formulating and testing (tentative) descriptive hypotheses. For example, when exploring the sense in which implicit memory might be said to be unconscious, psychologists have to operate with very precisely formulated expectations as to what kinds of data would speak in favor of which hypothetical possibility.

In other words, even when psychologists systematically vary experimental conditions, and even when they do so with no general explanatory theory in mind, they can still do so with the aim of testing competing descriptive hypotheses about their research objects. Thus, I argue, exploratory experimental research in the sense that I propose is not exhaustively described by the notion of *exploratory experimentation* as (mere) variation of experimental parameters. Rather, when researchers explore descriptive features of research objects, this can involve the systematic creation of experimental effects where these effects are treated as relevant to a specific research object (e.g., implicit memory, working memory) that the experiments aim to explore.[13] And, in doing so, they frequently conduct differential tests of descriptive hypotheses. Insofar as this type of research is both exploratory and experimental, there is no harm in calling it "exploratory experimentation" (as I have done in Feest [2012a]). However, for the sake of avoiding confusion, I now prefer the broader category of *exploratory research* to distinguish my analysis from existing accounts of *exploratory experiments*.

4.4.2. Exploratory vs. Theory Driven?

Having argued for a notion of *exploratory research* that highlights that psychological research typically explores objects of research and that such explorations can involve the testing of descriptive hypotheses, I now want to embed my analysis in the context of debates about what has surely been the most controversial aspect of Steinle's account of exploratory experimentation, that is, his claim that such experiments are not theory driven. Steinle acknowledges that the notion of *theory* is notoriously hard to pin down and, therefore, primarily characterizes exploratory experimentation in negative terms, that is, by saying that mere statements of empirical regularities do not qualify as theories. This leads

---

13. This raises the question of what effects can be relevant to objects of research. I discuss this question in detail in chap. 6 and my concluding remarks.

him to state that the goal of exploratory experiments—to "find . . . empirical rules" and formulate "classificatory and conceptual frameworks" (Steinle 1997, S71)—is pretty uncontroversially not a theoretical goal.

As we saw, this analysis of exploratory experimentation does not fit my case studies since—as I just argued—researchers explore objects in a way that goes beyond establishing empirical regularities. I take this to show not that Steinle's account of exploratory experimentation is wrong but rather that the kind of experimental exploration he describes does not fully capture the kind of exploration I try to analyze here. Other authors have formulated more fundamental objections to his claims about the atheoretical nature of exploratory experimentation. While, at least in his original formulation, Steinle mostly said that researchers who conduct exploratory experiments do not have any theory-driven expectations about the outcomes, he has at times been taken to say something stronger, namely, that exploratory experiments rely on no theoretical presuppositions at all. This latter assertion has been challenged on the ground that recent exploratory research in the biomedical sciences (by virtue of using instruments) makes heavy use of background theories (e.g., L. Franklin 2005) or even relies on theoretical knowledge about the subject matter under exploration (Colaço 2018a).

Laura Franklin-Hall has distinguished between two ways in which theories can be involved in experiments. They can function either as "theoretical background" or as "local theory," that is, the theory or hypothesis actually being tested (L. Franklin 2005). She argues that experiments can be described as exploratory when there is no local theory being tested even if they are directed or constrained by a background theory. The example she provides is taken from experiments aimed at studying the regulation of gene expression during the cell cycle. Specifically, she looks at the technique of *microarrays*, an instrumentation that allows the experimenter to make twenty-five thousand measurements of cellular mRNAs at once, and she discusses research that used this technique to create "a comprehensive catalog of yeast genes whose transcript levels vary periodically within the cell cycle" (L. Franklin 2005, 892–93). It is clear that the very goal of this research presupposes a great amount of theoretical knowledge concerning yeast genes, transcript levels, and cell cycles. However, her point is that the goal itself is not to test a specific hypothesis but to arrive at classificatory schemes.

I suspect that Steinle would not disagree with the idea that there is often theoretical background knowledge involved in exploratory experiments (Feest and Steinle 2016; Steinle 2016). But Franklin-Hall is saying not merely that there is theoretical background knowledge involved in exploratory experimentation but that it can direct and constrain the ways in which exploratory experiments are conducted. This contention potentially challenges the notion that, by definition, exploratory experiments allow for "completely unpreconceived outcomes" (Steinle 1997, S70). Franklin-Hall's case, for example, takes as a given a particular object of research—the gene, under a particular theoretical conceptualization—thus raising the question of how much conceptual openness there really is. (In other words, even if the aim of this research is classification, it seems to be already determined what kind of entities are being classified.) David Colaço (2018a) makes this point even more explicit in that he argues that exploratory experiments in neuroscience in fact often presuppose "local theories" (i.e., theories of the subject matter under exploration). In other words (to put this in my terminology), they presuppose theoretical assumptions of what kind of research objects they are exploring. As Colaço puts it: "Contemporary exploratory experiments are often performed on systems where researchers already have a local theory and need to use this theory in tandem with their techniques to take the next step to explore more details of the system" (2018a, 4/17). Accordingly, he argues that local theories of a given subject matter can function as auxiliary hypotheses in experiments designed to explore further features of that very subject matter. This is compatible with Steinle's notion of *exploratory experiments* as not making specific predictions about expected experimental outcomes while making it clear that the possible outcomes are both enabled and constrained by theory.

The notion of experiments, even when they are in some sense exploratory, requiring background and local theory is well taken, though it does appear to narrow down the degree of conceptual openness of the situations thus described. As such, the accounts discussed in the previous paragraphs come closer to my contention that, in psychological experiments, the conceptual space within which exploratory research occurs is shaped and constrained by existing assumptions and knowledge of the objects that are being explored. I argue that, even in the absence of a well-defined local theory of (say) implicit memory or working memory, scientists see themselves as experimentally exploring those research objects, and they are able to do so precisely because of the operational definitions they employ.[14] In such situations, operational definitions play the role Colaço (2018a) attributes to local theory. They function as auxiliary hypotheses relative to which the experimental data

---

14. In turn, as I have argued in previous chapters, those operational definitions derive at least part of their plausibility from folk psychology, but this is not what most philosophers of science mean by "theory."

can be treated as relevant to the subject matter, and it is only relative to these auxiliary assumptions that the experimental data can be appealed to when trying to explore the subject matter. Notice that, by saying that operational definitions function as auxiliary hypotheses, I am not saying that they are the only auxiliary hypotheses that matter here. Indeed, for any given experiment, there is actually an open-ended number of additional auxiliary assumptions pertaining to the design and implementation of the experiment as a whole.[15]

One more point we can take away from the case studies discussed in section 4.3 above is that, while the distinction between local theory and theoretical background is heuristically useful, it is not as clear-cut as it might seem at first, just like there is no clear-cut way to separate specific concepts and the larger conceptual space that is presupposed when we use them. For example, as we saw in the case of working memory, the exploration of working memory capacity has to operate with the (at first sight seemingly descriptive) notion of a *chunk*. But the task of controlling for chunking requires some theoretical understanding of how chunking occurs, which prompts researchers to turn to theories of long-term memory. Likewise, the description of implicit memory as unconscious might seem simple at first, but it soon becomes clear that its precise meaning is underdetermined by the experimental data, and, moreover, it also raises questions about its contrast class, explicit memory, that is, it raises a wider set of conceptual and theoretical questions. There is, perhaps, a sense in which accounts of long-term memory and explicit memory function as background knowledge in the exploration of working memory and implicit memory. But our case studies have also revealed that, in psychology, such background knowledge is itself on shaky grounds.

4.4.3. Exploratory Research as a Process

One way of summing up the argument of the previous subsection might be to say that, when it comes to the exploration of objects of psychological research, the very distinction between exploratory experiments and theory-directed experiments is of limited value. Once we have abandoned the idea of exploratory experimentation in psychology merely collecting empirical regularities, we necessarily need to bring auxiliary hypotheses into play, that is, background assumptions (provided by operational definitions and local theories as well as more general

---

15. I return to this in chap. 7, where I develop a fuller account of experimental inferences.

theoretical background assumptions) relative to which the experimental data can be treated as relevant to the subject matter at hand.

An advocate of the notion of *exploratory experimentation* might, however, make the following objection. Even if it is the case that we are unlikely to find experiments that are free of conceptual and theoretical assumptions, it does not follow that all theory-driven experimentation is exploratory. This is a valid point, and, thus, there is still a meaningful distinction between experiments that aim at exploration and experiments that aim at theory testing. In response, one might raise the question of how these two modes of research are related in practice. It is this question that Maureen O'Malley seems to be addressing when she argues that, when it comes to specific research fields, we typically do not find that they employ only one or the other kind of experimentation. Rather, as she points out: "A continuum of practices, with TDE [theory-directed experimentation] at one end and EE [exploratory experimentation] at the other, might be a better way of describing the relationship between these two scientific approaches" (O'Malley 2007, 349). One might be tempted to read this as saying that researchers can go back and forth between these two modes/aims of experimentation. However, on closer inspection, O'Malley says something much stronger, namely, that hypothesis-testing research can also be exploratory. Accordingly, she conceives of exploratory experimentation as a process "that moves back and forth as various groups engage in exploration and theory evaluation in relation to ... newly emerging phenomena" (349). With this analysis she wants to do justice to the fact that "as new concepts and classificatory frameworks are being developed ... older theoretical frameworks ... are being called into service and examined for their adequacy in relation to this new knowledge" (348). She argues that, within specific research agendas, we find exploratory research that attempts to evaluate competing theoretical frameworks and might even involve hypothesis testing.[16]

As should be clear from my earlier argument, according to which the experimental exploration of descriptive features of research objects can involve something like hypothesis testing, I am very sympathetic to this analysis. However, as I have tried to argue here, we should be clear that it might be stretching the original intended meaning of *exploratory experimentation* too far. My systematic interest here is, therefore, not in how to use the term "exploratory experimentation" correctly but rather in how to describe the exploratory research practices we encounter in

---

16. In this regard, O'Malley clearly goes beyond the authors discussed in sec. 4.4.2 above, who merely argued that exploratory experiments typically rely on theories.

psychology. While O'Malley is not looking at psychology, her focus on the ways in which competing theoretical/conceptual frameworks can be evaluated, compared, and brought to bear on specific research agendas has similarities with my aim to gain an analytic grip on how researchers explore their objects of research.[17] Such explorations are, I argue, best understood as iterative and dynamic and as engaging with theory in various different ways (including by testing hypotheses). Finally, let me also emphasize (again) that, while many philosophers of science use the term "theory" as synonymous with "explanation," the kind of exploratory research I have covered in this chapter is better described as *descriptive research*. To reiterate, questions about the capacity limits of working memory or the conscious/unconscious status of implicit memory are descriptive questions, not explanatory questions. But this does not mean that the answers provided are atheoretical. Exploring descriptive characteristics of objects of psychological research is, thus, a theoretical task even if it proceeds by experimental means.

4.5. Conclusion

In this chapter, I picked up on the main point of the previous chapter, that operational definitions figure as tools in the research process, and elaborated on the question of what such tools are for. The main thesis of this chapter was that they are tools for the exploration of objects of research and that the need for such tools arises because objects of research are typically ill understood or, as I put it, "epistemically blurry." Accordingly, the chapter aimed to elucidate the notions of *object of research* and *exploratory research*. Section 4.2 laid out my conception of an object of psychological research, explaining that epistemic blurriness is not restricted to a lack of theoretical understanding but extends to descriptive and taxonomic questions. Section 4.4 situated my analysis of exploratory research in the context of existing work about exploratory experimentation, emphasizing the dynamic nature of exploratory research. Sandwiched between those two sections, section 4.3 provided detailed analyses of the intricacies of exploring specific objects of research using implicit and working memory as examples. It illustrated my conception of epistemic blurriness and highlighted the ways in which taxonomic, descriptive, and theoretical issues are intermingled in the process of exploring objects of psychological research.

17. In chap. 7, I argue that hypothesis testing (in the course of so-called converging operations) is, indeed, an important component of the exploration of phenomena.

Importantly, this chapter explained that the epistemic blurriness of psychological research objects is a crucial motivator of exploratory research. Contrary to the ideas that empirical descriptions of objects of research are easily constructed and that research then mainly focuses on providing their theoretical explanations, I argued that much experimental research in psychology is concerned with providing empirical descriptions of the relevant objects of research and that the very question of what these objects are and what their distinctive empirical features are is often at the foreground of the investigation. I also showed that the descriptive characterizations of implicit memory as unconscious and of working memory as having a particular capacity are much less straightforward than one might think. Such characterizations require appeals to other background assumptions, revealing also that the conceptualization of a specific object of research is often closely tied to others (e.g., short-term memory and long-term memory). I argued that empirical descriptions of the objects in question can be generated from task descriptions associated with specific measurement procedures (e.g., the task of priming as not involving conscious recollection), but I also pointed out that such descriptions can be ambiguous, requiring further theoretical and empirical work to disambiguate them. Thus, I argued that the exploratory process as such consists not simply in the variation of experimental parameters but also in considering the meaning and adequacy of the descriptions derived from the experimental paradigm, which will inevitably require some engagement with theoretical questions.

I adopted from Steinle (1997) and Burian (2013) the important insight that there is a lot of experimental research that takes place in conceptually open situations, and I argued that in psychology such research is often directed at specific objects (e.g., working memory, implicit memory). This means that there can be conceptually open situations that nonetheless operate with conceptual assumptions that guide and structure exploratory research. Thus, one upshot of my analysis was that conceptual presuppositions and operational definitions allow researchers to ask, and operationalize, questions about object of research, knowing full well that the very notion of what the object is may shift in the process. My analysis allows for the exploration of psychological objects of research to not only be theoretically guided and constrained, but also to allow for hypothesis testing.

There are several loose ends at the end of this chapter that need to be picked up and developed further. One that is taken up in the following chapter concerns an issue I have mentioned in passing in this chapter, namely, the relationship between objects of research and phenomena.

The following chapter addresses this topic and provides an analysis of the ways in which I see objects of psychological research to be (both constitutively and epistemically) related to phenomena. I argue (*a*) that objects of psychological research are typically composed of clusters of phenomena, (*b*) that phenomena come on a continuum between the context specific and the context transcendent, and (*c*) that researchers often use context-specific phenomena (i.e., experimental effects) in their efforts to investigate objects of research. This is followed (in chap. 6) by an analysis of the relationship between objects of research (as mind dependent) and psychological kinds (as potentially mind independent). Chapter 7, finally, returns to a fuller analysis of (the role of) operational analysis in the formation of psychological concepts.

CHAPTER FIVE

Phenomena and Objects of Research

5.1. Introduction

One upshot of the discussion in the previous chapter was that objects of research are epistemically blurry not merely because their unobservable properties are ill understood but because the very conditions of their empirical individuation and description are difficult to nail down. Indeed, one major contention of the previous chapter has been that this is why a lot of research in psychology is concerned with questions of taxonomy and description. I have also suggested in passing that objects of research can (putatively) involve multiple distinct phenomena. In this chapter, I further explain and defend the distinction between phenomena and objects of research. I thereby (*a*) disambiguate different usages of the word "phenomenon" in the existing philosophical literature and (*b*) spell out why I believe the distinction between phenomena and objects of research to be helpful for an adequate descriptive and normative analysis of research practices in cognitive psychology.

The overarching contention of this chapter is that researchers in psychology often conceptualize their objects of research as being composed of clusters of phenomena rather than as isolated phenomena. For example, we might think of memory as being composed of encoding mechanisms, phenomenological recollection, retrieval mechanisms, typical behavioral patterns, etc. All these phenomena can reasonably be construed as having to do with memory, but none of them, considered alone, is *the* object of memory research. Some of these phenomena (e.g., behavioral regularities that occur in response to specific experimental manipulations) are more accessible and salient than others, or so I argue. Others (e.g., cognitive or neural processes) are removed or hidden from view (for the distinction between surface phenomena and hidden phenomena, see Feest [2011b]). My claim is that researchers in psychology operate with the (not always explicated) understanding that the phenomena in question (both surface phenomena and hidden phenomena) jointly form the overarching objects at which the epistemic efforts of scientists are directed. It is precisely this conceptual presupposition that underwrites the rationale for using operational definitions as tools, that is, the practice of experimentally creating instances of behavioral phenomena as evidence for descriptive claims about specific objects of research (see Feest 2010; see also chap. 3 above).[1]

The main thesis of this chapter is, thus, that experimentally generated phenomena can function as evidence for claims about objects of psychological research, which are composed of clusters of phenomena. Experimentally generated phenomena can perform this function by virtue of the fact that they are (however preliminarily and defeasibly) taken to be typical indicators of the object in question. This thesis about the nature and function of phenomena seemingly puts me at odds with two important and influential currents in recent philosophy of science. First, Bogen and Woodward (1988) have famously provided an analysis of the notion of *phenomenon* according to which phenomena are stable and general features of the world, whereas data are local, contingent, and idiosyncratic. On the face of it, this seems to imply that the notion of an *experimentally generated phenomenon*—which I am proposing here—is an oxymoron. Second, within the mechanistic approach to explanation (and also in the explanation literature more generally), the term "phenomenon" is often used to mean *explanandum* phenomenon.[2] Again, this is very different from my contention that phenomena are features of objects of research and that some phenomena can function as evidence for hypotheses about objects of research.

In this chapter, I motivate and defend the theses articulated above, and I also show that they are compatible with extant accounts of phenomena and mechanistic explanation. Section 5.2 begins by providing a more detailed exposition of the key concepts of my account (*phenomena, data, evidence,* and *objects of research*). I argue that a close reading of Bogen and Woodward (1988) can accommodate my thesis insofar as phenomena, qua stable regularities, can occur on a sliding scale between

---

1. I suggest but do not argue here that such conceptual presuppositions also underwrite the rationality of intervening at either the level of behavioral phenomena or the level of neural phenomena to investigate a given object of research (e.g., Craver 2007).

2. Kästner refers to the notion "that phenomena are the *explananda* of mechanistic explanations" as a "mainstream assumption" within the approach of new mechanists (2021, 338).

the idiosyncratic and context specific, on the one hand, and the general and context transcendent, on the other. After explaining my analysis of the relationship between phenomena and objects of research (understood as clusters of phenomena), in section 5.3 I further explain my own focus on exploratory research in psychology by juxtaposing it with existing philosophical work on mechanistic explanations. I remind the reader of my claim (in the previous chapter) that, in cognitive science and psychology, not all investigative efforts are geared toward explanation and that some also include the task of exploring (delineating and describing) objects of research. In turn, this draws our attention to questions about the kind of evidence that is required for such exploratory efforts. I argue that experimentally generated phenomena can function as evidence for descriptive claims about objects of research by virtue of being conceptually tied to them. Section 5.4 turns to mechanistic analyses of discovery, arguing that, while existing accounts have focused on the discovery of mechanisms for specific phenomena, my analysis brings to the fore the larger conceptual assumptions (concerning objects of research) that explain why researchers take an interest in specific phenomena (including mechanisms) in the first place and what prompts them to treat specific experimental effects as evidence.

5.2. Phenomena vs. Data and vs. Objects of Research? Conceptual Groundwork

Two prominent philosophical treatments of the notion of *phenomenon*—Bogen and Woodward (1988) and Hacking (1983)—seemingly contradict each other insofar as the former emphasizes that phenomena transcend the specifics of experiments and the latter emphasizes the ways in which phenomena are experimentally produced and stabilized. I argue (*a*) that the two analyses can be reconciled (phenomena can vary in their degree of generality vs. their context specificity), (*b*) that this can help us appreciate how experimentally generated phenomena can function as evidence, and (*c*) that experimentally generated phenomena in psychology are, in fact, typically treated as evidence for descriptive claims about objects of research.

5.2.1. Phenomena and Data

In their influential "Saving the Phenomena" (1988), Jim Bogen and Jim Woodward characterized their notion of *phenomena* by distinguishing them from data: "Our general thesis . . . is that we need to distinguish between what theories explain (phenomena or facts about phenomena) from what is uncontroversially observable (data)" (1988, 314). Data are "idiosyncratic to specific contexts" (Bogen and Woodward 1988, 317), whereas phenomena are "relatively stable and general features of the world which are potential objects of explanation and prediction by general theory" (Woodward 1989, 393). Data "can be straightforwardly observed," but they "typically cannot be predicted or systematically explained by theory" (Bogen and Woodward 1988, 305–6). Phenomena, on the other hand, "are detected through the use of data, but in most cases are not observable in any interesting use of that term" (305–6). Their thesis was directed against the notion that theories enable predictions and explanations of observations. By contrast, they claimed, theories predict and explain phenomena, the existence of which, in turn, can be established by means of data. The ways in which data are produced, Bogen and Woodward argue, only rarely, if ever, rely on a theory of the phenomenon under investigation. This is, among other things, an ingenious move of dealing with the problem of theory-ladenness. If the theory of the phenomenon does not determine what kinds of observable data we gather in support of the theory, then those observable data cannot be theory-laden in any vicious sense.

Bogen and Woodward's points are well taken, especially in the context of the debates at which they were initially leveled, which still operated with the simple dichotomy between theory and observations. Bogen and Woodward are right to highlight that experimental evidence is created under the local circumstances of specific experiments. They are also right on the mark with their claim that scientific investigations are typically not aimed at explaining experimental data, that is, that experimental data typically function as evidence, not as explananda. And, finally, they are right to point out that data are often used as evidence not for some abstract theory but rather for something predicted by theory or even for something posited in the absence of a fully formed theory (Woodward 1989). All this is compatible with my contention that data can function as evidence in the investigation of objects of research. Indeed, I think that my account can explain why specific experimental data are taken to be able to play this role in the first place, that is, by virtue of the conceptual ties presumed to exist between objects and (types of) experimentally generated data.

However, while my analysis is inspired by Bogen and Woodward's, it also departs from theirs in important ways. Specifically, I argue for a notion of *object of research* as distinct from that of *phenomenon*. This distinction allows for the possibility that objects of research can be composed of clusters of phenomena. In addition, I argue that experimental effects (Bogen and Woodward's "data") are, in fact, phenomena. In the case of psychology, this allows us to appreciate that, even though researchers frequently turn their attention to specific effects (e.g., chunking, priming), there are also more overarching taxonomic categories in play (e.g., working memory and implicit memory) that enable researchers to treat those effects as instantiating phenomena that are relevant to the object in question. My claim, moreover, is that those overarching taxonomic categories are typically tied to specific, paradigmatic measurement procedures that provide what I have called "operational definitions." When used in specific experiments, those measurement procedures are used to generate experimental effects that are relevant to the research object by virtue of already being conceptually linked to it. Thus, experimental effects are (context-specific and idiosyncratic) phenomena, and they can figure as evidence for claims about objects of research (conceived of as clusters of phenomena) by virtue of being conceptually tied to them.

To motivate my claim that experimental effects are phenomena, let us review Bogen and Woodward's argument for why they are not. Their argument hangs on the idea that data are directly observable and phenomena are not. In response, I claim that, while it is true that individual experimental outcomes can be (in some sense) observed, they can serve as evidence only by virtue of exhibiting an empirical regularity or pattern (or at least by virtue of the assumption that they do). Such patterns are not directly observable. Hence, data can serve as evidence only by virtue of the assumption that they instantiate phenomena. To back up my case, let me briefly examine why Bogen and Woodward think that phenomena (unlike data) are unobservable. Bogen and Woodward (implicitly) provide two distinct reasons for this claim that ultimately point to two types of phenomena (both of which are, however, covered by their definition of *phenomenon* as stable regularity). The first reason is that individual measurements are always subject to errors and noise. For example, Bogen and Woodward argue that, while it is correct that scientists attempt to explain why the true melting point of lead is 327 degrees Fahrenheit, only very few, if any, measurements of the points at which samples of lead melt will actually produce exactly this thermometer reading. Hence, the melting point of lead is at best an average result of many thermometer readings. A different way of putting this might be to say that different thermometer readings instantiate a particular phenomenon (the value that can be read of the thermometer), but, because of the imperfections and idiosyncrasies in the devices taking those reading, they never instantiate it perfectly. Add to this the fact that it is possible to measure instances of the melting point of lead in different labs and with different thermometers, and it emerges that the true phenomenon (thermometer reading) can be inferred only on the basis of multiple measurements.

Notice that the true phenomenon at stake here is essentially an averaged-out empirical finding, that is, an empirical regularity. The expression "the melting point of lead" refers to such an empirical regularity. The phenomenon as such is unobservable because averages cannot be observed, unlike individual instances (thermometer readings). It is important to recognize that thermometer readings can be classified as instantiating a particular phenomenon only by virtue of the assumption that they are part of a distribution around a mean and, hence, instantiate an unobservable phenomenon, however imperfectly. I argue in a moment that it is precisely on this assumption that instances of phenomena can function as data in an experimental context. However, there are other places where Bogen and Woodward make it quite clear that they intend their notion of *phenomenon* to encompass more than averaged-out empirical measurements. For example, they mention weak neutral currents and proton decay as examples of phenomena. To these we might add things like retrieval mechanisms or cognitive chunking in the mind/brain. This suggests another reason why phenomena are, according to Bogen and Woodward, unobservable: because they are tied to what philosophers of science sometimes call "unobservable entities" (either unobservable right now or unobservable in principle). In other words, even instances of them are not directly accessible to the human eye or other measurement instruments. Take, for example, the phenomenon of proton decay. If protons are not directly observable, then neither is their decay. In a similar vein, memory decay is not directly observable because memory is not. Does this mean that it is not possible to conduct empirical research on such phenomena? Surely not. But the connection between the phenomena in question and the empirical data used to investigate them is more mediated. Phenomena like memory decay, chunk formation, or priming are inferred not just by establishing regularities in the empirical data but by making abductive inferences from empirical regularities to some more observationally removed regularities.

5.2.2. Discernible Regularities as Evidence: The Relational Framework

This brief discussion of different arguments for the unobservability of phenomena suggests that, while different kinds of phenomena are unobservable for different reasons, there is also something they have in common: they are repeatable regularities. In the one case, they are regularities in the data, that is, regularities in the behavior of observable entities. In the other case, they are regularities in the behavior of some (perhaps unobservable or currently unobserved) entity. To avoid the fraught language of observability, I have in the past referred to them as "surface phenomena" and "hidden phenomena," respectively (Feest 2011b). The question might be raised whether they have enough in common to be described by the same concept. My claims are (*a*) that they do and (*b*) that it is helpful to understand that research objects in psychology, such as implicit memory or working memory, are composed of both surface phenomena (behavioral regularities) and hidden phenomena (regularities displayed by entities not accessible to the naked eye). It is precisely because of this that experimental effects are presumed to provide epistemic access to the objects in question.[3]

Having just argued that Bogen and Woodward's conception of phenomenon covers both what I call "surface phenomena" and what I call "hidden phenomena," let me point out that Ian Hacking uses the term "phenomenon" to describe "noteworthy discernible regularities" (1983, 225); that is, he appears to want to restrict its usage to surface regularities. He is especially interested in those regularities that can be produced experimentally, such as the Hall effect. More provocatively (or at least seemingly so), he has famously argued that many phenomena exist only in specific experimental contexts, contexts in which they have been laboriously produced. By contrast, Bogen and Woodward downplay the importance of experimentally produced phenomena, suggesting instead that phenomena are general features of the world, not idiosyncratic features of data production in experiments (i.e., Hacking's experimental effects). This prompts them to argue: "[The] features which Hacking ascribes to phenomena are more characteristic of data" (Bogen and Woodward 1988, 306n6). Given what I have just argued, it should be clear that I agree with Hacking that experimentally produced effects are phenomena. I would add that the fact that such effects do not exist outside the laboratory is not surprising and does not stand in the way of them serving the purpose for which they are generated. In other words, the point I want to highlight here is not just that experimental effects are phenomena but also that they play a specific epistemic role, to function as *evidence*. The follow-up question is by virtue of what they can play this role.

---

3. Notice that here I am talking only about the conceptual assumptions that need to be in place in order to treat experimentally generated effects as evidence for claims about objects of research. This is separate from the ontological question of whether specific psychological objects do indeed have specific surface phenomena that ineliminably belong to them. This latter question is addressed in chap. 6 below.

In addressing this question, let me start by repeating that, even though I agree with Bogen and Woodward's contention that data are highly local and idiosyncratic to specific contexts, they still have to exhibit a regularity in order to be able to figure as evidence (see also Feest 2011b). That is to say that, even if we have only one experiment (and, consequently, one set of data), the resulting data can be treated as evidence only on the assumption that they are instantiating a pattern, that is, that repeated experiments will yield similar data. But, clearly, this is not the only prerequisite. Evidence is a deeply relational notion in that it is always evidence *for* something. In fact, this is a point that Sabina Leonelli (2015, 2016, 2020) has stressed, arguing for what she calls a "relational" view of data. Within such a relational framework, she argues, the notion of *data* collapses into that of *evidence* (Leonelli 2020, sec. 5). As she puts it: "It is meaningless to ask what objects count as data in the abstract, because data are defined in terms of their function within specific processes of inquiry, rather than in terms of intrinsic properties" (Leonelli 2015, 818). I agree with this contention, though I would stress that it is not meaningless to ask what necessary requirements have to be met in order for measurement outputs to be candidates for data. (As stated above, I contend that they have to exhibit a stable regularity.)

So, given the assumption that data are defined in terms of their function within specific processes of inquiry (i.e., relative to their purpose), two questions arise. The first is a descriptive question regarding the processes by which specific marks (Hacking 1992) or traces (Rheinberger 2011) become data/evidence, that is, are invested with the power to speak to a specific inquiry. To put it differently, what background assumptions need to be in place in order for scientists to treat specific experimental outputs as evidence? The second question is a normative one in that it asks what the criteria of adequacy for treating outputs as evidence are. Surely, on a relational notion of *evidence*, there must be constraints on what constitutes *good* evidence. With regard to the first question, my answer is plain and simple. Experimental marks are treated as evidence for claims about a given object of research relative to, first and foremost, an operational definition of the corresponding concept plus the assumption that the operational definition is adequately implemented and there are no other potentially confounding factors in the experiment (see also Feest 2022). As I have argued in previous chapters, these conceptual assumptions are built into the very design that informs data production. I take this to be a descriptive fact, but I think the normative implications are obvious. If it should turn out that the operational definition of a given concept is inadequately implemented, other factors are not controlled for, or the concept has been abandoned for other reasons, the purported evidence in support of claims about the object thought to be in the extension of the concept will retrospectively have to be deemed inadequate or even meaningless. This implies that a crucial part of the investigative processes consists (or ought to consist) in evaluating the evidence that is presented in support of specific claims about a given object of research.

In chapter 7, I return to the important normative question of how data quality can be evaluated. I argue there (*a*) that to treat the output of an experiment as evidence for a specific claim is to make specific (implicit or explicit) assumptions about the causal history through which the output came about and (*b*) that the process of probing these assumptions is or ought to be a central part of the intertwined process of exploratory research and concept formation. I also address there the question of whether the locality and idiosyncrasy of experimentally produced phenomena challenge the generality of the conclusions we can draw from them, an issue that has variously been discussed under headings like "extrapolation" (e.g., Steel 2008) and "external validity" (Guala 2005; Jimenez-Buedo 2011; Sullivan 2009). In a nutshell, I argue that, once we appreciate that experimentally generated phenomena are intended as evidence, this helps us reframe the question of generalizability and extrapolation as being about the question of whether specific experimentally generated phenomena meet the standards required to serve as evidence for any specific claim at the intended level of generality and scope.

Leaving the normative issue of data quality aside for now, let me restate the main aim of this chapter, which is to argue that inferences from experimentally produced (surface) phenomena in psychological research are typically not to isolated (hidden) phenomena. Or, rather, insofar as researchers make inferences from surface phenomena (experimental effects) to specific hidden phenomena, this is often in the service of describing features of a larger object of research. For example, evidence of experimental chunking is typically used not simply as evidence for chunking in the mind/brain in and of itself (though it can be used that way, too) but as evidence for specific descriptive assumptions about working memory (e.g., the claim that working memory is characterized by chunking and that it has a particular capacity). Likewise, in the investigative contexts I covered in chapters 3 and 4 above, psychologists use experimentally generated priming effects not simply to make an inference about some hidden priming process in the mind/brain but rather to make an inference about a particular object of research (implicit memory) that they assume to have specific descriptive features (being unconscious or automatic).[4]

---

4. Notice that, at this point, my aim is not to evaluate these inferences but merely to make a descriptive claim about scientific practice.

5.2.3. Phenomena, Clusters of Phenomena, and Objects of Research

The discussion in the previous subsection illustrates why it is analytically useful to distinguish between phenomena and objects of research. Such a distinction is not currently drawn in the literature. Instead, as already mentioned, many philosophers use the term "phenomena" more broadly to refer to the targets of scientific investigations that are assumed to be in need of explanation. Let me emphasize that I do not mean to suggest that there is anything inherently wrong with referring to targets of research and explanation as "phenomena." I would like to suggest, however, that this common philosophical parlance sometimes glosses over an important difference, namely, that between phenomena simpliciter and phenomena as components of something else where the exact nature and identity conditions of this "something else" is to be determined.

Again, Bogen and Woodward's analysis is instructive here since they are a little ambiguous when it comes to those two usages. According to them, phenomena are "relatively stable and general features of the world" (Woodward 1989, 393) that "have stable, repeatable characteristics" (Bogen and Woodward 1988, 317). The latter statement is somewhat loosely formulated, however. Reading it inattentively, one might suppose that phenomena are entities in the world that have certain repeatable characteristics. On this reading (as applied to one of my examples), working memory might be considered a phenomenon that has certain repeatable characteristics, such as that of chunking. One might then suppose that it has other repeatable characteristics as well, for example, those having to do with duration and capacity. However, when we look at the examples provided, it is clear that Bogen and Woodward also regard the chunking effect *itself* as a phenomenon, not simply as a characteristic of some other phenomenon. So it seems that they use the term "phenomenon" in two distinct senses: (*a*) as a kind of thing (e.g., working memory) that has repeatable characteristics and (*b*) as simply a repeatable regularity (e.g., chunking).

In the philosophical literature (as in ordinary language), the term "phenomenon" is frequently used in the former sense in that it is often used when we mean not just isolated empirical regularities but more complex (and frequently not fully understood) occurrences of the natural and social world that we associate with specific salient empirical regularities. Accordingly, global warming or racism might be called "phenomena" even though, clearly, neither of them can be reduced to any one empirical regularity. Most of us have no doubts that these phenomena exist and that they are stable (though, it is to be hoped, changeable) features of the world, but it is equally clear that we think of them as complex systems that are composed of multiple regularities. This is, obviously, compatible with the notion that sometimes we want to explain specific regularities as resulting from the behavior of the complex system. There has been important work about the question of how multiple models of a complex system might be integrated with one another (e.g., Mitchell 2003) and about the question of how different models of complex systems might be said to embody different perspectives (Massimi 2022; Mitchell 2020). (See also Kästner [2018], which specifically applies the perspectives metaphor to levels of mechanistic models of cognition.) Here, my aim is more modest: suggesting that it is analytically useful to draw a conceptual distinction between *phenomena* (simple regularities) and *clusters of phenomena* (systems of regularities) and, furthermore, that objects of psychological research are often clusters of phenomena that are (by definition) not yet well understood. This is helpful for my purposes because it allows us to understand why chunking can be both a phenomenon (in the sense of an empirical regularity) and a feature of an object of psychological research (in the sense of an overarching object that is composed of a cluster of phenomena, such as behavioral phenomena, cognitive phenomena, experiential phenomena, or neurological phenomena).

We can then make sense of the objects of psychological research as typically consisting of clusters of phenomena that are treated as somehow belonging to the same overarching, albeit for the time being, epistemically blurry object. For example, when studying global warming, racism, or memory, we assume these to be viable objects of research, albeit objects that consist of multiple phenomena. We know or assume them to be real but are unsure about their precise descriptive contours and features or about the mechanisms that cause and sustain them. Often the question of what kind of a thing we are dealing with and what phenomena (stable regularities) belong to it is precisely what is at stake. Treating a specific phenomenon (e.g., a specific behavioral regularity) as part of a cluster of phenomena, giving the cluster a name (e.g., "memory"), and setting out to develop a concept to describe the class of objects in the extension of that concept is tantamount to treating memory as an *object of research*. For this reason, I suggest a terminological distinction between simple phenomena and clusters of phenomena, on the one hand, and objects of research, on the other, and I argue that, insofar as they occur (or are perceived to occur) as part of a cluster, individual (simple) phenomena are often viewed as features of an overarching object. It is because of this assumption that individual phenomena can be treated as explanatorily, descriptively, and evidentially relevant to the object.

The bottom line is that, by distinguishing between phenomena and objects of research, we get a better analytic grip on the research process as it presents itself to practicing scientists. I therefore suggest restricting (for analytic purposes) the use of the term "phenomenon" to things like the chunking effect, the priming effect, the melting point of lead, and other events that occur (or are assumed to occur) regularly under specified circumstances in order to distinguish them from objects that are characterized as clusters of phenomena. Both simple phenomena and clusters of phenomena can become objects of research once we turn our epistemic attention to them, thereby acknowledging that they present themselves as epistemically blurry and, thus, as conceptually underdetermined. Thus, by distinguishing between simple phenomena and clusters of phenomena, on the one hand, and objects of research, on the other, we can appreciate that both simple and complex phenomena can become objects of research. It is entirely conceivable that researchers might simply study the chunking effect their entire careers without taking any interest in the larger context that might give this phenomenon meaning. Indeed, I do not intend to deny that a lot of psychological research concerns itself with collecting evidence for specific experimental effects. However, I argue that, in psychological research, we frequently see experimental effects being generated with the explicit purpose of collecting data about an overarching research object such as working memory or implicit memory.[5]

5.3. Objects of Psychological Research as Explanandum Phenomena?

The account I have been developing thus far states that experimentally generated effects (local and idiosyncratic phenomena) can figure as evidence in relation to specific taxonomic and descriptive accounts of objects of research in psychology by virtue of the fact that they are treated as conceptually relevant to the research object and, qua operational definitions, already built into the experimental design. I now turn to the question of how my analysis of phenomena bears on talk about explanandum phenomena we encounter within the tradition of the new mechanism (e.g., Bechtel 2008b; Craver 2007; Glennan 2017; Kästner 2017; Krickel 2018). Clearly, the mechanistic approach to cognitive

---

5. The fact that evidence in psychology is often created on the basis of its conceptual connection with a specific object of research has recently given rise to worries about the concept ladenness of evidence (Dubova and Goldstone 2023). I address this concern in sec. 5.5 as well as in chap. 7 and my concluding remarks.

neuroscience is relevant here since it analyzes research practices that are, apparently, directed at subject matter similar to that of cognitive psychology.[6]

5.3.1. Phenomena as Explananda? The Case of Spatial Memory

The mechanistic tradition hardly needs an introduction here, so I keep my review to the bare minimum, referring readers instead to excellent recent discussions and state-of-the-art overviews of epistemological and metaphysical aspects of mechanisms (e.g., Kästner 2017; Krickel 2018; and Glennan, Illari, and Weber 2022). One central distinction found in that literature is that between two kinds of mechanistic explanations: etiological explanations and constitutive explanations (Craver 2007).[7] While the former explain phenomena by reference to mechanisms that precede them, the latter explain them by being constituted by or constitutively relevant to them. While it is a matter of some debate what "constitution" and "constitutive relevance" mean precisely (e.g., Krickel 2018), there seems to be some consensus that constitutive mechanistic explanations are not causal explanations even though they involve reference to causal mechanisms, namely, those that explain component phenomena of the phenomenon of interest. Mechanisms, in general, are taken to be organized assemblies of entities that are productive of regular change and, thereby, explain that change. Another important feature of mechanistic explanations is that they can be "interlevel." Minimally, this means that such explanations link items in different domains, such as the neural and the cognitive domains (Kästner 2017, 66). In the case of constitutive mechanistic explanations, this interlevel characteristic plays out in terms of the idea that the explanandum phenomenon (e.g., the formation of a specific memory trace in the hippocampus) is at the top of a mechanistic hierarchy, that is, is constituted by mechanisms that can in turn be broken down into lower-level mechanisms (e.g., specific molecular processes that occur when a memory trace is formed in the hippocampus).

Since both etiological and constitutive explanations are taken to be explanations of explanandum phenomena, this is a good moment to ask

---

6. The literature on this topic has exploded in the last twenty years, and I certainly cannot give a complete overview here. The main purpose of what follows is (*a*) to argue further for my analysis of phenomena by contrasting and comparing it with those of some prominent figures in the mechanistic literature and (*b*) to highlight what I take to be underlying agreements between my approach and that of mechanists in philosophy of cognitive science.

7. In the literature, one also finds reference to maintaining mechanisms (Craver and Darden 2013; Kästner 2018), but I ignore those here.

what mechanists mean by the term "phenomenon." There is relatively little sustained discussion of this question by mechanists, though both Bechtel (2008b) and Craver and Darden (2001) state that their usage of the term in line with Bogen and Woodward's (1988). To get a clearer sense of how their usage compares with mine, I now take a closer look at a case that is frequently invoked: that of spatial memory. My aim here is to argue (*a*) that there is a relevant distinction between the explanation of specific (spatial memory) phenomena and the explanation of *the phenomenon* of spatial memory and (2) that, while accounts of mechanistic explanation typically work with the idealizing assumption that explanandum phenomena are already known and merely await explanations, the reality of the investigative process in psychology is more complicated, highlighting the usefulness of my notion of an *object of research*.

In the history of psychology, spatial memory was famously put on the agenda by Tolman's "Cognitive Maps in Rats and Men" (1948). In this paper, Tolman conceptualized spatial memory in terms of the ability to form and then utilize spatial maps, where this ability (and, thus, the concept of *spatial memory*) was operationally defined in terms of successful search behavior (for more details, see chap. 1). Tolman reports on research conducted in his laboratory in which hungry rats placed at the entrance of a maze found their way toward the food that was located somewhere in the maze. This process was repeated a number of times on different days, and it turned out that after a few trials the rats ran straight to the food. Tolman suggested the hypothesis that "in the course of learning something like a field map of the environment gets established in the rat's brain" (1948, 192). Later on, the so-called Morris Water Maze was devised in which mice performed spatial navigation tasks in mazes that were filled with water, thereby making it impossible for them to use their sense of smell to find their way (Morris et al. 1982).

Intuitively, there are many phenomena involved here. This is acknowledged by Craver and Darden (2001) in an article that is, as far as I know, the first instance of spatial memory being used as a case study. After noting that they follow Bogen and Woodward (1988) in defining "phenomena" as "relatively stable and repeatable properties or activities that can be produced, manipulated or detected in a variety of experimental arrangements," Craver and Darden name a few examples of such phenomena—"the acquisition, storage, and retrieval of spatial memories; the release of neurotransmitters, and the generation of action potentials"—and proceed to present a multilevel mechanistic explanation of the phenomenon "mouse navigating Morris Water Maze" (Craver and Darden 2001, 114, 118; see also Craver 2007). However, a few pages later they identify *spatial memory* as "the phenomenon to be explained" (Craver and Darden 2001, 121).

From the perspective of the analysis developed earlier in this chapter, it should be clear that in the investigative context, in which scientists experimentally created the phenomenon of "mouse navigating Morris Water Maze," this phenomenon was intended as *evidence*, that is, as evidence (loosely speaking) for the ability to navigate in space.[8] In turn, this phenomenon could be the explanandum of (*a*) an etiological explanation, which spells out the mechanism that causally leads up to this particular instance of navigational behavior (e.g., mouse feeling hungry, which triggers the retrieval of a spatial map, which in turn prompts the initiation of navigational behavior), or (*b*) a constitutive explanation, which refers to the mechanisms that are active while the mouse engages in navigational behavior.[9]

However even if we grant (as we should) that the navigational behavior of mice (or navigational behavior more generally) can be *an* explanandum phenomenon of spatial memory research, the claim that it is *the* explanandum phenomenon of spatial memory research seems obviously false, and it is unlikely that this is what Craver and Darden had in mind (I propose a more plausible construal in sec. 5.3.2). Other relevant explanandum phenomena might be, for example, the formation, maintenance, decay, and retrieval of spatial maps. But how are these various spatial memory phenomena related to spatial memory? Krickel provides an interesting hint when she writes that "the hippocampus's generation of spatial maps" is a "component in the mechanism for spatial memory" (2018, 95). This formulation suggests that there is a mechanism that explains spatial memory as a whole. We can speculate that such a mechanism (if it exists) is multilevel and spatiotemporally extended and that it integrates and explains the totality of spatial memory phenomena. However, since the concept of *spatial memory* is much broader and more diffuse than *mouse navigating Morris Water Maze* or *hippocampus generating spatial map*, it is not clear that the former refers to a phenomenon in the same sense (stable, repeatable regularity) as the latter. I take this to illustrate my thesis that here, too (as in the literature about phenomena

---

8. I return in chap. 7 to the intricacies of determining the exact scope of an inference from the experimental evidence.

9. Here I am following Kästner's (2021) analysis, according to which etiological and constitutive explanations have different explananda (results vs. processes). Note, however, that (*a*) the result of the mechanism (e.g., navigational behavior) can also be a process and thus also receive a constitutive explanation, and (*b*) the process leading up to the result is itself a phenomenon that can receive a constitutive explanation.

discussed above), we should distinguish between two uses of the term "phenomenon." Sometimes it is used to refer to specific regularities (encoding, retrieval, spatial navigation), and sometimes it is used to refer to something broader (spatial memory) that displays or is constituted by specific regularities but is not easily reduced to any one of them.

I argue that these two usages correspond to my distinction between phenomena and objects of psychological research, where the former are regularities and the latter are clusters of regularities. Now, treating a cluster of phenomena as a specific kind of object is not the same as committing to the idea that there is a mechanism *for* this object (or, indeed, that there is *one* object corresponding to a given concept). Inquiring into the existence of such a mechanism for spatial memory (or for other psychological kinds) amounts to asking whether there is a designated system for the kind in question (Michaelian 2011). I return to the question of psychological kinds in chapter 6. My point here is merely to highlight that psychologists individuate and explore their objects prior to having a firm understanding of the existence, boundaries, and mechanisms of those objects. This follows from the epistemic blurriness of research objects, as explained in chapter 4 above. In turn, this means that researchers are not in a position to delineate and describe *the* explanandum phenomenon (of, e.g., memory research) very clearly. Nor can they be sure whether specific, more confined phenomena belong to a specific object of research.

If I am right in my contention that there is a lot of epistemic blurriness surrounding the nature and descriptive outlines of research objects like spatial memory, it seems that we need to distinguish between two kinds of questions with regard to mechanistic explanation in cognitive neuroscience. The first is what such an account should look like given the idealizing assumption that we already have our explananda clearly delineated. The second is how the explananda of cognitive neuroscience present themselves to scientists in the investigative process, precisely which phenomena are being explained, and how they relate to objects of research like spatial memory. While the former issue has received a lot of sustained attention in philosophy of neuroscience, Jacqueline Sullivan (2010) has rightly noted that it is not very clear what precisely the phenomenon circumscribed by the Morris Water Maze data is. Candidate explananda range from a cognitive function (spatial navigation) and representational changes (cognitive map formation) to observable changes in behavior. I argue that, in a diffuse kind of way, the concept of *spatial memory* is a catchall for all the phenomena (and probably more) mentioned by Sullivan. While it is surely correct that cognitive neuroscientists are interested in explanatory mechanisms, my analysis highlights the fact that the very existence, shape, and descriptive features of explanandum phenomena such as spatial memory are epistemically blurry and require sustained research. My notion of an *object of research* tries to capture that situation, and my analysis of the conceptual and epistemic relationship between objects of research and their phenomena offers a way of thinking philosophically about the taxonomic and descriptive efforts of psychologists.

5.3.2. Constitutive Explanations as Delineating Heterogeneous Sets of Capacities

Having used the case of spatial memory (as analyzed by proponents of mechanistic explanations) to further articulate my distinction between phenomena and objects of research, I now turn to the question of whether the project of delineating and describing psychological objects (which I am attributing to psychologists) and that of constructing constitutive explanations (which mechanists have attributed to neuroscientists) might be said to be two sides of the same coin.

Let me start by returning to the question of what the explanandum phenomenon of the search behavior researchers observe in the Morris Water Maze is. One suggestion is that this experimentally generated phenomenon instantiates a behavioral capacity that is common to mice and humans (which is why mice can be used as models to study it) and that it is, in fact, that capacity that spatial memory research ultimately wants to explain. The notion that the explananda of psychological research are capacities was first articulated by Robert Cummins (1983). The analysis offered above, according to which experimentally generated phenomena are not the explananda of psychological research but rather function as evidence, directly mirrors Cummins (2000) and owes a lot to his account.[10] If we think of psychological explananda in that way, and if we follow Cummins in arguing that explanations of capacities are provided by functional analysis, then this suggests a plausible construal of what constitutive mechanistic explanations are really trying to accomplish: to explain capacities. While Craver (2007) initially distanced himself from Cummins's view that explanations in psychology are explanations of capacities, this notion seems to have become accepted in more recent writings. This opens up the suggestion that something like Cummins-style functional analysis can provide sketches of mental

---

10. The notion that the explananda of psychological research are capacities has also been argued for in recent work coming out of cognitive science (e.g., van Rooij and Baggio, 2020, 2021).

mechanisms that might ultimately lead to neuroscientific explanations of the capacities in question (e.g., Piccinini and Craver 2011).

I take this to be on the right track, but I think that the proposal needs some amending to capture psychological research practices as I construed them above. The first amendment is that behavioral phenomena are not the only phenomena making up objects of psychological research (and, therefore, that behavioral capacities are not the only capacities of interest). The second amendment is that in psychology not all research efforts are aimed at constructing explanations. Let me begin with the first point. There is a way in which we can construe Tolman's positing of spatial maps (along with demands) as positing a mental entity that explains our behavioral capacity to navigate in space. This may have been good enough to satisfy Tolman's behaviorist sensibilities. However, I do not think it captures spatial memory research today. Surely, once spatial maps are on the table, other explanandum phenomena are added to the agenda of spatial memory research, such as the capacities to generate, sustain, represent, retrieve, and use spatial maps. While the capacities just mentioned are explanatorily relevant to the behavioral capacity of navigating in space, they are not themselves behavioral capacities but better understood as cognitive capacities.[11] To this I would like to add the capacity to experience cognitive representations in a certain way. Given this, I argue that the objects of research in cognitive psychology (beyond being clusters of phenomena) are best understood as sets of heterogeneous capacities. I suggest referring to these as "behavioral," "cognitive," and "experiential" capacities (see also Feest 2023).

This analysis is compatible with my claim that macro-level objects of psychological research (such as spatial memory) present themselves as clusters of phenomena since it explicates these phenomena as instances of the behavioral, cognitive, and experiential capacities just mentioned. It is also compatible with the idea that these phenomena and their integration can receive mechanistic explanation (by virtue of unpacking the mechanisms that enable the capacities in question). Going back to the question posed above of what a mechanistic explanation of spatial memory would look like, I suggest the following. Spatial memory is constitutively explained by a model that explains the totality of phenomena that make up this object, and it does so by explaining the capacities that,

---

11. This is reflected in the recent philosophical literature about cognitive neuroscience. For example, as mentioned above, Kästner (2018) argues that the point is to explain cognitive phenomena, and Sullivan states that cognitive neuroscience attempts to construct "mechanistic explanations of cognitive capacities" (Sullivan 2016b, 47).

when instantiated, give rise to the phenomena in question. If that is the case, then the effort to formulate a constitutive explanation for a given object of research, such as spatial memory, may well contribute to efforts adequately to delineate this object in theoretical and descriptive terms. In other words, building a mechanistic model that constitutively explains a set of capacities amounts to delineating what the corresponding psychological kind is.

On the analysis just provided, the explanatory project of constructing constitutive mechanistic explanations for complex capacities converges on the descriptive and taxonomic project of exploring them. However, I argue that this convergence is not well captured by the notion that the epistemic project of psychologists mostly consists in providing sketches of mechanisms that neuroscientists fill in (Piccinini and Craver 2011). The reason for this—to reiterate—is that a lot of psychological research is concerned not with the explanation of phenomena but rather with the classification and description (i.e., the exploration) of objects of research. This is particularly evident when we look at the experiential aspects of psychological objects of research. Think, for example, of the capacity to navigate space in accordance with a mentally represented spatial map. From the point of view of psychological exploration, one question of descriptive interest here is how spatial maps are phenomenologically represented so as to provide a tool for efficient navigation. For example, we can ask questions such as whether spatial maps are more like images or like architectural models, whether they are accurate, whether they are consistent, whether they encode perspective, whether they represent a bird's-eye view, etc. (Tversky 1993; see also Feest 2017a).

These are descriptive questions that characterize experiential capacities insofar as they pertain to spatial memory. I am not suggesting that these capacities cannot ultimately be explained mechanistically but merely making the point that often the search for mechanisms is not what psychologists are interested in. For example, as I argued in the previous chapter, one of the guiding questions of psychological research into the phenomenon of chunking is not what the mechanisms that produce it are but what this phenomenon can teach us about the storage capacity of working memory. Likewise, one of the guiding questions of psychological research into the phenomenon of priming is not what the mechanism that produces it is but whether it can be used to tell us something about the nature of implicit memory (e.g., about the respects in which implicit memory processes are unconscious).

Summing up, then, in this section I have tried to sharpen my distinction between phenomena and objects of research by contrasting my analysis of the research process in psychology with influential philosophical accounts of mechanistic explanations. My analysis has revealed the two approaches to be mutually compatible, thereby bringing into clear view the idea that objects of psychological research, such as memory, can be understood as (epistemically blurry) sets of behavioral, cognitive, and experiential capacities that for the purposes of research are subsumed under a single concept. I have also emphasized that, despite this, there is more to the descriptive aims of psychology than the mere construction of constitutive explanatory models.

5.4. Psychological Discovery as Phenomenal Decomposition?

Mechanist philosophers have put not only explanation but also discovery on the agenda of philosophy of science (e.g., Bechtel and Richardson 1993; see also Bechtel 2008a, 2008b; Craver and Darden 2013; Darden 2006; and Kästner 2017). The focus in this literature, which has been heavily influenced by interventionism (Woodward 2005), is on the discovery of mechanisms. Still, the notion of *phenomenon* is also invoked frequently, either as specifying the level at which a given experimental intervention is aimed or as delineating the observable correlate of the to-be-discovered mechanism. In this section, I submit some of the underlying assumptions of this literature to critical analysis and argue that my distinction between phenomena and objects of research can provide a helpful supplement.

5.4.1. Decomposition, Localization, and the Reconstitution of Phenomena

A widely held assumption in the literature about mechanisms is not only that phenomena are the explananda of mechanistic explanations but also that the discovery of mechanisms typically proceeds by way of constructing higher-level ("phenomenological") descriptions of the relevant explanandum phenomena with the aim of later supplying the mechanistic explanation, which in turn can reconstitute the explanandum phenomenon (e.g., Bechtel and Richardson 1993). In a similar vein, Craver and Darden (2001, 2013) describe the process of discovery as a process of "going down one level" to search for the mechanisms and "going up one level" to try to redescribe precisely what the mechanisms do (i.e., to redescribe the phenomenon). Craver (2007, chap. 4) remarks that, during the course of categorizing the relevant phenomena, two kinds of error can occur—lumping errors and splitting errors—where in each case the identity conditions of the phenomena are, ultimately, determined by the identity conditions of mechanisms.

One early important work that analyzed the process of discovery in fields like neuroscience was Bechtel and Richardson's 1993 *Discovering Complexity*. The spirit of the book is somewhat similar to the project pursued here in that Bechtel and Richardson set out to analyze the rationality of the actual research process in fields that investigate complex systems. Their basic premise differs from mine, however, insofar as they maintain that these research fields are characterized by a search for explanatory mechanisms, whereas I am highlighting that task of describing the overarching objects of research. They identify two heuristic strategies—decomposition and localization—as guiding this search. Roughly, on this view, phenomena are identified functionally on the basis of the system's behavior, and the search for the relevant mechanism proceeds by functional decomposition of the relevant behavior as a way of breaking down the explanandum into subunits for which specific mechanisms can be localized. In turn, Bechtel and Richardson argue that the phenomenon can be "reconstituted" as a result of the mechanistic decomposition: "A mechanistic approach is not limited to explaining phenomena that are taken as simply 'given,' but can mandate a revision of the way the phenomena are to be conceptualized" (1993, 196). This basic contention is, of course, highly congenial to my approach in that it acknowledges that the ways in which a research object is conceptualized can change in the course of the investigation. In this section, I begin by reviewing one of Bechtel and Richardson's examples before turning to Bechtel's more recent analysis of memory phenomena. I argue that, while Bechtel and Richardson's account provides some valuable insights, the discovery process in psychology is better understood if we place the scientific investigation of phenomena within the larger context of the scientific aim to explore overarching objects of research.

One way of paraphrasing Bechtel and Richardson's account is to say that phenomena are identified in a top-down fashion (i.e., in terms of specific causal/environmental conditions) but that their conditions of identity are also constrained from the bottom up (i.e., in terms of what we know about the mechanisms). The implication is that, once we find out more about the mechanism, we might have to redescribe the phenomenon. One example of a phenomenon Bechtel and Richardson discuss is the fact that there is a particular likelihood of being born with, say, blue eyes given the genetic makeup of the parents. In a similar vein, we might say that the fact that, in general, performance on explicit tests is impaired in amnesiacs is a phenomenon. Phenomena are obvious targets of experimental investigation since the very idea of experimentation consists in manipulating conditions and observing the effect as a function of those conditions. I argue that to fully understand why such experimental manipulations are performed we frequently need to take a slightly wider perspective and ask how scientists see the phenomena in question related to their overarching objects of research. In what follows, I explain what I mean by looking at Bechtel and Richardson's notion of the *reconstitution of phenomena*.

One of Bechtel and Richardson's examples of the reconstitution of a phenomenon concerns the synthesis between classical genetics and biochemistry as giving rise to a new understanding of the phenomena at stake. Briefly, the story is that Mendel's experimental paradigm for studying inheritance identified specific phenotypic traits as resulting from specific experimental interventions. This suggested to many that there are specific mechanisms that bring about specific traits and that these can be localized in specific genes (Bechtel and Richardson refer to this as a "heuristic identity thesis" regarding the relationship between a mechanism and a given phenotype). However, there was soon mounting evidence that (with just a few exceptions) phenotypic traits are the products of many genes and that the ways in which they are expressed depends crucially on environmental factors. Bechtel and Richardson argue that research into the biochemical mechanisms of inheritance revealed there to be a one-to-one correspondence after all. But this correspondence was not between genes and phenotypic traits but between genes and enzymes. From this they conclude not that the phenomenon, as originally conceived, requires a more complicated explanation but that the explanandum phenomenon was initially conceived at the wrong level: "We simply had to reconceptualize the relevant traits at a lower level of analysis. . . . It was no longer eye colors, but the enzymes that produced them, which become the proper units for a Mendelian analysis" (Bechtel and Richardson 1993, 195).[12]

It seems eminently plausible that explanandum phenomena can be reconceptualized in this way. At the same time, there is something vaguely unsatisfactory about Bechtel and Richardson's analysis if we take it as an analysis of the search for mechanistic explanations of a particular phenomenon. One has the sneaking suspicion that, instead of the intended explanandum determining the search for strategies, the strategies of decomposition and localization determine what phenomena can

---

12. Notice that this account of reconstitution at a different level is different from the notion that a phenomenon can also be reconstituted at the same level, e.g., when the identification of a mechanism can lead to lumping or splitting in the descriptions of a phenomenon, as discussed in Craver (2007). (For the argument that the identification of a distinct explanatory mechanism need not give rise to kind splitting, see Colaço [2020].)

be explained. In reconstituting the phenomenon, scientists are no longer explaining what they initially wanted to explain (i.e., the percentage of blue eyes in a given population). In other words, the worry is that the initial explanandum phenomenon (phenotypic trait) is not reconceptualized, that it has simply been replaced by a different phenomenon (enzyme production). The suspicion arises that this amounts to a change of subject (see also Kronfeldner 2015).

Kronfeldner argues that this kind of "redefining of phenomena" is common in the investigation of complex phenomena, where scientists sometimes go up "a level of abstraction" or down "a level of composition" in accordance with specific explanatory aims (2015, 180). She suggests that this is quite compatible with a pluralist and pragmatic explanatory stance. While I agree with this thesis in general, it raises the important question of what the common denominator of such pluralist explanatory practices is. Surely, pluralism requires not only that we use different explanatory strategies or methods but also that those strategies or methods are roughly directed at the same question or object. But how can this be ensured if the very explanandum phenomenon has shifted? Once again, I argue that this seeming paradox goes away if we understand that often the aim of research is not to explain individual phenomena but rather to explore an overarching object of research. One candidate object of research in this particular case is *heritability*. It is only when we recognize the two phenomena as being related to the same overarching object of research that we understand that a shift in explanandum phenomenon is not necessarily a change of subject. It is merely a shift to a different phenomenon among the cluster of phenomena regarded as relevant to the overarching object of research.

This brings us back to the question of how Bechtel and Richardson conceptualize the epistemic function of identifying a phenomenon. As we saw, descriptions of phenomena are for them convenient ways of empirically identifying putative mechanisms that in turn explain the phenomena. While it is plausible that phenomena are caused or realized by mechanisms (or complex sets of mechanisms), this does not explain why researchers are interested in the phenomena to begin with. Nor, I claim, does this characterization exhaustively describe the role phenomena play in the research process. Phenomena can be explananda, but not every phenomenal description of what a given mechanism does is, thereby, of interest as an explanandum, let alone as an object of research. However, phenomena can figure as parts of a descriptive characterization of overarching objects of research. For example, phenomena that are deemed relevant to implicit memory can include unconscious/automatic learning and retrieval mechanisms as well as experimentally generated priming effects. The latter can, thus, be used to investigate the former. Crucially, such investigations involve both explanatory and descriptive questions. Thus, phenomena not only figure as explananda but are also utilized as providing epistemic access to a given object of research. This is possible because objects of research are conceptualized as clusters of phenomena that range from the context specific to the context transcendent.

Summing up, given my own guiding assumption that much scientific research in cognitive science is aimed at the exploration of objects of research, which are composed of multiple interconnected phenomena, I argue that the exploration of objects of research importantly involves both the investigation and the utilization of the phenomena thought to be relevant to the object in question. For example, it is quite conceivable that the experimentally individuated phenomena of classical genetics remain epistemically important as operationally defining heritability as an object of research. Various procedures of intervening at the biochemical level can be used and their effect on the phenotype observed. Notice that this is precisely how Craver has analyzed interlevel experiments (Craver 2007; see also Craver and Darden 2001). However, while Craver and Darden have viewed such experiments as supporting a mechanistic account of explanation, one in which the explananda are specific phenomena, I wish to highlight the role of the experiments in exploring a given object of research (e.g., heritability, implicit memory, spatial memory), something made possible, in part, because of the ways in which multiple phenomena are conceptually tied up with the object under investigation.

5.4.2. William Bechtel: Phenomenal vs. Mechanistic Decomposition of Memory

In the previous subsection, I provided a brief discussion of a particular account of scientific discovery, that is, one that views experimental research on complex systems as mainly focused on discovering explanatory mechanisms. One of the authors discussed above, William Bechtel, has extended his analysis to psychology, focusing specifically on the example of memory research. I now argue that his account makes an important point, though my framing of this point is different than his.

In *Mental Mechanisms*, as in his previous work, Bechtel starts out by suggesting that mechanisms can be identified "in terms of phenomena" (2008b, 14), where by "phenomena" he appears to mean regular behavior of organisms under specific conditions. Of course, given what I said earlier, it is clear that I agree with the contention that the regular behavior of organisms under specific conditions (e.g., conditions of experimental manipulation) should count as a phenomenon, though (as I have argued above) I do not think they necessarily function as explanandum phenomena in an investigative context. Bechtel addresses the question of whether the functional identification of behavioral phenomena (and the corresponding construction of explanatory mechanisms) can guide scientific constructions of psychological taxonomies specifically: taxonomies of such types of long-term memory as episodic memory, declarative memory, procedural memory, and implicit memory. In other words, the question is whether, for example, the priming effect can be used to identify the mechanism of implicit memory, understood as some underlying brain system. On the basis of the analysis provided in Bechtel and Richardson (1993), one might expect that the natural strategy would now be one of decomposition and localization, that is, the search for a realizer of the types of memory identified as being at the phenomenal level. Such a strategy, which Bechtel refers to a "phenomenal decomposition," would then simultaneously explain the phenomena in question (by identifying the realizing mechanism) and guide the discovery process (by providing evidence that the taxonomic distinctions provided on the basis of behavioral phenomena in fact got it right, i.e., correspond to a robust underlying mechanism). In other words, one would search for distinct systems/mechanisms that are responsible for the various types of experimentally generated phenomena.

However, Bechtel's argument then takes an unexpected turn. Contrary to the decomposition and localization strategy, Bechtel (2008b) no longer thinks that phenomenal decomposition is likely to carve nature at the relevant mechanistic joints of memory types. Thus, he argues that, whereas phenomenal decomposition does a decent job of delineating empirical phenomena, it may not be the best way to identify explanatory mechanisms for those phenomena. Another way of putting this is to say that, while the phenomenon of priming can (presumably) be explained mechanistically, the explanatory mechanism might not belong to a specific type of memory (implicit memory). What Bechtel therefore calls for is a decomposition of the system into the kinds of mechanisms that actually explain the empirical regularities yielded by phenomenal decomposition even if those explanatory mechanisms do not constitute memory types. He refers to this other kind of decomposition as "mechanistic decomposition" (Bechtel 2008b, chap. 2).

This turn of events is significant because it may well turn out that the neural mechanisms in question cut across the different kinds of memory identified as the phenomenal level. And, indeed, this is precisely what Bechtel suggests with regard to implicit memory. Rather than behavioral priming phenomena mapping on a priming mechanism, which is, in turn, part of the mechanism or system of implicit memory, an alternative explanation of the behavioral phenomenon of priming suggests that it is due to a process (or "operation" [Bechtel 2008b]) that is also operative in different kinds of memory phenomena. A competing explanatory hypothesis is provided by the so-called transfer appropriate processing (TAP) (Roediger, Weldon, and Challis 1989), according to which priming occurs when the processes involved in the encoding of a memory are similar to those involved in its retrieval. This means that, to explain priming, we do not need to posit the existence of a type of memory system (implicit memory). In a similar vein, effects that indicate working memory capacity do not need to be attributed to a type of memory system (working memory) but can be explained by means of mechanisms of attention (see the discussion in the previous chapter), which might, in turn, also feature in numerous other objects of research. Methodologically, the point is that these putative explanations are derived not simply by phenomenal decomposition and localization but by turning to other accounts of cognitive mechanisms (in this case, encoding, retrieval, and attention) as potentially explanatory of the empirical regularities generated by working memory paradigms.[13]

If Bechtel is right about this, there are two possible conclusions we can draw. The first is that there is no such thing as implicit memory (or, if there is, priming is not explained by or related to it). The second is that implicit memory can be a worthy object of research even if it cannot be identified with a specific brain system but is, rather, a cluster of phenomena many of which may also feature in other clusters of phenomena (which can become objects of research in their own right).[14] I entertain here the latter possibility. Consequently, I maintain that, even if it turns out that behavioral phenomena like priming and chunking cannot be localized in designated brain systems that are responsible for implicit memory and working memory, it does not follow that implicit memory and working memory cannot be objects of research. It only follows that such objects are more complex than a simple localization account might suggest. To put it in my terms, they are composed of

---

13. Catherine Stinson gets at a similar point when she remarks: "Early cognitive models of memory, rather than aiding discovery by providing a mechanism sketch, were rather misleading guides to the discovery of neural mechanisms" (2016, 587).

14. I take this view to be compatible with a recently popular approach in philosophy of cognitive neuroscience that has argued against the notion of specialized neural mechanisms for specific cognitive functions, highlighting instead the idea of neural use and reuse (e.g., Anderson 2014, 2016).

multiple phenomena—such as (in the TAP approach) phenomena of encoding and retrieval—that are orchestrated in such a way as to enable cognitive, behavioral, and experiential capacities. Such capacities and the clusters of phenomena we observe when the capacities are instantiated are, I argue, the objects of psychological research. Looking at psychological research in this way, we can—once again—appreciate the basic point of this chapter, that is, that behavioral effects such as priming and chunking are not best understood as the explananda of psychological research. They are, rather, empirical phenomena that are considered conceptually connected to specific objects of research and can, thus, figure as evidence in experimental investigations designed to delineate and explore them.

Summing up, I argue that, while it is not incorrect that memory researchers try to explain empirical phenomena and search for mechanisms, we have to analyze these research activities within the broader framework of attempts to identify and explore objects of research, which are typically conceptualized as clusters of interconnected phenomena. I also want to highlight that the dichotomy between phenomena and mechanisms misleadingly suggests that they are two distinct ontological categories. But, clearly, if we follow Bogen and Woodward's (1988) definition, mechanisms (qua regularly occurring events) *are* phenomena. The issue I am focused on here concerns the conceptual assumptions that need to be in place when researchers deem specific classes of phenomena to be relevant to the same object of research. Such conceptual assumptions are a crucial component of the research process, but they can also be revised as a result of that research.

5.5. Toward an Analysis of Norms of Exploration in Psychology

The analysis presented in this chapter, with its distinction between phenomena and objects of research, aims to provide a descriptively accurate picture of the process of exploration in psychology. However, I am also pursuing a normative aim here, that is, arguing that the investigative strategies I am analyzing are, in fact, good strategies. This normative pronouncement is important as it can potentially block counterexamples to my analysis. In other words, even if it were the case that some psychologists study only isolated experimental effects with no explicit mention of overarching research objects, such as working memory, implicit memory, or spatial memory, I would like to argue for a vision of psychology that takes clusters of phenomena, *such as they are described by macro-level psychological concepts*, rather than isolated regularities as the objects of research and that treats specific—experimentally generated—phenomena as evidence for claims about these objects.

In light of what was argued in the previous section (where we encountered the thesis that there is no dedicated mechanism or system for implicit memory), this normative vision might appear to be undermined. If we take it to be a norm of inquiry that our concepts should respect the mechanistic structure of the world (Kästner and Haueis 2021), then that might initially imply the normative requirement that there can be a science of implicit memory only if there is a mechanism for implicit memory. But, if Poldrack and Yarkoni are right that "it is exceedingly unlikely that there is any single brain region, cluster, or network that corresponds neatly to high-level psychological concepts such as episodic recall, working memory, or phonological rehearsal" (2016, 599), what would be the point of using such macro-level concepts to individuate objects of research and of formulating a normative recommendation about how to study such objects?

In response, let me emphasize the following four points. First, I am obviously in agreement with the assertion that psychological concepts ultimately cannot violate any underlying (brain) mechanisms. But that is a much weaker requirement than searching for a dedicated mechanism or system for every psychological kind. Second, one of the main objectives of this chapter (and the previous one) has been to argue for a notion of *object of research* as epistemically blurry. If we take this seriously, it follows that the very question of what the underlying mechanisms are and how they are relevant to a given object of research cannot always be readily answered by researchers. Exploring this question is part of the package I am outlining here. This means that we cannot establish that there is a mechanism for a given object prior to investigating the object. Let me also emphasize here that I am fully on board with the vision of cognitive neuroscience as a unified field of research (e.g., Boone and Piccinini 2016). Thus, my aim is not to argue for some kind of irreducible autonomy of psychology. Rather, I wish to highlight the epistemic value of conceptualizing objects like working memory and implicit memory on the macro level as inspired by folk-psychological language, and I believe that the academic discipline of psychology has valuable tools to offer when it comes to studying the phenomena associated with those objects. A third, related reply is that macro-level psychological concepts are "compact and psychologically interpretable" (Poldrack and Yarkoni 2016, 600) in a way that neurobiological concepts are not, thereby allowing us to parse the subject matter at a level that connects with our everyday (folk-psychological) practices. I recognize that this reply might seem unconvincing since one might argue that psychological interpretability alone cannot be a good reason to adopt a concept. After all, our everyday practices might simply be mistaken. This gets me to my fourth (and last) point, namely, that existing scientific projects of finding mechanisms for cognitive functions presuppose (implicitly or explicitly) some folk-psychological vocabulary. This has been remarked on in recent years by various authors (e.g., Burnston 2021; Francken and Slors 2014, 2018; and Poldrack and Yarkoni 2016). In a move similar to the one I have been arguing for in this book up to this point, Burnston (2022) argues that the usage of folk-psychological concepts in cognitive neuroscience should be construed as providing not explanations but rather heuristics. Such heuristics can guide the investigative process, but there is no reason to suppose that they cannot be adjusted or discarded.[15]

In arguing for these points, I adopt (and adapt) Norwood Russell Hanson's thesis that scientific inquiry can be described as a search for patterns where scientific discovery consists in conceptual insights as to what specific patterns mean, that is, what precisely is being observed when patterns are detected (see Hanson 1958, chap. 4). Kästner and Haueis make a similar point by drawing on John Haugeland's (1998) notion of *patterns* and by distinguishing between patterns as things that "persist from below" and things that are "salient from above" (Kästner and Haueis 2021, 1644). The crucial point here is not only that the "above" from which patterns are salient is relative to our perceptual abilities, but also that we tend to classify patterns by kind, that is, as belonging to specific kinds of objects (see Hanson 1958, chap. 1).[16] Salient patterns are the same as what I have previously referred to as "surface phenomena" (Feest 2011b), of which experimentally created phenomena are a subset. The normative claim I wish to make here is that, if we do not conceptualize such patterns/phenomena as belonging to a specific object of research, it is not clear what the research that draws on these patterns is about. Also, once we have accepted that our ability to discern salient patterns from above is relative to our perceptual and conceptual prerequisites, there is no obvious reason to rule out specific perceptual and conceptual prerequisites, that is, for example, those that

---

15. Needless to say, this raises the question of what standards ought to force us to adjust (or discard) a given macro-level concept. In other words, what is the nature of the ontic constraints on good psychological concepts? I return to this question in chap. 7 below.

16. The way Hanson puts this is to say that, when we observe something, we observe it *as* something. To illustrate this with an example used in this chapter, investigators involved in spatial memory research observe navigational behavior *as* a spatial memory phenomenon.

individuate objects of research at the macro level of complex capacities and by means of our folk-psychological resources.

Having thus provided a justification for using macro-level psychological concepts to single out objects of psychological research, I want to add that the strategy of creating experimental evidence in accordance with operational definitions is rational as well. Macro-level psychological concepts envision the objects of psychological research as being composed of clusters of phenomena, thereby bringing with them assumptions about the types of behavioral (and experimentally generated) phenomena that will be relevant to the objects. Thus, in the spirit of using existing concepts as heuristics, we can make sense of the epistemic practice of deriving from them operational definitions that function as tools in the exploration of the purported objects. It should be clear, however, that these normative recommendations need to be supplemented with an analysis of the criteria that determine whether a given experimentally created phenomenon is, in fact, relevant. I will provide such an analysis in the following two chapters, in which I probe further into the ontological status of psychological objects (chap. 6) and the issue of how high-quality evidence for claims about such objects can be ensured (chap. 7).

5.6. Conclusion

In this chapter, I have further elaborated my notion of an *object of research*, contrasting it in particular with the notion of a *phenomenon*. My account is inspired by Bogen and Woodward's but also departs from and extends theirs in a crucial way. While Bogen and Woodward conceptualize phenomena by contrasting them with data, I have argued that local and idiosyncratic patterns or effects can also be phenomena. Drawing on Leonelli's (2015) relational account of data, I have argued that such locally and idiosyncratically instantiated phenomena are often best understood as evidence that is produced for specific epistemic purposes.

With regard to the question of what experimentally created phenomena are evidence *for*, I argued that they are evidence for claims about objects of research, that is, the ill-understood (epistemically blurry) targets at which scientists direct their attention. Picking up on the argument of the previous chapter, I argued that anything can be an object of research provided that researchers direct their epistemic attention at it. This includes midsize things, mechanisms, or experimental effects. However, I have claimed that basic psychological research is organized around a class of concepts that delineate objects like memory, attention, or perception that typically are (but do not necessarily have to be) taken or derived from folk psychology and that are composed of clusters of phenomena, including behavioral regularities. When researchers create experimental effects, they essentially produce local and idiosyncratic instances of the behavioral regularities that are conceptually tied to the object in question. This is what it is to provide an operational definition of the corresponding concept.

I developed my argument (in secs. 5.3 and 5.4) by contrasting my notion of experimentally created phenomena as evidence with the way phenomena are commonly treated in philosophy of neuroscience (specifically within the mechanistic approach to explanation), that is, (*a*) as explananda or (*b*) as providing phenomenal descriptions that guide in the discovery of mechanisms. My argument suggested that an exclusive focus on individual phenomena and explanatory mechanisms not only fails to get into clear view the evidential role that experimentally created phenomena play in the research process in psychology but also misconstrues the research efforts of psychology as directed at specific (as it were isolated) phenomena when they are, in fact, often directed at objects of research that are composed of clusters of phenomena.

While the first four sections developed a novel descriptive account of the investigative process in psychology, section 5.5 raised a more explicitly normative question in that I asked whether my analysis can be at the foundation of an analysis of norms of inquiry for psychology. I argued that, if we take seriously the notion that the objects of psychological research are clusters of phenomena that are brought about by complex cognitive-behavioral capacities, then the practice of operationally defining the relevant concepts in an effort to gather evidence for descriptive claims about the corresponding objects of research is a useful normative guideline, though it is presumably not sufficient. In order to further develop this thread, chapter 6 addresses the following two desiderata. The first is to strengthen my thesis that the objects of psychological research are complex cognitive-behavioral capacities, thereby underwriting my claim about folk-psychological concepts providing adequate tools of inquiry. The second is to explore the extent to which objects of psychological research pose ontic constraints on the norms of psychological inquiry. Chapter 7 picks up the notion of *operational analysis* (first introduced in chap. 2) in order to further articulate the normative picture for which I wish to argue.

CHAPTER SIX

What Kinds of Things Are Psychological Kinds?

6.1. Introduction

One upshot of the discussion in chapter 4 was that objects of research are epistemically blurry not merely because their unobservable properties are ill understood but because the conditions of their empirical individuation and description are difficult to nail down. Indeed, a major contention of the previous two chapters has been that this is why a lot of research in psychology is concerned with seemingly simple questions of taxonomy and description. By highlighting the blurriness of research objects, I have focused on the epistemic predicament of researchers, and I have provided an account of methodological responses to this predicament. However, an obvious question now is, what is the ontological status of the objects of psychological research? A different way of phrasing the question is that, while the previous chapters have looked at the formation of psychological kind concepts, the ontological question raised here seems to ask whether there are psychological kinds that correspond to such concepts.

One quick and easy answer is that, from the perspective of practicing scientists, this question makes little sense. Objects of research, as I have conceptualized them, are, by definition, ill understood, and, hence, their ontological status is not well-known. If we stick with our methodological commitment to honor the epistemic perspectives of practicing scientists, it might, therefore, be argued, we cannot inquire about the true status of their objects. This would be a misunderstanding of the question I am asking here. My question does not concern the true status of, say, implicit memory or working memory. Rather, the question is what the nature of psychological kinds in general. On what conception of *psychological kind* can we make sense of the possibility that scientists can be right or wrong about specific kinds, such as implicit memory or working memory? As I argued in chapter 2, scientists are generally committed to the idea that there is something that corresponds to their concepts, and, as I showed in chapters 4 and 5, scientists expend considerable effort trying to get things right. I maintain that an adequate philosophical analysis of the investigative practices of psychologists needs to include a story of what this means.[1] This chapter develops such a story. As such, it has the descriptive aim of teasing out the implicit ontological commitments that inform scientific practice and the normative aim of providing a justification for those commitments.

One way of phrasing the question this chapter asks is in terms of whether there are real kinds out there that correspond to our macro-level psychological concepts (see sec. 5.5 above) and, if there are, what the ontic underpinnings are that determine or constrain them. My answer, briefly, is (*a*) that there are real psychological kinds but (*b*) that the ontic underpinnings are not (just) located at the neural level. Instead, I argue for a realism about psychological kinds at (what I call) the "whole-organism level." According to the analysis I propose, the units that are singled out by psychological kind terms are specific patterns of phenomena that are salient at the level of the (sensing, behaving, and cognizing) individual organism as distinct from but also functionally embedded in the (social and natural) environment. Regarding the ontological status of psychological kinds, the thesis of this chapter is that psychological kinds are relational. They come into being and are sustained at the interface between specific features/processes distinctive to human subjects and specific features of the world. The former has to do with our perceptual, conceptual, and instrumental apparatuses but also with the aims that explain why specific patterns are picked up and grouped into kinds. The latter has to do with specific material, social, and cultural contexts in which the bearers of psychological kinds (i.e., whole organisms) are embedded. The basic intuition is, thus, similar to Massimi's (2022) contention that perspectivalism and realism about kinds need not be mutually exclusive.[2]

At first glance, the ontic and the perspectival aspects of my account appear to be in tension if we assume that the reality of something hinges on the idea that it can exist independently of our minds and/or human practices (we can call this the "realism objection"). To address and resolve this tension, section 6.2 shows that the realism objection can be broken down into several distinct worries. They are (*a*) the worry that, on a relational account, kinds cannot be real (we can call this the "mind-dependence objection"), (*b*) the worry that, on a

---

1. I am very grateful to Katie Kendig for pushing me to address this point (in conversations).

2. While Massimi aims at a general account of natural kinds, my aim in this chapter is more modest, i.e., to develop an account of psychological kinds.

relational account, we end up with a pluralism that speaks against the reality of kinds (we can call this the "conventionalism objection"), and (*c*) the worry that only microstructural entities can be really real (we can call this the "reductionism objection"). I address all these concerns by arguing that conventionalism about macro-level kinds is compatible with realism about them. In section 6.3, I argue for the stronger claim that a macro-level (or whole-organism) perspective on psychological kinds is, indeed, the way in which we ought to individuate psychological kinds. I develop my argument by (*a*) showing that there are good reasons to suppose that neural structures cannot provide us with objective criteria as to the one correct way of individuating psychological kinds, (*b*) explaining why the possibility of a certain degree of ontological pluralism is not philosophically threatening, and (*c*) situating that ontological pluralism within the landscape of philosophical discussions about natural kinds.

While sections 6.2 and 6.3 are mostly concerned with laying out my account and defusing challenges to it, section 6.4 attempts to provide additional arguments in favor of my analysis. Specifically, I consider some underpinnings of my claim that psychological kinds are created and sustained by similarity relations and by causal and conceptual practices at the levels of both folk psychology and scientific psychology. I also point to a natural fit between my analysis and early to mid-twentieth-century ecological approaches, which partially influenced the emergence of operationism in psychology. Section 6.5 attempts to elucidate my whole-organism take on psychological kinds further by juxtaposing it with recent discussions of cognitive ontology, which are concerned with the question of whether and how cognitive functions can be localized in brain regions. I argue (*a*) that, while my analysis of psychological kinds is obviously compatible with attempts to localize the component functions of such kinds, it is situated at a different level and (*b*) that my argument for the reality of whole-organism kinds is somewhat independent of issues of localization. I note that, while functional analyses of psychological kinds can certainly contribute to revisions of those kinds at the whole-organism level, attempts to localize mental functions in turn rely on psychological concepts pitched at the whole-organism level. Section 6.6, finally, offers several answers to the question of how to capture the notion that, in developing their concepts of psychological kinds, psychologists try to get it right.[3]

---

3. I treat concepts of psychologcal kind distinct from concepts of human kind, which describe kinds of people (see Godman [2021] for an excellent overview of the latter topic).

6.2. (Natural) Kinds: Setting the Stage

In this section I begin by providing a sketch of some influential twentieth-century positions about natural kinds in order to articulate and situate my account of psychological kinds. While I align my views with specific philosophical writers and positions about natural kinds in general, let me emphasize at the outset that I do not aim at such a general account myself but am, rather, motivated by the more specific goal of articulating a notion of *psychological kinds*. According to this notion, psychological kinds are *relational, macro level,* and *real*.

6.2.1. Similarity vs. Deep Structure as Determining Kinds

An early influential treatment of natural kinds was provided by W. V. Quine (1969), who approached the topic by working with the assumption that things are sorted into kinds by relations of resemblance. Quine notes that, while the concept of *similarity* is notoriously hard to define, this ultimately need not worry us because, once we have achieved a good scientific understanding of "the cosmic machinery," we will be able to dispose of kind terms: "I shall suggest that it is a mark of maturity of a branch of sciences that the notion of similarity or kind finally dissolves, so far as it is relevant to that branch of science" (Quine 1969, 121).

Quine describes our conceptual progression as one from pretheoretic to scientific kinds in order to argue that mature sciences will, ultimately, be able to dispose of kind concepts altogether. Examples of pretheoretical kinds are color concepts that are rooted in our physiological makeup, which is adapted to the environment. In the course of scientific research and the emergence of a more theoretical understanding of the world, Quine suggests, there is "a development away from the immediate, subjective, animal sense of similarity to the remoter objectivity of a similarity determined by scientific hypotheses and posits and constructs" (Quine 1969, 133). The cosmic machinery has a structure, but, once we have discovered it, Quine suggests, we no longer need the notion of kind since, in that case, we are talking about not how we judge things to be but how they are. For example, once we have understood the chemistry of solubility in theoretical terms, we no longer need the kind category of soluble as it is based on a pretheoretical judgment of similarity among the behaviors of various substances.

Quine's account nicely exemplifies a particular way of thinking about kinds as individuated relative to a judging person or community. Two or more entities belong to the same kind if they seem similar to creatures like us with a particular physiological makeup and sociomaterial context. This also means that we individuate kinds at macro levels specific to our contingent conceptual and perceptual abilities. This is an aspect of Quine's analysis to which I am sympathetic. However, for Quine, this contingency implies (*a*) that kinds are, at the end of the day, not real and (*b*), therefore, that they will have no place in science in the long run. The two points need to be separated since one could hold on to kind concepts for pragmatic reasons even if kinds do not really exist. Thus, one could accept implication *a* but reject implication *b*, arguing that, even though there are no kinds out there, kind concepts play important instrumental roles in science and in everyday explanatory practices. However, such an instrumentalism is not the line of attack I wish to pursue. Rather, I think that, if we want to capture the notion of scientists being able to be right or wrong about their objects, we should do so on the grounds that kinds are real, and this means that we need an analysis of how they can acquire such a robust status despite being based in mere judgments of similarity. For Quine, kind judgments were transitory epistemic vehicles in investigating the world rather than singling out parts of the furniture of the world. His approach, moreover, privileged the smallest possible (physical) level of grain as adequate for an understanding of reality. By contrast, I argue in this chapter that, while psychological kind judgments are, indeed, epistemic vehicles, they can also contribute to the practices that confer reality on the corresponding kinds, individuated at the macro level.

The notion that our superficial criteria of sorting things into kinds can be distinguished from the real structural features of the world is also prominent in the other important philosophical tradition that has had an impact on debates about natural kinds, namely, the sort of essentialist thinking that has come to be associated with Hilary Putnam and Saul Kripke. According to this tradition, there are deep structures that settle kindhood by providing the essence of what a given kind really is. The best-known example of this is water as a natural kind that is specified by a specific molecular structure (though, for complications of this simple picture, see Chang [2012b] and Havstadt [2018]). As I laid out in chapter 3 above, I do not subscribe to a theory of concepts that takes any part of a concept's meaning to be singling out an essential or intrinsic property of its referent. Moreover, especially when it comes to properties that have resulted from evolutionary processes, this strong form of essentialism seems implausible. Finally, as indicated above, this account, too, assumes that the true structure of the world will, ultimately, be revealed by turning to microstructures that are objectively given in the world. Nonetheless, the realistic intuition behind this kind of microstructuralist essentialism is worth preserving if we want some standards by which to judge whether any given kind judgment is getting it right or wrong. This is relevant to the case of psychological kinds because a version of essentialism (albeit in a naturalized form) has long been prominent in the idea that psychological kindhood is settled by neuroscientific facts. (I return to this in sec. 6.3.)

Roughly, the two classic positions just outlined (kinds as similarity judgments and kinds as microstructural essences) seem to leave us with two possibilities. Either kinds are relational but not real, or kinds are real but only by virtue of their microstructural essences. By contrast, on the position I am developing here for the case of psychological kinds, (*a*) such kinds are real by virtue of being relational, and (*b*) the relevant relations can be forged and maintained at the macro level. With these two theses I wish to push back against the intuitive idea that, in order for something to be real, it has to have a mind- or practice-independent existence as well as against the implicit (or explicit) reductionism in the background of many discussions about natural kinds (and about psychological kinds, for that matter). With regard to the realism point, the assumption that reality requires a mind- or practice-independent existence is clearly false since artifacts made by humans are obviously real even though their existence depends on humans. Likewise, to return to an example already mentioned, the existence of colors depends (among other things) on the observer, but this does not mean that colors are not real. Indeed, the fact that they are central to, for example, the way in which we regulate traffic suggests that they are capable of making a causal difference in the world by virtue of being integrated in our practices.

Notice that I have not yet presented a positive account of psychological kinds. Thus, my procedure has been mainly negative; that is, I have argued that specific arguments countering a notion of kind as relational, macro level, and real need not convince us. One other objection that needs to be addressed, however, is the following. If kinds depend on similarity judgments, this seems to open the door to a pluralism with respect to kinds since there are obviously many ways in which things can be similar. Applied to psychology, for example, pluralism might mean that there could be a plurality of different taxonomies in that past taxonomies do not map onto present ones or even that there are cultural differences in how we carve up the psychological domain at any given point in time (e.g., Danziger 1997) but that, nonetheless, the kinds within these various contexts should be considered real. In considering this possibility, we should distinguish between (*a*) the notion that it is possible to sort the world into kinds in more than one way (*pluralism$_1$*) and (*b*) the notion that there are in fact multiple (perhaps even cross-cutting) real kinds in the world (*pluralism₂*). The mere possibility of alternative ways of kinding (pluralism₁) does not result in a rampant plurality of kinds (pluralism₂). (For the notion of *kinding*, see Kendig 2016a.) This means that we need an account of how possible psychological kinds can become real and that we need an answer to the question of whether a plurality of real kinds should be regarded as problematic. I conclude this section by addressing the first question. (I address the second question in sec. 6.2.2.) According to the account I am gesturing at here, kinds become real if they are components of robust conceptual and causal practices (Mallon 2018), where such practices can be operative at the folk-psychological level or be more specific to investigative contexts. By "conceptual practices," I mean practices or reasoning; by "causal practices," I mean practices of causally intervening in the world, both in the laboratory and in the real world.

The notion of *conceptual practices* that I am using here is similar to Sullivan's (2016a), but, in addition to practices of classification, I include practices of drawing inferences (e.g., Brigandt 2010), which can include inferences about the likely success of specific causal interventions. In this way, I view causal practices as closely intertwined with conceptual practices. For example, if I classify a set of symptoms as a migraine, this will lead me to infer that taking Sumatriptan is an effective causal intervention. Likewise, if I classify someone as having an attention deficit disorder, I might infer that Ritalin is an effective causal intervention. The claim I wish to make here with regard to psychological kinds (e.g., attention) is that the very conceptual and causal practices that are enabled by the relevant concepts (insofar as they are successful) contribute to the consolidation of the kinds in question.[4]

6.2.2. Conventionalism vs. Realism about (Psychological) Kinds: A False Dichotomy

Having briefly outlined the contours of a position that allows macro-level kinds to be real, let me now start to bring it into sharper relief by situating it vis-à-vis some of the more recent existing literature. In doing so, I want specifically to address the concern that my account is, ultimately, a form of conventionalism that will commit me either to an anti-realism about psychological kinds or to a rampant pluralism about (real)

---

4. This basic idea is, of course, similar to Ian Hacking's (1995) analysis of human kinds, though my account of the malleability of kinds does not require a looping effect since it is not restricted to kinds of people (like people with autism) but also includes kinds of psychological properties (such as memory or attention).

psychological kinds. This section argues that conventionalism does not imply antirealism and that, while it does imply a form of ontological pluralism, we need not worry about it (at least not in the case of psychological kinds). So where is the conventionalism worry coming from?

Bird and Tobin (2018) suggest that there are two broad ways of moving forward after rejecting a strong metaphysical realism about kinds. One is to formulate a weak realism that tries to salvage the idea of real structures and entities in the world as underwriting natural kinds without buying into notions of the metaphysical necessity or intrinsic nature of their properties. One popular version of this approach is sometimes referred to as a "naturalized essentialism" (e.g., Griffiths 1997) and can be associated with Richard Boyd (e.g., Boyd 1991). The other response is to endorse a more or less full-blown conventionalism, suggesting that we classify the world into kinds of things according to conventional criteria in order to serve specific (epistemic or practical) needs and that the world offers relatively little structure of its own to constrain the classifications in question. Such a position is sometimes attributed to John Dupre (e.g., Dupre 1999, 2002). Despite the fact that Boyd's and Dupre's positions are seemingly very different, it is worth considering them together since (as Craver 2009 remarks), it is not clear whether Boyd's position ultimately succeeds in steering a middle ground between (strong) realism and conventionalism. Hence, if we find that Boyd's position collapses into a conventionalist one, we need a better understanding of what that entails.

Richard Boyd's homeostatic property cluster (HPC) theory (Boyd 1991, 1999) is potentially useful to my analysis for two reasons. First, there are obvious parallels between it and my claim (put forth in the previous chapter) that psychological concepts refer to clusters of phenomena. Second, the HPC account tries to reconcile the fact that kind determinations are made from specific perspectives with a realist intuition. The HPC account tries to achieve such a reconciliation by arguing that natural kinds have both an epistemic component (i.e., they are individuated relative to specific epistemic interests) and some kind of ontic underpinning. Roughly, according to Boyd, natural kinds are clusters of properties (criterion a) that are explained by a common structure (criterion b), provided that the corresponding concepts are projectible (i.e., figure in our best theories) (criterion c) and accommodate the real causal structure of the world (criterion d) (see Boyd 1999).[5] Let us briefly consider how these criteria are intended to hang together. There is a sense in which it is trivially true that, if we understand *cluster* as statistical correlation, there

---

5. I am following Craver's (2009) very lucid explication of HPC theory here.

is likely to be a causal fact that explains it. However, many correlations are spurious; that is, even if we can explain them, it does not follow that the correlated properties form kinds. This is how criterion c comes in, narrowing kinds down to those clusters of properties that do some work in our epistemic practices. But a kind concept might give us the illusion of epistemic work while, in fact, not really latching onto the world. This is how criterion d comes in, demanding that our kind terms in fact accommodate the world. The bottom line is that kinds are simultaneously constrained from the epistemic side and from the ontic side, as laid down by criteria c and d. (For a similar move, see also Khalidi [2013].)

I am well aware that the HPC account has been (perhaps rightly) criticized as a general theory of natural kinds (e.g., Ereshefsky and Reydon 2015). But here my concern is merely whether it can give us insights into the nature of psychological kinds. I argue (*a*) that, with a few tweaks, it can but (*b*) that those tweaks do indeed put the HPC account on a slippery slope toward a more thoroughgoing conventionalism than Boyd perhaps initially intended, thus raising the issue of rampant pluralism (already remarked on at the end of the previous section). The two adjustments I want to make are (*a*) that the interests that shape the way we individuate psychological kinds need not be epistemic and (*b*) that the causal and conceptual practices that individuate and sustain kinds need not be scientific.

As far as I can see, Boyd and Khalidi privilege epistemic aims because nonepistemic aims cannot be trusted to identify the more fundamental explanatory structures of the world. Presumably, this is also the reasoning behind appealing to criterion c, which not only demands that kind terms be projectible but also ties this idea of projectability to our best scientific theories. By contrast, I am inclined to point out that concepts that are tied to nonepistemic and folk-scientific aims are frequently projectible too. For example, if we have a macro-level concept of *spatial memory* and reliable criteria that allow us to apply it to specific people, this enables us to make predictions about those people's behavior, to recommend them for specific jobs, etc. I thus argue that successfully meeting epistemic aims can contribute to the reality of kinds, but this is not because there is anything special about epistemic (let alone scientific) aims per se. Rather, it is because epistemic practices exemplify the more general point emphasized by my account, namely, that kinds can come into being at the interface of our similarity judgments (which are enabled by perceptual and conceptual prerequisites and typically tied to specific interests and aims) and the successful worldly practices in which these similarity judgments allow us to engage. The point is that, while macro-level kinds clearly cannot violate microstructural facts, what makes them real qua macro-level kinds is not those microstructural facts but macro-level conceptual and causal practices, epistemic and nonepistemic alike. Let me emphasize that I do not wish to deny that the ways in which we relate to the world scientifically enjoy a special status. My account merely suggests that the simple picture by which this status derives from science's ability to uncover the true structure of the world is mistaken.[6] I recognize, however, that there is a potential danger attached to my move of widening the range of interests that can contribute to the creation of kinds: The move seems to rob science of the ability of correcting people when they operate with problematic kinds (an obvious example that comes to mind is that of race). I return to this question in the final section of this chapter, where I disambiguate this question and rebut the worry.

Boyd himself came to endorse something that sounds a lot like my relational account of kinds. For example, he writes that "natural kinds are always, in an important sense, social constructions and practice relative" (Boyd 2000, 54), and that kinds are "historically situated" as well as "relationally and historically defined" (69). Since he took this to be compatible with "a broadly realist and naturalist conception of natural kinds, causation, truth and knowledge" (54), the question is what "broadly realist" means here. Boyd cashed this out in terms of "the (causal structure of the) world [playing] a heavy legislative role" in the construction of kinds (66). While "legislative" is a strong term, I argue that, in the case of psychological kinds (but probably also for a variety of other kinds), the causal structure of the world at most constrains what kinds are possible but cannot by itself determine which kinds are, in fact, real. The reason for this can be found in Boyd's own formulation as quoted above. Kinds are historically situated and, thus, in an important way, contingent. This means that the criterion of accommodating the world is necessary for kindhood but does not explain why specific kinds are real at specific points in time in specific epistemic and practical contexts. Massimi addresses a similar issue in formulating the need to have an account of natural kinds that allows them "to be flexible and malleable enough to accommodate contingentism without necessarily opening the door to conventionalism" (2022, 277).[7] When Massimi uses the term "conventionalism," she presumably has in mind a position according to which kind determinations

---

6. Reydon (2016) critically refers to this as a "zooming-in" notion of the discovery of kinds.

7. Let me emphasize again that, whereas Massimi aims at a general account of natural kinds, I am merely aiming for an account of psychological kinds. That said, I think there are overlaps between our approaches, particularly with regard to her definition of natural kinds as "historically identified and open-ended groupings of modally robust phenomena" (Massimi 2022, 16).

are arbitrary and, thus, the kinds in their extension are not real. By contrast, on my analysis (which I further flesh out in the following section), psychological kind determinations are (in a certain sense) conventional insofar as the world does not dictate one right way of kinding. However, the historical contingency of kinds does not imply arbitrariness or antirealism. I thus argue that my relational account of kinds can have it both ways. Psychological kinds can be conventional and real at the same time.

Given such a radically relational notion of kinds, we might wonder whether the term "kind" is even appropriate since it seems to conjure a notion of *naturalness*. Ingo Brigandt (2022), for example, argues that we should drop the term "natural" but continue to use the term "kinds" so as not to cede the field to essentialists. Weisskopf (2020) takes a slightly different approach in distinguishing genuine natural kinds from what he calls "anthropic kinds," where the latter have some of the relational features I describe above. For my purposes, it does not matter whether psychological kinds are natural. What I have been trying to argue here is merely that their relationality does not stand in the way of their potentially being real. The more important issue at this point is what exactly I have in mind when I refer to psychological concepts as "macro-level concepts," what it means to pick psychological kinds at that level, and what kind of pluralism we should be expecting as a result of my account. I address these questions in the following sections.

6.3. Pluralism, Mechanisms, and the Whole Organism

In the previous section, I argued that, if psychological kinds are constituted by similarity judgments, on the one hand, and conceptual and causal practices, on the other, then this can potentially give rise to a pluralism of psychological kinds. In section 6.3.1, I push back against the idea that such pluralism can be resolved by identifying mechanism kinds, while, in section 6.3.2, I start fleshing out my notion of *psychological kinds* as whole-organism kinds, contrasting it with other existing accounts that also try to resist reductionism about macro-level kinds.

6.3.1. Epistemic Aims and Mechanistic Structure

As we saw above, one seductive way of thinking about natural kinds has it that we individuate things at the macro level on the basis of similarity judgments but that it is ultimately at the level of underlying structures that the world as it really is reveals itself. Within philosophy of neuroscience and philosophy of psychology as well as within much of twentieth-century philosophy of mind, something like this idea is often adopted in terms of the notion that an adequate understanding of the neural foundations of mental functions will help us carve up the mental domain in the right way. As already indicated in the previous chapter, I am not discounting the value of an integrated cognitive neuroscience. However, we should not give in to the temptation of thinking that there is a fact of the matter as to the kinds of mechanisms at the neural level. In this regard, I follow Craver's (2009) analysis, according to which there are no objective criteria for the delineation of mechanism kinds. If Craver is right, it follows that we cannot rely on mechanism kinds to help us identify psychological kinds. Let me briefly summarize Craver's argument so that I can address the question of what conclusions we should draw from it regarding the issues of realism and pluralism.

Craver takes up the suggestion that scientists split (and also sometimes lump) their classificatory categories in accordance with what they find out about the world.[8] In order to address the question of how this might play out in the classification of mechanism kinds, he starts out with his own (now-canonical) account of mechanisms as composed of entities and activities (Machamer, Darden, and Craver 2000) and specifies that relevant mechanisms might be "brain regions, cells, proteins, and organisms" (Craver 2009, 582). He then remarks that a putative mechanism kind can participate in—and be composed of—many mechanisms. This leads him to raise the question of how the HPC account can be applied here to fix the identity conditions of a mechanism kind. In other words, following Boyd's idea that there has to be a common structure that explains why the properties that belong to the putative mechanism kind (e.g., the hippocampus) are correlated, is there a fact of the matter as to its underlying structure? Craver points out that there are many mechanisms involved in any given mechanism and that, depending on which of these mechanisms one picks as providing the underlying structure for a mechanism kind, one might end up with different accounts of the shape of the mechanism kind in question. "The simple point," Craver states, "is that most putative kinds are entangled with myriad mechanisms. By attending to some of these, one is led to lump. By attending to others, one is led to split" (2009, 584). As he notes, this introduces a perspectival and pluralist element into the notion of *mechanistic kinds*: "The HPC view allows that they can each be right for their own purposes, so long as we include perspectival and pragmatic considerations" (585).

This situation gets exacerbated once we recognize that the very question of how to individuate a given mechanism is also far from clear. In an

---

8. This is a suggestion he had made in an earlier article (see Craver 2004).

argument reminiscent of Bechtel and Mundale's (1999), Craver points out that one brain mechanism (e.g., the hippocampus) can be described at different levels of abstraction, which will lead us (again) to look for different component mechanisms. Each of these descriptions serves a unique epistemic purpose, and, thus, it is unclear how one might argue that any one of them is objectively pitched at the appropriate level of abstraction (i.e., the level that maps to some ideal mechanism for a particular kind). Noting that there might be "different, yet partially overlapping kinds of mechanism" (Craver 2009, 588), Craver concludes: "Human perspectives and conventions enter into judgments about how mechanisms should be typed and individuated. This raises a challenge to defenders of the HPC account either to find an objective basis for taxonomizing the mechanistic structure of the world, or to argue that these perspectival intrusions into the accommodation process do not threaten the realist objectives that motivate the belief in natural kinds in the first place" (591). A similar point is made by Danks (2015), who argues that there is no neutral standard that determines which scientific ontology we should choose.

Craver himself extends the conclusion of the argument to psychological kinds, stating: "The same could . . . be said for taxonomies of memory, emotions, and other mental modules" (2009, 585). Eric Hochstein picks up this idea and maintains: "The assumption that there is a single ideally correct way of classifying or categorizing mental phenomena, and that neuroscience and psychology should adhere to this correct scheme, runs counter to productive scientific practices in these domains" (2016, 746). He argues for this claim by showing (*a*) that the idealized picture according to which some essential structure uniquely fixes the reference for a kind term does not even hold for seemingly simple cases such as jade and (*b*) that it also does not hold for cognitive science.

While Craver and Hochstein both still assume a relatively traditional picture of cognitive science, one according to which (even under multiple descriptions) the relevant kinds are individuated and explained by mechanisms within the human skull, recent debates in philosophy of cognitive science suggest that we can delineate memory in radically different ways, depending on interest and theoretical commitments. Importantly, each comes with specific views about relevant mechanisms, and, thus, each makes the case that all of them are individuating a natural kind. Pöyhönen, for example, compares three ways of classifying memory kinds for which he coins the concepts of "intercranialist memory," "transactive memory," and "exogram memory" (2016, 147). The first situates memory within the skull and, hence, assumes the relevant unifying mechanism to be neural. The second takes the dyad as a basic unit of memory in a social context, thus including unifying mechanisms of social interaction. The third makes more radical assumptions about extended cognition, including in its account extracranial memory stores such as clay, paper, and computers. Pöyhönen argues that, if (and insofar as) all three ways of classifying memory are underwritten by causal mechanisms, it is not clear that there is an intrinsically right way of individuating memory.

There are several different ways in which we can interpret the discussion offered above. First, if we were hoping that within the HPC account of natural kinds neural mechanisms would deliver the one correct way of identifying psychological kinds, this hope is clearly doomed. In addition, the discussion of Pöyhönen's argument also raises the question of why we should assume that the answer should be sought at the level of brain mechanisms in the first place. Which underlying structure or mechanism we deem relevant is determined by how we conceptualize the kind in question. But, as we just saw, mechanisms do not legislate how we *should* conceptualize kinds. They constrain only how we *can* conceptualize them. It follows that, if we are inquiring about the reality of psychological kinds as they currently exist, the availability of underlying mechanisms is at best a necessary condition.

Through the remainder of this chapter, I push the argument that the entities in the extension of psychological kind concepts such as *memory* are best understood as complex (cognitive, behavioral, and experiential) whole-organism capacities of humans and animals. I also argue that these complex whole-organism capacities are real and that they acquire their status as real kinds from the causal and conceptual practices of folk psychology. By this, I do not mean to deny reality to neuroscientific kinds; I merely claim that folk-psychological concepts do not refer to neuroscientific kinds. This means that, even though research at the level of scientific psychology and neuroscience can certainly contribute to conceptual revisions at the folk-psychological level, there are real psychological kinds out there, prior to scientific investigations, that are sustained at the interface of our psychological similarity judgments and the conceptual and causal practices they enable. To a first approximation, it is precisely the kinds of folk psychology that we expect the science of psychology to investigate.[9]

Before pressing on with this analysis, let me quickly circle back to the issue of pluralism. If there are (as we just saw) different ways of conceptualizing memory, on what grounds am I privileging a conceptualization at the whole-organism level? The answer is, simply, that this level best

---

9. In sec. 6.6 below, I return to the question of the ways in which science can correct people with regard to psychological kinds.

fits with our folk-psychological usages of psychological concepts. When people use the term "memory" with respect to humans and animals, they are referring to a complex set of capacities that are bundled within the behaving, cognizing, and experiencing individual organism. They are not referring to chunks of the brain, nor are they referring to specific mental states, and they are not (just) referring to behavioral patterns. They are also not referring to extended cognition (even if the latter usages can be backed up by underlying mechanisms). In other words, while there are different ways of conceptualizing memory (pluralism$_1$), I argue that our widely shared folk usage is, in fact, the dominant one (thereby restricting pluralism$_2$). (See sec. 6.2.1, where I introduce the distinction.) I want to be sure, however, to indicate that my account is compatible with other arguments for the plurality of kinds in psychology and cognitive science. First, even if folk-psychological kinds are individuated and sustained at the whole-organism level, it is entirely conceivable that different people at different times and places arrive at different taxonomies and incorporate them in their practices, making them equally real in those contexts. This has been argued by Kurt Danziger in a number of publications (e.g., Danziger 1997, 2002, 2003), and I am on board with this general idea (see also Feest 2023). Second, we can be realists about psychological kinds at the whole-organism level while also acknowledging a plurality of other kinds (e.g., at the level of neuroscience) that are relevant to our scientific understanding of these macro-level kinds.[10]

6.3.2. Reductionism and the Reality of Whole-Organism Kinds

The issue just touched on opens up a larger topic, however, namely, that of reductionism. The question is whether psychological kinds can have stand-alone reality distinct from the kinds that make up their physical basis. This issue, among others, is taken up by Muhammad Khalidi in *Natural Categories and Human Kinds* (2013). Khalidi is interested in these questions in the context of his own theory of natural kinds, according to which kinds are individuated from an epistemic standpoint and are justified if and when they map onto a unique causal-mechanistic profile in the world.[11] In the previous section, I argued that there are

10. I return to this in secs. 6.4 and 6.5.
11. This is essentially a development of Boyd's HPC account. Khalidi refers to it as an "epistemic" account because it privileges those kind determinations that serve epistemic needs, but he takes them to be underwritten by ontic/mechanistic structures. By contrast, other recent accounts try to steer clear of ontic commitments altogether (e.g., Slater 2015). (For a critique of the notion that epistemic considerations can be divorced from underlying metaphysical commitments, see Kendig and Grey [2019].)

problems with the notion of a *unique causal-mechanical profile*. Nonetheless, Khalidi's analysis offers some insights into the issues at the heart of the question of antireductionism (i.e., the question of whether psychological kinds have a stand-alone reality at a macro level). I briefly outline some of his arguments here to bring my own position into clearer view.

The thesis at the heart of reductionism is that higher-order kinds might not pass the test that is required if we want to say of a property that it really exists qua kind. These types of concerns are, of course, well-known from the extensive literature about higher-order (or emergent) properties. The worry is that, if such properties can be reductively explained by the causal mechanisms that realize them at a "lower" level, then they are not genuine kinds. Khalidi explains this point by citing Kim's (1992) example of viscosity as realized by multiple physical mechanisms, depending on the physical substance at hand (viscosity as realized by liquids, viscosity as realized by solids, etc.), suggesting that viscosity is not a genuine kind but rather a disjunction of lower-level kinds. Khalidi counters this argument by stating that properties like viscosity "earn their keep in science by virtue of their projectibility" and that "projectibility is an indication that the categories track properties and kinds that enter into real causal relations" (2013, 90). Notice that projectibility is here seen as an indicator of causal relations that are similar across different instances of viscosity, which, in turn, is seen as underwriting kindhood. The argument is, then, that, as long as there are causal relations that can be picked out only on the basis of a specific macro-level (theoretical and observational) vocabulary (i.e., in this case, the vocabulary of fluid mechanics), this legitimates the use of such descriptive terms as individuating natural kinds.

This argument is, of course, very familiar in its appeal to the (potential) multiple realizability of kinds, though multiple realizability arguments have also been contested (e.g., Polger and Shapiro 2016). I return to some aspects of this debate in section 6.5. Here, I merely want to gesture at the position I develop there, namely, that the question of identity vs. multiple realizability is, in a certain sense, beside the point, at least when it comes to psychological kinds, because traditional debates about identity theory vs. functionalism are typically pitched as debates about the identity conditions of mental states or cognitive functions, not psychological kinds as I conceive of them. In a similar way, Khalidi (2013) ultimately appears to find neither reductionism nor multiple realization to be very helpful in thinking about the kinds in different scientific domains. Both these positions presuppose that higher-level descriptions can be mapped onto lower-level descriptions in terms of either an identity or a realization relation. One reason why such mappings are unlikely to be successful, Khalidi argues, is that the kinds of different domains are individuated at different spatiotemporal scales and in different ways. For example, Khalidi notes, "in many cases, the macrolevel entities are individuated extrinsically, functionally, or even etiologically, and cannot be identified with simple aggregates of microlevel entities" (2013, 220). This seems exactly right for the case of psychological kinds. The point, as I would put it, is that, while various phenomena that constitute a psychological kind may well be identical with kinds that figure prominently in other disciplines, such as neuroscience, the psychological kind *itself* is individuated at a different spatiotemporal scale than those of other disciplines. This can easily be illustrated with the memory case discussed in previous chapters. Memory as a folk-psychological kind is, I argue, a set of (cognitive, behavioral, and experiential) capacities, that is, the capacity to have experiences, the capacity to engage in cognitive processes, and the capacity to display specific behavioral patterns. The instantiation of those capacities spans events from triggering conditions to behaviors. The mechanisms that figure in the instantiation of this capacity are individuated at different scales, ranging from the molecular mechanisms to the movement of limbs. Obviously, two mechanisms at such different scales cannot be mapped onto each other as a relation of identity, though they all figure in the overall set of capacities that are exhibited by the whole organism.

One possible response to this observation might be to say that, by individuating kinds at the whole-organism level rather than at the level of underlying mechanisms, we might find ourselves lumping behaviors together as instantiating kinds by virtue of superficial similarities rather than meaningful explanatory differences. Paul Griffiths (1997) presents an argument to this effect when arguing against individuating emotions functionally, appealing instead to evolutionary lineage as individuating kinds of emotions at the most natural level. He thus argues that it is deep structure, not superficial similarity, that determines kindhood. Griffiths is a proponent of a naturalized essentialism, which tries to tie the reality of kinds to specific explanatory mechanisms. In this way, his motivation is similar to Khalidi's desire to make sure that the kinds categories of the special sciences "demarcate genuine aspects of reality as opposed to mere subjective perspectives of reality" (Khalidi 2013, 121). By contrast, my strategy in this chapter is to tie the reality of psychological kinds to judgments of similarity and to corresponding successful practices involving the kinds in question. By this, I do not mean to deny that there are contexts in which it will be epistemically justified to individuate kinds by appeal to specific explanatory mechanisms. My point is merely to reject the notion that explanatory mechanisms provide an objective foundation for the correct description of real kinds.

And, moreover, I would like to resist the dichotomy between genuine aspects of reality and mere subjective perspectives on reality.[12] If, as I have argued, kinds come into existence at the interface between our cognitive/perceptual/instrumental preconditions and the world, their reality does not necessarily stand in any tension with their perspectival nature. In response to Griffith's worry that functional individuations of kinds are too superficial to capture genuine kinds, I would argue that this is an empirical question that hinges on whether the kinds do the work they are supposed to do.

6.4. Similarity Judgments at the Whole-Organism Level: Echoes from Ecological Psychology

In the previous sections, I have argued that there is no reason why we could not individuate psychological kinds on the macroscopic, whole-organism level and grant them reality at that level. They are (potentially) as real as the entities we get if we identify kinds at more microscopic levels like cells or brain mechanisms. I argued that psychological kinds are relational in the sense that they come into being at the intersection between our perceptual and conceptual prerequisites, on the one hand, and the physical and social world, on the other, which allows for certain conceptual and causal practices. The crucial work done by our perceptual and conceptual prerequisites is that they enable us to sort things into similarity classes. But what does this mean, exactly? How do we judge instances of psychological kinds as similar to one another? How is similarity between behaviors determined so that it allows instances of behavior to be grouped into kinds of behavior (e.g., behaviors associated with memory)? And, finally, what are the causal and conceptual practices that sustain the kinds? In this section, I focus on the issue of how instances of psychological kinds are recognized.

One seemingly obvious answer is that kinds of behavior are classified as similar on the basis of their falling under the same concept, where this concept, in turn, is entrenched in a network of reasonably successful folk-psychological assumptions and practices that confer a robust reality to the kinds. There is a sense in which this idea is central to the entire argument of this book. Existing conceptual frameworks allow us to parse the world into kinds of things, and those conceptual frameworks themselves provide the operational definitions that make it possible to

---

12. See also Hatfield (2020), who points out that there is a legitimate usage of the term "subjective" as "pertaining to the subject" (79). Hatfield contrasts this with the common idea that the notion of *subjective* implies "idiosyncratic variation" (86).

investigate the kinds in question. But this answer begs the question of how the concepts (and the kinds they help enable) are possible to begin with. Given my relational theory of kinds, according to which kinds come into being as a result of successful similarity judgments, we need an account of how we come to judge observable token behaviors (and conceivably also introspectively assessable token mental states) into similarity classes that jointly form psychological kinds.

A full analysis of this question would clearly be beyond the scope of this book, but let me briefly outline a suggestion that is inspired by work of Corinne Bloch-Mullins (2018, 2019), who has articulated what she calls a "similarity-based analysis of concepts," which relies on a particular area of research in the psychology of perception: the structure-alignment view of similarity (e.g., Markman and Gentner 1993). The basic two components of Bloch-Mullins's account are (*a*) that similarity judgments are heavily context and foil dependent and (*b*) that there are lawful regularities in the ways in which features of the foil determine similarity judgments. The former suggests that similarity is flexible (by virtue of depending on context), while the latter suggests that similarity is not arbitrary (by virtue of being underwritten by perceptual laws) and can, therefore, be explanatory. To phrase this in the terminology developed in this and the previous chapter, the ontic facts of the world do not uniquely determine kinds, but it does not follow that kind judgments are arbitrary. Rather, kinds depend on similarity judgments, where similarity is constituted by a triadic relation between the judging individual, the stimulus, and the foil/context. According to this line of thinking, similarity is a context-dependent process of dynamic alignment rather than a static and decontextualized relation between features of the environment. Similarity judgments are made at the convergence of structural and contextual features as judged by individuals with specific biological and social makeups. Consider, for example, psychological kindhood. It is not the mere behavior of an individual that prompts us to apply a kind concept but rather the behavior in the context of specific contextual/environmental features. Accordingly, a behavior can be classified as instantiating a variety of different kinds, depending on context.

While the psychological literature cited by Bloch-Mullins does not draw this connection, the thesis about the dynamics of similarity judgments (and perception in general) as being determined by foil/context comes remarkably close to the basic ideas of the Gestalt principles of perception (for a systematic overview and historical analysis of the emergence of Gestalt psychology, see Ash [1995]), which has also emphasized that the way things are perceived and, thus, the conditions under which things are perceived as similar depend not on (the similarity of) their parts but on how the parts are arranged, thus forming similar structures. There are some obvious parallels between Gestalt perception and ecological accounts of perception, according to which perception enables the perceiving organisms to pick out invariant structures in the environment, thereby allowing for adaptive behaviors in a complex world. The parallel between the structure-alignment view of similarity and these Gestaltist/ecological approaches of perception is welcome here as it also highlights the holistic character of our percepts. This gives further credence to my claim that (insofar as similarity ratings can give rise to kinds at all) there is no reason why kinds cannot be individuated at the macro level. Moreover, the affinity of this line of thinking to ecological approaches reminds us that we move through the world as whole organisms and perceive others on a whole-organism level.

If I am right in my contention that kinds are relational properties (which come into being at the interface between mind and world) and can acquire reality insofar as they work (i.e., figure in our practices and make a causal difference), then the question is how judgments of similarity can inform the concepts that allow for psychological kinds to come into existence. Drawing on Bloch-Mullins's work, I argue that the structure-alignment theory is a good contender in that it can account both for the robustness and for the malleability of psychological kinds. Moreover, connecting this insight with Gestalt and ecological accounts of perception, we can make sense of the fact that (as I have argued) folk-psychological kinds are typically described and identified at the level of the behavior of organisms since instances of psychological kinds are typically attributed to behaving organisms, not to brains in vats, and not to individual brain states or mental states (even if individual brains or mental states belong to the class of phenomena that make up the kind).[13]

To clarify, my claim is not, as some versions of behaviorism might have it, that psychological kinds are nothing but behavioral dispositions. (In fact, my talk of behavioral-cognitive-experiential capacities is supposed to suggest otherwise.) Rather, my claim is that, when identifying kinds in everyday life, we do so on the basis of behaviors (verbal or nonverbal)

---

13. My account of the similarity-based classification of psychological kinds seems compatible with philosophical analyses of the direct perception of psychological kinds like emotions, which argue that specific behavioral patterns are constitutive parts (rather than mere indicators) of the kinds in question (e.g., Newen, Welpinghus, and Juckl 2015). The structure-alignment account can potentially explain the role of contextual aspects in determining which psychological kind a given behavioral pattern is a part of in any given instance.

that are often triggered by external circumstances and seemingly driven by some internal states toward specific goals. To put this differently, we typically attribute the functions or activities of psychological kinds to organismic units that can exhibit behaviors in response to and directed toward physical and social environments. Indeed, Tolman's insight that behavior is goal directed highlighted the organism as the unit to which we attribute cognitive capacities (see chap. 1). For Tolman and some of his Harvard teachers, a crucial point was that it is impossible first to describe behavior neutrally as an explanandum and then to explain it in terms of cognitive states. Rather, descriptions of behavior are already infused with an ascription of purposes and hypotheses and, thus, with the assumption that the behaving organism instantiates psychological kinds. Both Tolman and Hull chose to focus on the study of what they called "molar" behavior, thereby acknowledging the whole organism as a natural unit of analysis.

Other prominent early and midcentury psychologists also highlighted that psychology should focus on the organism and the way it is functionally embedded in its environment. This is particularly clear in the work of proponents of (what came to be known as) the Chicago school of functionalism such as John Dewey (see, e.g., Green 2009; and O'Donnell 1985). By contrast with the prominent school of structuralism (usually associated with Titchener), which investigated the structure of the conscious mind, functionalists viewed the mind in a more biological manner as serving the function of allowing organisms to be oriented in complex environments. William James's work, with its emphasis on the dynamic and adaptive function of mental phenomena, can also be placed in that tradition. The focus on the organism as a whole was also taken up by other early to mid-twentieth-century psychologists, for example (in the immediate vicinity of Tolman), in Egon Brunswik's probabilistic functionalism (1934, 1943) as well as James Gibson's (1979) ecological approach to perception. In the work of these scholars, the focus on the organism was combined with the claim that no reference to cognitive processes was needed to explain the phenomena in question; instead the lawful manner in which psychological functions are dynamically adapted to environmental features was studied. While ecological approaches are sometimes juxtaposed with an information-processing account of cognition, Hatfield (2019) argues convincingly that such an ecological perspective is compatible with allowing for a study of processes that make adaptive behavior possible.

The organism-centered approaches in the history of psychology mentioned above are fairly marginal in mainstream psychology today. However, I claim that the practice of taking the behavior of organisms as the central unit of study is, in fact, deeply ingrained in psychological methodology. It is, after all, the whole organism that enters the experimental situation, is presented with stimuli and instructions, and is asked to respond. Psychologists do not observe random behaviors and then infer inner cognitive states or processes. Rather, behaviors are typically already observed as cognitive (e.g., as search behavior or, more generally, as problem-solving behavior) in accordance with folk-psychological assumptions, and it is precisely those assumptions that are exploited in psychological experiments.[14] It is with a similar sentiment that Hatfield argued that psychology cannot be replaced by neuroscience because the functions that neuroscience studies are individuated at the level of organisms: "The study of those brain functions of interest to cognitive science *is* the study of the psychological processes of organisms" (2002b, 229). The point here is that, while my account of psychological kinds as individuated at the whole-organism level places me in the company of ecological approaches, this does not commit me to the view that we cannot appeal to cognitive or neurological capacities as making up such kinds.

6.5. Psychological Kinds and Cognitive Ontology

Picking up on the idea that macro-level psychological kinds are composed of cognitive and neurological capacities but cannot be reduced to any one of them, I now review the relationship between my account of psychological kinds and existing discussions of cognitive ontology. I argue (*a*) that the reality of psychological kinds is compatible with the search for realizing functions but neutral with regard to the localizability of the latter and (*b*) that research in cognitive ontology relies on implicit ontological/conceptual assumptions about psychological kinds.

6.5.1. Cognitive Ontology

In the past twenty years, there has been a lot of interest in methodological questions that arise in the context of attempts to map mental functions to brain structures (e.g., Poldrack 2010; and Price and Friston

---

14. At first glance, it may seem counterintuitive that the mental state of experiencing tonal volume (as described in relation to Stevens's research discussed in chap. 1) is a whole-organism disposition. However, I claim that folk psychology very naturally thinks of auditory experience as an attribute that manifests itself as a result of environmental stimulation and can give rise to behavioral responses. As we saw in chap. 1, it is precisely this folk understanding on which psychophysical methods build.

2005). It is important to note here that, even though the aim of this literature is to devise an agreed-on cognitive "ontology" (by which authors typically mean a taxonomy of mental functions that is in accordance with brain functions), the questions they are raising are epistemological in nature in that they grapple with the issue of how such ontologies can legitimately be inferred on the basis of behavioral data and neuroscientific data such as those provided by brain imaging. This debate is instructive for my purposes because (*a*) it helps me articulate the way in which the exploration of objects of psychological research is compatible (and can be fruitfully combined) with that of studying brain functions and (*b*) authors within that literature are keenly aware of the fact that cognitive neuroscience relies on folk-psychological concepts both in their construction of ontologies and in the ways in which they devise experimental tasks to investigate the corresponding cognitive functions.[15]

One central epistemological issue in the literature on cognitive ontology concerns the question of how to individuate cognitive functions in a way that allows them to be mapped onto brain structures. This question was importantly put on the agenda by Price and Friston (2005), who noted that, even though there is evidence to suggest that some brain areas are involved in multiple different functions, it is a realistic goal to identify brain regions that realize them once we have succeeded in individuating them at the right level of description. Others have been less optimistic, arguing that the very project of locating specific mental functions in specific brain regions should be abandoned in favor of one that identifies functions in a more context-sensitive manner (Klein 2012). Trying to strike a middle ground, McCaffrey (2015) argues for a "functional heterogeneity hypothesis," according to which different brain regions are multifunctional in different ways. Some can realize different functions by means of different specialized components, allowing us to do function/component mappings, whereas others can realize multiple different functions by means of only one kind of component. While his point is that only the latter kinds of functions require a contextualist approach, Burnston (2016) argues that contextualism and localization need not be mutually exclusive (see also Burnston 2021; and Dewhurst 2019). (I return to this below.)

---

15. Khalidi (2022) distinguishes between cognitive kinds and neural kinds, stating that the former (in contrast to the latter) are often individuated relationally. Since his book came out right as I was completing my manuscript, I cannot give it the attention it deserves. However, I would like to suggest that Khalidi's notion of a *cognitive kind* captures aspects of what I want to draw attention to with my notion of a *psychological kind*.

Regardless of where one stands on this issue, it is clear that, in order even to ask how cognitive functions are realized in the brain, one needs already to have a descriptive vocabulary of cognitive functions to get the project off the ground (Poldrack 2010). This raises two interrelated questions, namely, (*a*) how to ensure that one has the right kind of psychological vocabulary for functions that one might actually find in the brain and (*b*) what an empirical method that settles the matter might look like (Poldrack and Yarkoni 2016). A seemingly straightforward strategy to attack these questions is, first, to present subjects with tasks whose execution requires the function of interest (e.g., to have them engage in an *N*-back task on the assumption that this task will engage working memory functions) and, second, to use brain-imaging data to then locate the function. The first suggestion essentially amounts to operationally defining one's concept of the function of interest. One problem with it is that, even if the chosen task does require experimental subjects to engage the function of interest, the subject is likely also going to be engaging in other functions when carrying out the task, making it hard to isolate the function in its pure form (thus hampering the inference from task performance to mental function). Needless to say, this problem also afflicts attempts to use brain-imaging techniques to make an inference from a task to a brain region where a function is localized. As Poldrack (2010) has pointed out, there is, in addition, the "problem of reverse inference." Even if we were sure that the brain activity that we observe under a specific task is, in fact, connected to a specific function (e.g., a function that contributes to working memory), this would not establish that the activity occurs only in connection with working memory. The activity we are observing might also be present in other mental functions, and, hence, we are not warranted in using the results of imaging studies to infer that working memory is active whenever the relevant brain region is active.[16]

In response to these problems, researchers have suggested the use of text-mining techniques to extract task ontologies (Figdor 2011) as well as reported activations and semantic annotations from published articles (Yarkoni et al. 2011), thereby hoping to effect some stable mappings between tasks, functions, and brain regions. Recognizing that high-level psychological concepts (such as *working memory* and *implicit memory*) most likely do not have a simple physical correlate, Poldrack and his colleagues set up an online, community-based repository with a large number of concepts, tasks, and phenotypic and

---

16. Notice that this is a version of a problem that I already touched on in the previous chapter, where it was found that phenomena psychologists have attributed to implicit memory might be due to mechanisms that are not unique to implicit memory.

behavioral traits with the aim of establishing a catalog of concepts and their hierarchical relations to one another as well as the tasks with which they are associated (Poldrack and Yarkoni 2016; see also http://www.cognitiveatlas.org).[17] However, Sullivan (2017a) has argued that such attempts are unlikely to be successful in their current form since researchers do not have a unified terminology for cognitive functions (i.e., they use different terms, and they also use the same terms in different ways) and it is not clear that any two implementations of a task really operationalize the same concept (see also Sullivan 2009). This raises the worry that current research practices might not be conducive to the aim of coordinating the plurality of approaches, methods, and concepts in cognitive neuroscience.

This very brief overview of some of the methodological and conceptual problems and debates in (the philosophy of) cognitive neuropsychology should suffice to make it clear that, as it is used in that literature, the term cognitive ontology is pitched at a different level than my term "psychological kinds." While I have argued that psychological kind terms single out complex sets of (cognitive, experiential, and behavioral) capacities of whole organisms, research on cognitive ontology has concerned itself with the question of whether specific functions that are required for the realization of some aspects of such capacities are appropriately described and whether they can be localized in specific regions of the brain. The two projects are mutually compatible. However, my claim about the reality of psychological kinds is not dependent on any specific answer to the localization of cognitive functions. Notice that, by this, I am not committing to the stronger claim that findings about brain mechanisms might not have an impact on the psychological kinds.

6.5.2. Psychological Concepts and the Question of Realism

Despite the differences between my conception of psychological kinds and the neuroscientific kinds that are taken up in discussions of cognitive ontology, there is one thing that the two approaches do share, namely, the usage of folk-psychological terms such as "memory." This gives rise to the question of what these terms refer to, precisely, and how we should square the fact that researchers use (and operationally define) them in very narrow and impoverished ways with the fact that research is, ultimately, expected to elucidate something about the real world

17. For an overview of this and related community-based attempts at building neuropsychological and psychiatric classification systems, see Sullivan (2017a).

beyond the laboratories of cognitive neuropsychologists. An interesting version of this worry has been articulated by Francken and Slors (2014, 2018), who argue that, when folk-psychological concepts (which they call "common cognitive concepts") get turned into "scientific cognitive concepts" (which are operationalized in terms of specific tasks), a lot of assumptions are made about the referents of the concepts. These assumptions remain implicit and unsystematic, resulting in a confusing multitude of operationalizations that, in turn, lead to problematic extrapolations about the object of interest. Francken and Slors argue that, since operationalizations of cognitive concepts embody impoverished and partial (and possibly false) assumptions about the referents of our folk-psychological concepts, we should not interpret the findings of neuropsychological experiments realistically, that is, we should not assume that terms like "working memory" refer to a brain region that causally explains behavior. Instead, they adopt a "Dennetian" proposal, according to which our folk-psychological concepts track behavioral patterns that can be interpreted by means of the folk-psychological language of beliefs and desires. They then suggest that this interpretivist stance can be put to practical use to inform discussions about how specific operationalizations and their data should be interpreted.

I would like to tease out a distinction between two questions here, namely, between (*a*) the question of whether mental states (qua mental states) as individuated by folk-psychological belief/desire terminology have explanatory efficacy and, hence, are real and (*b*) the question of whether folk-psychological kinds (qua complex whole-organism capacities) are real. Francken and Slors are concerned with the former question, arguing convincingly that, given the partial and impoverished ways in which cognitive concepts are operationalized, we should not expect scientific and folk-psychological language to be coextensive. My claim as to the reality of psychological kinds is concerned with the latter question, however. As I argued in the previous chapter, psychological research is better understood as investigating (aspects of) complex (behavioral, experiential, and cognitive) capacities rather than explaining individual instances of behavior by means of reference to mental states. Once we appreciate this distinction, it seems clear, first, that individual functions or mental states as investigated by neuroscientists can certainly be explanatorily or descriptively relevant to the folk-psychological kind as a whole and, second, that, even if folk psychology and cognitive neuroscience use the same words (e.g., "pain" and "memory"), their vocabularies are not going to be coextensive. For example, cognitive neuroscientists might take the term "spatial memory" to refer to a brain region that is active when rats navigate mazes, whereas folk psychology takes it to refer to a complex set of capacities at the whole-organism level, for example, the capacity to navigate in space, the capacity to store and retrieve knowledge about a spatial layout, the capacity to represent such knowledge in a particular way, etc.

Importantly, our folk-psychological understanding at the whole-organism level informs the more localized research of cognitive neuroscience, making the relationship between folk-psychological kinds and scientific concepts potentially tighter than Francken and Slors (2014, 2018) predict since it is our folk concepts that individuate the explananda of scientific efforts by psychologists and cognitive neuroscientists. In a similar vein, Burnston argues: "The best reading of the role of psychological constructs is as heuristics for investigation, rather than as explanatory kinds" (2021, 260). Elsewhere, he suggests: "Rather than serving as explanantia, we should view mental categories as helping us parse behaviors into rough, and revisable, similarity classes" (2022, 272). This sentiment is also echoed by Dewhurst, who states: "Even if folk psychology does not correctly identify the fine-grained functional structure of the brain, it can nonetheless correctly identify behavioural patterns and dispositions which are just as real as those described by neuroscience" (2021, 326). I would add that such investigations can certainly yield explanations but that such explanations will likely not make reference to folk-psychological kinds qua explanatory kinds. Folk-psychological kinds are, rather, the explananda of constitutive explanations (see sec. 5.3.2 in the previous chapter).

The argument just provided implies that realism about psychological kinds (at the whole-organism level) and realism about cognitive functions (at the level of the brain) need not be mutually exclusive. We can be realists both about whole-organism capacities and about the component functions that either contribute to their realizations or jointly constitute the complex capacities in question. Importantly, this is the case even if atomism about cognitive functions is false (i.e., even if there are no dedicated brain regions for component functions, regardless of context) since, as Burnston (2021) points out, the rejection of atomism does not imply that component functions (individuated in a context-sensitive manner) cannot contribute to the explanation of psychological kinds (understood at the whole-organism level). Context-sensitive localizations are localizations, too (Burnston 2021). Nor is it the case that, if we cannot construct a one-to-one mapping between folk-psychological concepts and neuroscientific concepts, the former has to be eliminated. This notion (well-known from eliminativism à la Churchland [1981]) is deeply rooted in a particular picture of folk-psychological explanation, according to which folk psychology is a theory that explains behavior by reference to kinds of mental states that figure as theoretical entities. It is also rooted in a picture of scientific psychology as trying either to vindicate or to disprove folk-psychological theories. In turn, this latter notion is rooted in the assumption that the vindication or refutation of folk psychology hangs on whether the mental states or functions it picks out can be mapped onto brain states or regions.

Another way of putting this is that eliminativism shares with traditional approaches such as identity theory (e.g., Place 1956; and Smart 1959) the following two ideas: (a) that psychological terms refer to mental states (rather than whole-organism capacities) and (b) that for a mental state to be real is to be type identical with a brain state. If we were to buy into this picture, we might indeed worry that we face eliminativist consequences if it should turn out that there are no simple one-to-one mappings between mental states/functions and brain regions.[18] This worry is also addressed by Dewhurst, who suggests that "current debates over cognitive ontology revision pose a novel eliminativist threat, in a similar manner to the original eliminative materialism of the 1980s" (2021, 214). Dewhurst argues that we should resist this threat by proposing that folk-psychological and scientific concepts serve entirely different functions, the former being conceived as coarse-grained descriptions of behavior and the latter being directed at fine-grained mechanisms. As should be clear from the discussion above, I concur with this assessment, adding, however, that psychological concepts are not merely coarse-grained ways of referring to individual cognitive/neural functions. They are, rather, descriptions of whole-organism capacities. This move diffuses the eliminativist threat by pointing out that (folk-)psychological kinds are individuated at a different level than are those of cognitive neuroscience and that their reality is not determined by whether there are one-to-one mappings between, say, mental states and brain mechanisms. As explained above, I contend that their reality is created by virtue of the conceptual and causal practices they enable.

---

18. Though functionalism in the philosophy of mind rejects the notion that mental states are type identical with brain states, it shares with identity theory the focus on mental states as the referents of psychological constructs (at least in the classical formulations of Putnam and Fodor). An interesting recent critique of functionalism (Polger and Shapiro 2016) subjects the notion that functions can be multiply realized to critical analysis. While a discussion of this would go beyond the scope of this chapter, it suffices to point out that Polger and Shapiro, too, are interested in mere component parts of what I regard as the referents of psychological kind terms, thereby not really touching on the question of how psychological kinds are sustained at the whole-organism level.

My analysis does not deny that there might be research that restricts the scope of specific psychological kind terms, clears up confusions, or results in kind splitting (e.g., by pointing out that a specific term or theory is used to refer to or explain distinct phenomena). Within philosophy of psychology and cognitive science, versions of this have been argued for with regard to emotions (Griffiths 1997), memory (Craver 2004), concepts (Machery 2009), and consciousness (Irvine 2014). However, I have argued in this chapter that there is an important difference between this claim and the stronger claim that nonpsychological facts settle what is the right way of carving the world into psychological kinds. It is this latter claim that I reject (see also McCaffrey and Wright 2020). More importantly, in the context of this book, the research that has the potential to deconstruct some of our folk kinds does so by drawing on those very kind concepts. For example, as we saw, even if we end up with experimental findings that suggest that memory ought to be split into different kinds, this research still relies on an overarching folk-psychological understanding of memory as involving the capacity to store and retrieve previously learned information. It is precisely this folk-psychological understanding that is utilized in operational definitions.

I argue that this is relevant because it allows us to view experiments in psychology and cognitive neuroscience as implicitly acknowledging the complex whole-organism character of the kinds they ultimately want to explain. Outside any folk-psychological context, there is nothing in the brain that says "memory." It is only by putting whole organisms in specific experimental environments (exposing them to specific tasks and measuring the outcomes in specific ways) that we can interpret experimental results as pertaining to specific psychological kinds. To put this in the terminology developed in previous chapters of this book, when scientists design experiments about a complex capacity such as memory, they are able to take advantage of certain assumptions about the kind in question that are built into the (whole-organism) folk-psychological concept of *memory*. Specifically, they might assume that memory is something that can be manipulated in certain ways (e.g., by learning procedures) or that specific memories can be retrieved by particular kinds of prompts, that memory involves the storage of remembered material, that it can have phenomenological aspects, and that specific resulting behaviors are memory behaviors. Such assumptions are precisely what is being put to use when memory gets operationally defined (in both the wide sense and the narrow sense) as an object of research. Such assumptions are defeasible, as is the very concept of *memory* (or kinds of memory) as a viable scientific kind, but it is only on the basis of such assumptions that research can proceed at all.

6.6. Conclusion

It is time to return to the question with which I started this chapter, namely, how we can reconcile the notion of a psychological kind as relational with the idea that scientists are trying to get it right. In other words, how can we reconcile the notion that psychological kinds depend on *us* with the notion that we can be wrong about them, which presupposes some kind of epistemic gap between us and the psychological kinds we deal with and investigate?

Let me begin my response by briefly summarizing my account of the ways in which psychological kinds depend on us. I have argued that (folk-)psychological kinds are created and sustained by our context-sensitive ability to group behaviors into similarity classes and by virtue of the causal and conceptual practices such classifications allow us to engage in. I have, moreover, argued for a potential pluralism of psychological kinds, thereby pushing back on the idea that there is only one correct macro-level taxonomy of psychological kinds that is somehow uniquely determined by micro-level (nonpsychological) facts. The macro level at which the relevant psychological kinds are individuated is, I claimed, not the level of mental states or cognitive functions but rather the level of the complex capacity of whole organisms to display purposeful behavior, engage in cognitive processes, and have experiences. This is the level at which folk psychology individuates its kinds, and these are, therefore, the kinds investigated by scientific psychology. If we accept this premise, there are three ways in which psychological research can potentially get it wrong. Understanding these three ways will help us unpack what is involved in trying to get it right.

The first way in which researchers can fail to get it right is by providing inaccurate functional analyses of the macro-level kinds under investigation. To put this differently, while the reality of psychological kinds is settled at the macro level, this does not mean that we cannot be wrong at the level of the contributing phenomena (including behavioral phenomena). Given the epistemic blurriness of psychological objects of research, it should not come as a surprise that we can make descriptively false (or at the very least ambiguous) assumptions about the kinds in question. For example, even though the notions of various forms of memory (short-term, long-term, implicit, etc.) are captured by a more general folk-psychological concept of *memory*, and even though there are good empirical reasons for splitting memory into subkinds in ways that have already looped back into our folk psychology, the case studies provided in previous chapters have shown that we can still have misconceptions about how these kinds are realized and what their descriptive features are. As we saw, the behavioral phenomena associated with implicit memory and working memory may not be due to specialized systems (see chap. 5 above). Likewise, the notion that implicit memory is unconscious is compatible with different conceptions of unconscious, which means that we can be wrong about the sense in which implicit memory is unconscious (see chap. 4 above).

There is another important sense in which psychologists can get it wrong, namely, regarding what I call the "shape" of specific kinds, which includes the generalizability of specific findings about them. The phenomena associated with kinds of memory may be robust under some circumstances but not others; they may be robust for some populations but not others. In other words, the generalizations psychologists draw from their experimental findings may get it wrong because they are working with incorrect assumptions about the representativeness about their samples, experimental settings, and measurement instruments (Feest 2022). Thus, while a kind may be real in the sense that it is firmly anchored in conceptual and causal practices at the societal level, it may not describe each individual or group of individuals equally well. I return to a detailed analysis of this problem in chapter 7, where I argue that a significant amount of psychological research should be devoted to determining the exact ways in which units of psychological analysis (individuals) are sensitive to varying circumstances and populations, which directly corresponds to specific descriptive facts about the shape of specific objects of research. I show that the method of operational analysis, which we already encountered in chapter 2, can play a valuable and so far underappreciated role in helping with the fine-tuning of our psychological kind concepts, thereby contributing to the elimination of false assumptions about their referents.

While not at the center of the questions I am addressing in this book, I want to point to one other, more meta-level way in which concepts can be determined to be getting it wrong. This has to do not with empirical inaccuracies or overgeneralizations but with the values and aims that are transported by the ways in which specific kinds are conceptualized (Brigandt 2022). If we take this into account, the evaluation of a kind concept will involve analyzing questions like whether we are on board with the aims and values and whether the aims and values are well served by the kind concept. I therefore suggest that, when looking at the kinds posited by our folk-psychological ontology, we need to distinguish between the reality of the kind and the *goodness* of the kind, that is, between the kinds we have and the kinds we might want. By distinguishing between real kinds and good kinds, I pick up on recent discussions of social scientific categories that have pointed out that the social constructedness of categories like *gender* and *race* does not change the fact that they have a powerful social reality (see sec. 6.2 above). The basic point is that, even though the kinds are real, they rely on conceptual and evaluative frameworks that can be critically reflected. While such considerations are especially obvious in relation to concepts like *race* and *gender* (i.e., concepts that classify kinds of people), other psychological entities can also potentially acquire reality when the corresponding concepts become ingrained in our folk psychology. They are taken to refer to psychological entities, but they are also connected to practices such as meditation, hiring policies, or specific training programs that are adopted with the aim of making a difference in people's lives. This means that the question of whether these concepts get it right can be evaluated by asking not only whether the causal and conceptual practices associated with them are in fact effective but also whether they reflect our values, are biased against (or in favor of) specific social groups, etc. Pursuing the question of the value-laden nature of psychological kind terms and their operationalizations is an obvious follow-up for the account presented in this book. But it would go beyond the scope of the book to pursue it here.

CHAPTER SEVEN

Operational Analysis and Converging Operations

7.1. Introduction

While objects of psychological research do not correspond to some independently fixed psychological kinds that the scientific concepts of psychology try to accommodate, it does not follow that the material world does not put constraints on the ways in which psychological kinds are constructed, both by folk-psychological practices and scientific practices. This is also why experimental efforts to probe the material world by way of causal manipulation and measurement are promising ways of exploring psychological kinds. As the previous chapters have argued, such efforts are best described as gradual processes of conceptual development.

In this chapter, I offer a closer analysis of the ways in which psychological experiments can be utilized to contribute to the formation of concepts pertaining to objects of psychological research, where those concepts transcend the specifics of any given experimental context. I develop my account of concept formation and conceptual development on the basis of a generalized account of experimental inference in psychology. This account is connected to the notion of *converging operations* (Garner, Hake, and Eriksen 1956), which we already encountered in chapter 2, to argue that experimental inferences can be regarded as supporting the intended conclusion to the extent that alternative explanations of the experimental data have been ruled out. With this analysis, I integrate three recent bodies of literature in philosophy of science, namely, (*a*) debates about experimental inferences in the social sciences (e.g., Guala 2005), (*b*) debates about experimental errors (Hon 1989; Mayo 1996; Schickore 2005), and (*c*) debates about what constitutes good (or reliable) data (Woodward 1989, 2000).

Picking up the chapter 5 argument that the quality of evidence has to be determined contextually (Canali 2020; Feest 2022) as well as the chapter 6 argument that the units of psychological analysis (whole organisms) are context sensitive, I argue that these two appeals to context bear an important systematic connection in the process of concept formation in psychology. Experimental data can reliably support specific inferences to the extent that they are not distorted by contextual factors. At the same time, the very context sensitivity (or reactivity) of the psychological subject matter is something that needs to be explored in its own right as it provides insights into (what I call) the "shape" of a given object of research and the scope of the corresponding concept. Thus, an overarching thesis of this chapter is that questions about the quality of evidence required for a specific experimental inference have to be addressed in tandem with descriptive and conceptual questions about the object under investigation. The flip side of generating reliable evidence for a particular experimental inference by controlling for distorting contextual factors (confounders) is asking how context moderates the object about which we wish to make inferences.

Section 7.2 begins by arguing that, when psychologists make inferences from experimental data to specific conclusions, they have to be mindful of three sets of challenges. I refer to those challenges as the "individuation challenge," the "manipulation challenge," and the "measurement challenge." The individuation challenge pertains to the precise description under which an experimental manipulation can be said to be causally responsible for the data (such that the data can be taken as evidence for a specific effect). The manipulation and measurement challenges pertain to the question(s) of (*a*) whether a given experimental intervention is manipulating the entity of interest and (*b*) whether the resulting data can be treated as reflecting the effect of that intervention (such that the data can be taken as evidence for a claim about that entity). Section 7.3 puts the insights of section 7.2 into a slightly more schematic mold to explicate the concept of *experimental result* and its counterpart, the concept of *experimental artifact*. An experimental result is, on my construal, the conclusion of an argument that takes specific assumptions as auxiliary hypotheses. Specifically, it assumes that the three challenges outlined in section 7.2 have been met. Whenever one of these premises is false, the conclusion of the argument is an experimental artifact. This means that experimental data are reliable vis-à-vis a specific intended result only if the three challenges have been met, which is the same as saying that potential confounders have been controlled for. I suggest, however, that epistemic attention to confounders is warranted not only in order to control for artifacts but also in order to explore the shape of the object under investigation. A contextual factor can be a confounder relative to one goal (obtaining clean data in support of a specific hypothesis about an object of research) at the same time as it is a moderator relative to a different goal (gaining an understanding of the shape of an object of research).

Section 7.4 argues that operational analysis, as introduced in chapter 2, can aid in the simultaneous task of weeding out confounders and identifying moderators in the process of conceptual development. I advocate for the method of converging operations as a specific instance of operational analysis and provide an example from implicit memory research to illustrate my thesis that the method of converging operations can lead to conceptual fine-tuning. In section 7.5, I raise a problem that arises if (*a*) we think of psychological research as attempting to converge on the context-dependent shape of psychological objects at the same time as (*b*) we recognize that there can be no acontextual data, giving rise to an issue I dub the "new problem of nomic measurement."

7.2. Inferences in Psychological Experiments

Throughout this book, I have argued that psychologists make inferences from experimental data to descriptive claims about their objects, where operational definitions figure as vital (if fallible) premises. As we saw in chapter 2, operationists early on understood the obvious point that an operational definition alone would not guarantee the success of the intended inferences. In this section, I outline three challenges that researchers have to tackle when making such inferences.

7.2.1. The Individuation Challenge: Identifying Effects qua Kinds of Effects

Let us begin by looking at the (seemingly) simple task of establishing the existence of causal effects by conducting an experimental manipulation and observing the result. We have already encountered such effects in previous chapters, for example, the priming effect and the chunking effect. But we can also think of effects that are named after their presumed causes, such as the power-posing effect or the Mozart effect.[1] In this section, I identify three problems that arise in relation to the

---

1. These are the alleged effects of specific types of manipulations (having subjects engaging in empowering behavior and exposing subjects to Mozart's music, respectively) on specific types of outcomes (job performance and outcome on intelligence tests, respectively).

individuation of causal effects in psychology. They concern (*a*) the challenge of correctly identifying the experimental cause, (*b*) the challenge of correctly identifying the population for which the effect holds, and (*c*) the challenge of correctly identifying the sociocultural and material circumstances in which the effect occurs. I subsume all three under what I call the "individuation challenge." Ultimately, they all concern the question of what kind of effect we are looking at, that is, what kind of effect is instantiated by a specific experimental cause/effect pair.

I begin with the challenge of correctly identifying the experimental cause. At first sight, this may seem like a trivial task. Clearly, if we conduct an experimental manipulation and observe an effect (compared to a control group), it seems easy enough to infer that the manipulation caused the effect. But this cannot be the whole story. As I argued in chapter 5, experimental effects are typically generated to provide data. And, in an experimental context, the point of data is to provide evidence for claims about phenomena, such as the claim that there is a regularly occurring causal relation that is assumed to hold under specific circumstances. Viewed this way, it is clear that when researchers make inferences from experimental effects to experimental causes, they do so *under specific descriptions* of those causes. It is only with specific descriptions of experimental causes in mind that the experimental data can be taken as evidence for a *specific type of effect*, that is, one that fits the description of the experimental cause.

Importantly, the correct answer to the question of how to describe an experimental cause is underdetermined by the experimental data. Take, for example, the case of semantic priming, where researchers present subjects with specific words (so-called primes) and later find that word recognition for another set of words is faster if the words in this second phase of the experiment are semantically related to those of the first (see Meyer and Schvaneveldt 1971). However, should it turn out that the semantically related words in the second phase are also more frequently used in ordinary language than the semantically unrelated word in the control group are, then this raises questions about whether the experimental effect really does instantiate a semantic priming effect as opposed to a word-frequency effect. If it does not, we have misdescribed the causally relevant aspect of the stimulus, and, consequently, the experimental data cannot be regarded as good evidence for the effect as we have described it. My claim here is that a similar argument can be made if we have reason to suspect that the experimental manipulation has not been described at the right level of abstraction. For example, let us assume that the semantically related words used in an experiment about semantic priming were all taken from a food-related vocabulary ("bread," "butter," "knife," etc.). Let us further assume that we describe the experimental manipulation as one where semantically related words are presented, that is, that we treat the food words as instantiating semantically related words more generally. But, of course, we might be mistaken in that the experimental effect might not instantiate semantic priming generally but only semantic priming for food-related words. Both these case studies describe hypothetical ways in which the causally relevant feature of the stimulus can be misdescribed, leading to mistaken assumptions about the effect that is instantiated by the experiment.[2]

The challenge of describing an experimental cause in the appropriate manner is one aspect of what I call the "individuation challenge," which pertains to the problem of describing the cause of an experimental effect in a way that accurately captures the causal relationship that the experimental effect is presumed to instantiate (see also Feest [2016, 2019], in which I adopt the notion of an *individual problem* from Soler [2011]). Solving the individuation problem for any given effect amounts to establishing facts about what I call the "shape" of the effect. The flip side of establishing descriptive facts about the shape of an effect consists in determining the scope of our concept of the effect. But there is more to the shape of an effect, having to do not merely with the kind of stimulus that prompts it but also with the (kinds of) populations in which it occurs and the (kinds of) larger, spatiotemporal conditions under which it occurs.

I argue that, when researchers make inferences from data to effects, they do so not only by presupposing a specific description of the experimental cause but also by presupposing specific assumptions about the populations and spatiotemporal circumstances that the experiment instantiates. In other words, experimental inferences rely on assumptions about shape (of the object) and scope (of the corresponding concept). This is particularly apparent when we look at the notion of an *experimental sample*. Experimental subjects are chosen with the understanding that they are—in relevant ways—representative of a population of interest. This implies that assumptions about the shape of the effect are already contained in the choice of the experimental sample: assumptions

2. My analysis is inspired by Cartwright's work on the problems researchers encounter when attempting to extrapolate from one context to another (e.g., Cartwright 2012; Rol and Cartwright 2012). The problem, this work suggests, is frequently not that the causal claim cannot be generalized but that it was not correctly characterized in the first place. As Jimenez-Buedo and Russo (2021) point out, a similar problem is sometimes referred to as the *identification problem* (e.g., Manski 1999). Here, I am discussing a particular instance of this problem, i.e., the problem of correctly identifying the causally efficacious feature of a manipulation.

about the demographic in which we expect the effect to hold.[3] The same, I suggest, goes for what Cronbach (1982) called the "experimental setting." Cronbach was specifically talking about field experiments. Therefore, for him the concept of an *experimental setting* referred to the sociocultural context in which an experiment was performed (e.g., a village in Guatemala), but it can also refer to lab experiments. Obviously, the latter kinds of experiments are going to leave out features of the real world. This means that, if we want to make inferences about the real world (past and present) at all, we can do so only on the assumption that the experiment is in relevant respects representative of the part of the world about which we want to make inferences (i.e., that the experiment abstracts away only those features of the world that are unlikely to make a difference to the effect). Given that researchers typically do want to make inferences about the real world, I argue that they assume (at least implicitly) that their experimental setting is—in relevant ways—similar to the spatiotemporal slice of the real world about which they want to make inferences.

Now, it should be obvious that none of the above-mentioned assumptions about the individuation conditions for a given effect of interest are unproblematic. It should also be obvious that the issues at stake touch on questions that have been discussed under the rubric of *external validity* in philosophy of economics (e.g., Guala 2005, 2012; and Jimenez-Buedo 2011) and philosophy of neuroscience (Sullivan 2007, 2009). Francesco Guala, in particular, has importantly distinguished external validity (as pertaining to inferences from experiments to the world) from internal validity (as pertaining to inferences within an experiment). He argues that internal validity is established by controlling for confounders within an experiment. With regard to dealing with the problem of external validity, Guala distinguishes between two approaches, of which he favors the latter, that is, (*a*) being mindful of whether the experimental subjects and the experimental design are representative of the populations and environments about which one wants to learn and (*b*) extrapolating from an experiment to a different context by way of analogical reasoning (Guala 2012, sec. 4). Guala himself makes some interesting remarks about the value of the former approach. For example, he refers to representative sampling as "the logic underlying the best-known external validity control" (2003, 1202). In addition, he acknowledges discussions about the need for "representative designs"

---

3. This does not rule out suspending these assumptions and asking what population is really being sampled. But my claim is that it is often simply assumed that the sample is representative of a specific population of interest.

(2012, sec. 4.1; see also Hogarth 2006). However, he argues that—at least for the case of experimental research in economics—simple and idealized experiments are often preferable to experiments that try to sample the complexity of the real world. This leads him *not* to adopt a representative design approach to external validity, opting instead for a different response, that is, establishing internal validity first and then extrapolating on the basis of analogical reasoning.

The idea that researchers operate with simplifying assumptions is, of course, familiar from the literature about scientific models. I do not want to dismiss the value of working with simplifying experimental designs. However, I am not sure that the problem of making inferences from the results generated by means of such experiments is well captured by a notion of *external validity* as distinct from that of *internal validity* (for a critical discussion of the distinction, see also Jimenez-Buedo 2011). More pointedly, within basic research, there is a temptation to treat "internally valid" experimental results as representing some kind of context-neutral facts. By contrast, the analysis of the individuation problem offered above implies that inferences from experimental data typically contain assumptions about the effect's shape (i.e., about the contexts in which it occurs). Such assumptions, as a rule, transcend the experiment. In addition, as indicated above, researchers typically assume that their sample represents a relevant population outside the lab and that their experimental setting as a whole represents a class of environments of interest. If I am right about this, this makes it hard to grasp the notion of a *purely internal inference*. And this means that, instead of distinguishing between internal validity and external validity, we should (at least in the context of laboratory experiments) simply worry about validity simpliciter. Thus, even though I agree with Guala and others that the question of how to extrapolate to novel contexts is important (especially in field experiments), I suggest that we not use the expression "external validity" for this problem of correct extrapolation because it misleadingly suggests that there can be such a thing as purely internal (acontextual) inference.

Terminology aside, the deeper issue is that the question of whether specific environmental or demographic features make a difference to the occurrence of the effect is directly relevant to addressing the individuation challenge. As such, I argue that, in the context of laboratory research, it should not be left to an after-the-fact extrapolation but should be systematically reflected on right from the outset. This draws our attention to theoretical considerations about relevant environmental features as well as to methodological considerations concerning representative experimental designs.

Summing up, this section has laid out the individuation challenge as a crucial issue faced by any attempts to make experimental inferences about effects. Addressing it amounts to acknowledging that assumptions about the scope of an intended inference are already built into the experiment and engaging in conceptual and empirical work involving questions like (*a*) whether the experimental manipulation is representative of the cause about which we want to make an inference, (*b*) whether the experimental subjects are representative of the population about which we want to make an inference, and (*c*) whether the experimental design is representative of a real-world context about which we want to make an inference.

7.2.2. Latent Variables: The Manipulation Challenge and the Measurement Challenge

The individuation challenge concerns, as discussed above, the issue of how to ensure that one has adequately individuated an empirical effect of interest, where by "effect" I mean a causal relationship between a stimulus and a response. However, as I have argued at length in previous chapters, much of psychological research is not concerned with establishing simple causal effects for their own sake but rather uses experimental effects as data with the aim of testing descriptive statements about objects of research, which are composed of multiple phenomena (chap. 5). For example, priming effects are used as data to make inferences about a feature of implicit memory where implicit memory has been operationally defined in terms of the effect.

This creates the following additional challenge. In the case of describing stimulus-response effects, we are confronted only with the individuation challenge. In the case of objects like implicit memory, things are more complicated. We cannot simply assume that the various (and context-dependent) manifestation conditions of the priming effect correspond exactly to the variations of some feature of implicit memory (i.e., of the psychological kind presumed to be indicated by priming effects). Doing so would beg the question of whether what we are detecting under any given condition is really (an effect of) implicit memory. This is, of course, exactly the reason why, in its original formulation, operationism was rejected (Garner, Hake, and Eriksen 1956), giving rise to the question of how it is determined whether the data generated in an experiment in fact allow the intended inference about a given object of research (see chap. 2).

When we treat a given experimental effect as providing evidence for some descriptive claim about an object of research like implicit memory, we are, thereby, acknowledging that the effect is mediated by processes internal to the behaving organism. That is, the inferences we are drawing on the basis of our data are inferences about internal processes that are not immediately observable or straightforwardly discernible. In psychology, such variables are called "latent variables" or "hypothetical constructs" (though I prefer to draw a conceptual distinction between the latent variables and the constructs that describe them). While it is often assumed that psychological entities (beliefs, desires, memory systems) are unitary entities, I argued in the previous chapter that psychological kinds are better understood as clusters of phenomena (such as learning mechanisms, storage mechanisms, retrieval mechanisms, behavioral regularities, presence or absence of consciousness, etc.). Either way (i.e., whether we conceive of implicit memory as one unitary entity or an entity that is composed of multiple phenomena), the notion of *latent variables* draws attention to the idea that experimental manipulations in psychology often target specific phenomena associated with an object of research that are assumed to reside beneath the behavioral surface of the psychological kind in question. This acknowledgment of internal processes as mediating behavioral effects brings to the fore two additional challenges for experimental researchers, which, following Danks and Eberhardt (2009), I call the "manipulation challenge" and the "measurement challenge." These two challenges are (*a*) the challenge of ensuring that the experimental manipulation is, indeed, causally affecting the latent variable of interest (e.g., implicit memory or some phenomenon thought to belong to the cluster of phenomena relevant to it) and (*b*) the challenge of ensuring that the experimental data are, indeed, caused by the latent variable of interest. Both these challenges have to be met if the experiment is to succeed in measuring the effect of the manipulation of the variable. And it is only when they are met that an inference to a claim about a latent variable is warranted. With regard to the Mozart effect, for example, the question of whether the effect (insofar as it is real) is really due to the piece being by Mozart concerns an aspect of the individuation challenge, whereas the question of whether there is evidence that the music has an effect on intelligence concerns the manipulation and measurement challenges (see Feest 2016). (For another illustration of the difference between the individuation challenge and the manipulation and measurement challenges, see Feest [2022].)

The general gist of the two challenges can also be illustrated with the example of a drug trial. Let us say that I give a painkiller to a group of experimental subjects who suffer from migraines and observe a positive effect (the patients report reduced pain compared to migraine patients who did not receive the drug). One possible explanation is that the drug causally acted on the nervous system to reduce the pain reported by the subjects. If this is the case, then the experimental data could be considered good evidence for my claim about the drug reducing pain. But this conclusion hangs on two assumptions, namely, (*a*) that the observed improvement was really due to the effect of the drug on the nervous system (rather than, say, on the patients' expectation to get better) and (*b*) that the patients' reported improvement accurately reflects their experienced pain (rather than, say, a desire to please the doctor). If either of these two assumptions is not met, this amounts to a failure to meet the manipulation challenge and the measurement challenge, respectively. And, clearly, in both cases the experimental data cannot be regarded as good evidence for the causal claim about the drug acting on the brain and resulting in pain reduction. It is common to control for confounders that are due to expectations or social desirability by adopting blind (or double-blind) designs, where the patients (and the researchers) do not know whether they are getting the drug. This illustrates that, in order to control for confounders, it helps to have specific hypotheses about what they might be (i.e., in this case, hypotheses about the effects of subject and experimenter expectations). The example also illustrates that, while the individuation challenge has to contend with confounding variables external to the organism, the manipulation and measurement challenges have to—in addition—contend with confounding variables that are internal to the organism.

It is easy to see that all the challenges outlined above get right to the heart of the question of the adequacy (and limits) of operational definitions in any particular experiment. Operational definitions draw on conceptual presuppositions about the subject matter to specify which experimental data are to be treated as relevant to a given question about the subject matter. However, given the three challenges discussed above, it is by no means clear that the data generated on the basis of a specific operational definition and other background assumptions are in fact good evidence for any given hypothesis. We now have a better sense of why the challenges discussed in this chapter are so crucial for experimental research. Meeting them is a prerequisite for having high-quality data that can serve as evidence for the intended conclusion about a given object of research, where that conclusion can be a descriptive claim about the shape of either the object or the specific variable that mediates the behavior of the object in question. In either case, and in line with the contextual theory of evidence (chap. 5 above), such conclusions are warranted only if the individuation challenge and the manipulation and measurement challenges have been met.

But, given the epistemic blurriness of objects of research and my claim that the quality of data cannot be determined in a decontextualized manner, how can evidence-based conceptual development proceed? In what follows, I present a slightly more schematic account of the analysis offered above, casting it in the form of a theory of experimental inference that elucidates the complementary notions of *experimental results* and *experimental artifacts*. This allows me to raise the question of how evidence-based conceptual development might proceed by way of a process of testing for artifacts.

7.3. Experimental Inferences as Constrained by Operational Analysis

In the previous section, I laid out several challenges that experimenters in psychology face when they generate data with the aim of drawing specific inferences from them. In this section, I spell out the general account of experimental inference contained in the discussion above, which will (*a*) pinpoint the nature of auxiliaries presupposed in conclusions psychologists draw from their experimental data, (*b*) allow for a principled analysis of the notion of an *experimental artifact*, and (*c*) provide a first understanding of the task of operational analysis.

7.3.1. Experimental Results and the Reliability of Data

The basic claim of this section is that experimental results are conclusions of deductive inferences, which, in order to be sound, require their premises to be true. In philosophy of science, such premises are also known as "auxiliary hypotheses." What I am suggesting here follows, of course, the basic mold of a hypothetico-deductive schema. Given the analysis in the previous section, I add to the hypothetico-deductive-schema a specific account of the auxiliaries in play, namely, (*a*) that the individuation challenge is met, (*b*) that the manipulation challenge is met, and (*c*) that the measurement challenge is met (P = premise; C = conclusion):[4]

> P1: If I conduct causal intervention I on a sample of population P in experimental context C with the intention of manipulating latent variable L and get data D, then this implies result R about latent

---

4. This is an expanded and modified version of a similar schema I developed in Feest (2022). However, there I focused on the manipulation and measurement challenges, whereas here I am also highlighting the individuation challenge.

variable L (pertaining to population P and real-world context C) if and only if
- P1a: the individuation challenge is met, that is,
  - P1a^i: the causal intervention is adequately described as I,
  - P1a^{ii}: the sample is representative of population P, or
  - P1a^{iii}: the experimental setup is representative of spatiotemporal context C;
- P1b: the manipulation challenge is met, that is,
  - the manipulation causally affects latent variable L;
- P1c: the measurement challenge is met, that is,
  - the measurement of the outcome of the intervention reflects changes in latent variable L that result from the causal intervention I.

P2: Conditions P1a–P1c are met.

P3: Causal intervention I is conducted on a sample of population P in experimental context C with the intention of manipulating latent variable L; data D are observed.

C: Result R about latent variable L pertaining to population P in real-world context C.

An experimental inference in psychology can be said to be sound (and, thus, guarantee not only the validity of the argument but also the truth of the conclusion) if the three challenges identified in the previous section have been adequately dealt with. Even though it is common in the philosophical and scientific literature to speak of validity as a mark of a good design or a good measurement instrument, I prefer to restrict the notion of *validity* to the formal character of an inference while adding the notion of *soundness* to highlight that the quality of scientific inferences also places requirements on the truth of their premises.[5] It is only when this latter requirement is met that the data generated in the experiment can be said to allow an inference to the intended conclusion, that is, the result of the experiment. For example, let us assume that I administer a particular manipulation (e.g., that of presenting subjects with specific words along with the instruction to form sentences with those words) and subsequently administer a specific priming test and an explicit test. And let us further assume that the manipulation differentially affects the data generated by the priming test and the explicit test. I can interpret

5. That said, I use the term "validity" when citing other authors' work.

this as offering support for the descriptive hypothesis that elaboration has an impact on implicit memory provided that all the premises contained in P2 are true. In other words, the inference in question can be regarded as sound if the manipulation (instructing subjects to form sentences) is accurately construed as requiring cognitive elaboration (individuation challenge P1a^i), the elaboration affected (or did not affect) implicit/explicit memory (manipulation challenge P1b), and the test results were, in fact, caused by implicit/explicit memory (measurement challenge P1c). In addition, the truth of the auxiliary premises (P1a^{ii} and P1a^{iii}) has to be presupposed relative to the scope of the inference I want to make. Conversely, by making these assumptions, I am also assuming that there are no confounders present. In other words, experimental results are propositions that are deductively derived from P1, P2, and P3.

Obviously, the burden of proof hangs on P2. But, clearly, there is absolutely no guarantee that P2 is true. This is why, despite the inference schema's deductive nature, the inference from the data to the result is an inductive one and a highly uncertain one at that. In other words, as already indicated above, there is a sense in which we are looking at a version of the underdetermination problem (see also Uygun Tunç and Tunç 2023). Specifically, we are looking at what Mayo (1996) refers to as "methodological underdetermination." Inductive inferences can go wrong. Indeed, in the case at hand, they are quite likely to go wrong. My claim here is that they can go wrong in systematic ways that are tied to the conditions stated under P1a–P1c.[6] If (contrary to the assumptions encapsulated in P2) any of the three challenges are not met in the experiment, the resulting data do not support the intended conclusion because they can be assumed to be distorted. For example, if I believe that my experimental data allow for an inference about semantic priming, this inference is warranted only if the data were, in fact, due to the semantic content of my verbal stimuli (rather than, say, to the frequency with which the words are used in natural language). And, if I believe that my experimental data allow for an inference to semantic priming in a specific population (say, all adults), this inference is warranted only if my experimental sample was, in fact, representative of that population.

One thing that is revealed by the analysis offered above is that the soundness of an experimental inference is closely related to the quality of the experimental data. In turn, the quality of the experimental data hangs on whether the material conditions presupposed by the auxiliary

---

6. And this means that such potential errors can also be addressed in a systematic fashion. I argue below that this is precisely what the method of operational analysis allows us to do.

assumptions of the argument in fact hold. This idea is also captured by James Woodward's (2000) analysis of *data reliability*, according to which data can be regarded as reliable evidence for a claim about a phenomenon if there is a pattern of counterfactual dependence between the phenomenon and the data.[7] Jacqueline Sullivan (2009) has argued that, in complex domains like experimental neurobiology, there is an inherent tension between reliability and validity because, on the one hand, we are more likely to get reliable data about a phenomenon under tightly controlled experimental conditions while, on the other, we are interested in finding out about phenomena in the real world (i.e., under conditions that are not tightly controlled). According to Sullivan, these two requirements pull in different directions: "Reliability prescribes simplifying measures in the context of the laboratory.... Validity, on the other hand,... prescribes that an investigator builds into an experimental design those dimensions of complexity that accompany the phenomenon of interest in the world" (2009, 523). While Sullivan's analysis is specifically directed to experiments in neurobiology, it also speaks to some of the issues with which I am concerned here. Specifically, the problem she is raising can be recast in terms of the difficulties of meeting the three challenges mentioned above.

Taking a step back, I would like to suggest that the root problem is the high degree of complexity and epistemic uncertainty of the experimental situation. The uncertainty can (seemingly) be reduced by reducing the complexity of the design. But the question is how to ensure that it is reduced in a meaningful way, that is, in a way that still samples the spatiotemporal context of interest and, thus, can give rise to data that speak to that context. While Sullivan's dilemma suggests that, in the case of experimental neurobiology, this cannot be done, I would like to argue that, in the case of psychology, the following two ways forward are worth considering. First, if we think of the objects of psychological research as individuated on the level of the whole organism, it is conceivable that there are macro-level regularities that emerge even within the complexities of the real world, which can be studied experimentally. This possibility is suggested by ecological approaches to psychology, on which I briefly touched in section 6.4 above. Second, epistemic uncertainty can be reduced by attempting to conduct operational analyses of the original complex experiments. The concept of *operational analysis*, as I introduced the term in chapter 2, refers to the process of analyzing the

---

7. I take this notion of *data reliability* to be compatible with Sullivan's (2007, 2009) understanding of *reliability* as a property of the entire data-production process (Sullivan 2009, 534).

experimental context with the aim of eliminating possible confounders that might stand in the way of reliable data. The need for such analysis is acknowledged by Bergmann and Spence (discussed in chap. 2 above) when they write that in psychology, the "complexity of the situation and insufficiency of knowledge tend to preclude successful segregation of all the variables necessary for the complete functional description attempted" (1941, 5).

This raises the question of how operational analysis should proceed. In the next subsection, I discuss various proposals for error elimination in experimental research, relating these to my own analysis of what constitutes an experimental artifact. Then, in section 7.4, I argue that, while the elimination of experimental error is important, it is not sufficient for a genuinely forward-looking account of conceptual development in psychology. I argue that it should be supplemented with the method of converging operations.

7.3.2. Experimental Artifacts and Learning from Error

Given my schematic analysis of experimental inferences in psychology, it seems clear that data reliability should improve as a result of error elimination and that this process will involve two steps, that is, first identifying sources of possible confounders and, second, controlling for them. The importance of experimental control to experimental research has been remarked on, and analyzed, in the recent literature. For example, Jutta Schickore (2019) points out that there are different notions of *experimental control* and that the corresponding scientific practices have slightly different functions. The first is that of introducing a control group to test whether a specific experimental intervention is responsible for the observed effect. This notion of *control* is familiar from randomized controlled studies. By contrast, the second notion of *control* consists in the systematic variation of the conditions of an experiment to probe for confounders (see Schickore 2019). Stephan Guttinger also points to this second function/notion of *control* when he argues that controls can "help to obtain a meaningful (i.e., interpretable) output of the experiment" (2019, 463). They can do so either by way of targeted experiments that control for background factors that might be confounding the data ("negative control") or by way of testing whether a given experimental design can detect an effect that is already known to exist ("positive control").[8]

---

8. Note the similarity between this idea and Allan Franklin's (1986) epistemological strategies allowing for both the elimination of error and the calibration of instruments.

Now, in science the idea of needing to control for confounders is closely linked to that of avoiding and eliminating experimental artifacts. Let me, therefore, briefly spell out how my account of the relationship between experimental inferences and experimental error elucidates the notion of an *experimental artifact*. The analysis presented here simultaneously captures two strong intuitions, namely, (a) that experimental artifacts are human-made entities (data, phenomena) that systematically mislead us into accepting certain propositions as experimental results and (b) that the very propositions we falsely accept as experimental results should also be understood as artifacts, namely, artifacts of mistaken assumptions about how the data came about. These two aspects of the notion of an *artifact* are systematically connected. The very assumptions that systematically mislead us about the evidentiary status of particular human-made entities (artifacts in the first sense) also feed into the inferential machinery that allows us to arrive at certain (flawed) experimental results (artifacts in the second sense).

Let me explain this by drawing the reader's attention once again to the inference scheme presented in the previous section. One way of reading it is to say that each of the premises P_{1a}–P_{1c} implies the assumption that potential confounding variables have been controlled for. Artifacts occur when the process of data production does not align with the background assumptions required for the intended interpretation of the data produced. This amounts to saying that the data are not suited for the purpose for which they are generated. This occurs when the experimenter fails to meet the individuation challenge, the manipulation challenge, or the measurement challenge. Notice that the data not being suited for the epistemic purposes for which they were generated can occur for two sets of reasons, namely, (a) that they do not have the causal history required (this is particularly evident in failures to meet the manipulation challenge and/or the measurement challenge) or (b) that the triggering conditions, contexts, or populations to which we want to infer have not been identified correctly (this happens when researchers fail to recognize some aspect of the individuation challenge).

My basic framework for thinking about the notion of an *experimental artifact* as juxtaposed to that of an *experimental result* synthesizes and explains several insights from the existing literature about experimental artifacts. For starters, it disambiguates and integrates two notions of *human-made*, a physical one and a cognitive one. First, experimenters physically create data that turn out to be inadequate for the evidential purpose at hand. Second, and as a result, they derive false conclusions from the data. It is the first notion at which Weber (2018) hints when he says that artifacts are "unreliable data." By contrast, Guala suggests something along the lines of the second notion when he states that artifacts are "interpretations, and mistaken ones, of a certain set of data" (Guala 2000, 49). My analysis can account for both ideas by clarifying what makes data unreliable (not standing in the right counterfactual relationship with the relevant phenomenon) and what underwrites a given interpretation of the data (specific background assumptions that are required in order to use data for an intended inference toward a specific result). It does so by explicating that data are produced for specific epistemic purposes in ways that inform the intended interpretation of the data, and it distinguishes between several components of this: choice of stimulus, choice of manipulation, and choice of measurement instrument. My analysis dovetails with Giora Hon's idea that the logic of experimentation can be cast in the form of an argument that takes as premises a number of background assumptions that, if they are false, give rise to various forms of error (Hon 1989, 1998). It also connects with Craver and Dan-Cohen's (2024) contention that the issue of experimental artifacts ought to direct our attention away from evidential justification and make us instead focus more on why and how specific data are produced and selected in the first place, an issue Craver and Dan-Cohen dub the "evidential selection problem."

This latter point is important in that it reiterates a basic insight from much of philosophy of experimentation, that is, that evidence is not simply given but rather selected and produced. In turn, this means that mistakes can be made at the stage of data selection and production. It also means that, whenever researchers have reason to suspect they made a mistake in the data selection/production process, they can produce additional data and do so in a manner that controls for the suspected source of the mistake. As mentioned above, this is similar to Deborah Mayo's idea that data can be made more reliable by way of a systematic process of error probing (Mayo 1996). What I wish to add to Mayo's analysis is the suggestion that the point of controlling for potential confounders is not simply to eliminate errors that may have made the data of previous experiments unreliable (though it can clearly do that, too). Rather, the production of new data with the aim of error probing can also be in the service of testing a more refined hypothesis. This would imply that error probing does not merely rule out errors but can also contribute to a dynamic process of conceptual development. In this regard, I pick up on a point made by Schickore (2005), who distinguishes between a process of eliminating errors and a process of learning from errors and argues that Mayo's notion of *error probing* applies only to the former (eliminate error). It is precisely this idea that I want to pick up here and develop further in section 7.4 to show that, properly understood, error probing can also contribute to the latter (learn something new from error).

In a nutshell, the argument goes as follows. On the relational theory of evidence, experimental data are generated with the aim of testing a specific hypothesis. When researchers design a new experiment to rule out a specific source of artifact, there is, indeed, (as Mayo holds) a sense in which they are trying to eliminate a possible error that might have stood in the way of drawing a sound conclusion to their intended experimental result. In that regard, if the experiment is successful, it can be seen as strengthening the reliability of the previous data and ruling out that the previous result was an artifact. However, there is also a sense in which they test a different hypothesis, namely, the hypothesis that a specific feature of the previous experiment created an artifact. If we understand the notion of an *artifact* in the way that I have explicated, learning that a specific conclusion was an artifact of a confounder can also be interpreted as giving rise to a new possibility, namely, that some feature of the object under investigation might be modified by the factor in question. In such a case, I claim, we might say that we have gained an insight about the shape of the object, as defined above. The following section provides a more rigorous account of this analysis by linking it to the method of converging operations.

7.4. Converging Operations

The conclusion I want to draw from the discussion offered above is that what researchers do when engaging in the simultaneous tasks of conceptual development and data evaluation is a form of error analysis. I have suggested, however, that the process of conceptual development requires more than mere error elimination and that what from the perspective of one descriptive hypothesis about an object of research is viewed as an artifact might from a different perspective reveal something about the object's shape. This raises the question of how Bridgman's notion of *operational analysis* might be fleshed out in a more systematic fashion, allowing us to understand not only what it takes to eliminate error but also how psychologists can gain knowledge about their research objects by way of error.

7.4.1. Converging Operations as Differential Hypothesis Testing

The aim of simultaneously exploring data quality and elaborating and refining existing concepts is addressed (albeit not in those words) by the method of converging operations proposed by Garner, Hake, and

Eriksen in "Operationism and the Concept of Perception" (1956), which has already been introduced in section 2.5 above. Their argument is specifically geared at addressing what I have referred to as the "manipulation challenge" and the "measurement challenge": the question in which they are interested is how to ensure that an experiment designed to investigate questions about perception in fact yields data that are informative about perception. Garner, Hake, and Eriksen start out by taking issue with the crudely operationalist notion that psychological concepts like that of *perception* can be understood purely in terms of discriminatory behavior in response to stimuli. Instead, they point out that the experimental data one gets in a perception experiment might be mediated by some processes other than perception. Let me briefly repeat how they illustrate this possibility. Let us assume that we wanted to investigate whether the normative valence of a word has an impact on the speed of perceptual processing. More specifically, we want to know whether vulgar words like "fuck" and "shit" have an impact on perceptual processing. We might study this question by exposing subjects to a selection of such words (in contrast with neutral words) and record the speed with which they respond to them. If we were to find such an effect, we might call it a "vulgar-word effect." As Garner, Hake, and Eriksen point out, such an effect (even if it existed) might well be due to something other than perception, and it is entirely conceivable that the behavioral effects are due to an inhibitory effect of the stimuli on the response to the perceived stimuli (i.e., after the perceptual processes have already taken place) rather than an inhibitory effect on the perception itself. If all we have to go by is the manipulation and the effect, we cannot decide which explanation is the correct one. To put this in the more traditional language of philosophy of science, the correct explanation is underdetermined by the data.

But, of course, we do not have only the manipulation and the effect to go by since it is possible to design additional experiments, that is, additional manipulations that yield new data. And this is exactly the line taken by Garner, Hake, and Eriksen, who attack the problem by noting that we are essentially looking at two competing explanations for the data (the discriminatory behavior that resulted from the experimental intervention). They might be due to the perceptual system, or they might be due to the response system.[9] The way forward, then, is to devise an experiment that specifically targets and differentially tests these two competing explanations. In the literature, a similar idea has been

---

9. Notice, however, that these two systems do not exhaust the available explanations of the data.

described in terms of experiments potentially providing "contrastive confirmation" (Grünbaum 2021). On my analysis, however, the point of such experiments (at least in the epistemic contexts under debate here) is not so much confirmatory as it is exploratory. Such experiments contribute to the exploration of the subject matter by means of the method Garner, Hake, and Eriksen call "converging operations," or "any set of two or more experimental operations which allow the selection or elimination of alternative hypotheses or concepts which could explain an experimental result" (1956, 150–51). In the case at hand, this is implemented by suggesting four experimental conditions (see table 7.1): one in which subjects are presented with a vulgar word and asked to give a vulgar response (condition A), one in which they are presented with a vulgar word and asked to give a nonvulgar response (condition B), one in which they are presented with a nonvulgar word and asked to give a vulgar response (condition C), and one in which they are presented with a nonvulgar word and asked to give a nonvulgar response (condition D). As Garner, Hake, and Eriksen put it: "A converging operation... would be to present the same stimuli as before, but to pair these stimuli with responses, such that vulgar responses are used for nonvulgar stimuli and vice versa" (151).

The rationale is as follows. It is known from the previous experiment that the threshold is high for condition A and low for condition D. If it turns out that it is high for condition C, this will be interpreted as showing that the previous results were due to the response system. If it turns out that the threshold is high for condition B, this will be interpreted as showing that the previous results were due to the perceptual system. And, if it turns out that the threshold is high for both condition B and condition C, this will be interpreted as showing that the previous results were due to the response system and the perceptual system (in which case experimenters would need to formulate and test a more specific hypothesis about their interaction). Notice that the results do not

Table 7.1: Illustration of the converging operations approach

	(III) VULGAR RESPONSE	(IV) NONVULGAR RESPONSE
(i) Vulgar stimulus	A. Vulgar stimulus, vulgar response	B. Vulgar stimulus, nonvulgar response
(ii) Nonvulgar stimulus	C. Nonvulgar stimulus, vulgar response	D. Nonvulgar stimulus, nonvulgar response

establish that there is such a thing as a *response system*. There might, after all, be other, as-of-yet unconceived of, competing explanations of the data. Nor do the results give us a very specific causal mechanism. They do, however, provide a more nuanced understanding of the shape of the phenomenon (discriminatory behavior in response to vulgar words). At the same time, they also narrow down the conceptual space of the mechanisms that might be mediating this behavior (perceptual mechanisms vs. response mechanisms) and of the object of research of which the behavioral effect and the mediating mechanisms are part. In this way, converging operations simultaneously probe the reliability of the data that were generated for a previous, cruder hypothesis and contribute to conceptual, forward-looking work.

Returning once again to the inference scheme presented in section 7.3.1 above, we can say that Garner, Hake, and Eriksen (1956) are concerned with a version of the measurement challenge. Thus, they presuppose that the effect is due to the vulgarity of the words (individuation challenge), and they presuppose that the words causally affect the perceptual system (manipulation challenge). The question they are asking is whether the response data are a good measure of the effect on perception or whether they are confounded by the response system (measurement challenge). If the latter were the case, then clearly the resulting data would not support an inference to a claim about how perception is affected by word meaning. I argue that, while the method they present here does not give conclusive answers, its basic logic provides a valuable tool with which to attack not just the measurement challenge but the other challenges as well. For example, in addition to asking whether the effect is really due to the vulgarity of the experimental stimuli ($P1a^i$), one can also ask whether the sample is representative of the population of interest ($P1a^{ii}$, e.g., English-speaking adults in the United States) and whether the experimental setting is representative of the spatiotemporal scope of interest ($P1a^{iii}$, e.g., contemporary society in the United States). Thus, I argue that the method of converging operations has the potential to be a powerful tool of conceptual development, one that can address questions of shape in addition to questions about how the effects of experimental interventions are mediated.

7.4.2. Applying the Method of Converging Operations to Implicit Memory

One important aspect of the method of converging operations is that it requires competing explanations of the experimental data so that something resembling a *crucial experiment* can be conducted. I hasten to add that such an experiment is, of course, not a crucial experiment in the sense of establishing the true explanation once and for all since (as noted above) it is obviously possible that there are competing explanations. What I mean by "crucial experiment" is merely that such an experiment pitches competing explanations of the data against each other. It thereby allows, simultaneously, for the investigation of data reliability and the exploration of the conceptual space of the relevant object of research.[10] As spelled out by the method of converging operations, if a researcher generates behavioral data on the basis of a particular operational definition within the context of a particular experimental design, the epistemically productive way forward is to generate alternative explanations of the data. Thus, in the language of confounders, we should rule out that the response system (should there be such a thing) systematically distorts the data in such the way that they are confounded by the response system and thus do not constitute reliable evidence for a claim about perception. Notice that, in Garner, Hake, and Eriksen's (fictional) example, the suspected alternative explanation is located *within* the organism in that it invokes a hypothetical internal process that might be mimicking the effect attributed to perception. But it should be clear that the method of converging explanations can equally be applied to alternative explanations that pertain to factors *outside* the organism and have to do with the correct *individuation* of the effect and, thus, with the shape of the object to which the effect is linked. Such external factors might concern the correct description of the experimental stimulus, sample, or experimental context. In each of these cases, having a specific competing explanation of the data will enable researchers to conduct a converging operation experiment.

Let me now turn to an example that illustrates how the method of converging operations can play out in memory research, a case discussed in section 5.4.2 in which I showed that there are competing explanations of the data generated by implicit memory experiments. Proponents of the implicit memory system hypothesis attribute the data to the existence of a separate memory system, whereas proponents of the transfer appropriate processing (TAP) hypothesis hold that they are due to a correspondence between encoding and retrieval processes.[11] Thus, ad-

---

10. For a related take on crucial experiments, see Weber (2009).

11. As we saw in sec. 5.4.2, Bechtel (2008b) appeals to this case to argue that phenomenal decomposition is not always the best way to identify the explanation for a phenomenon, contrasting phenomenal decomposition with mechanistic decomposition. By contrast, I treat this example as a case in which there are two competing explanations for a particular set of data. In order to figure out which of these two accounts in fact provide evidence for the data, something like the converging operations method must be employed.

Table 7.2: Application of the converging operations approach to implicit memory

	EXPLICIT MEMORY TEST	IMPLICIT MEMORY TEST
Perceptual mode of processing	A. *Graphemic cued-recall test*	B. Perceptual identification Word fragment completion
Concept-driven mode of processing	C. Free recall, recognition	D. *General knowledge question*

vocates of this latter explanatory approach claim that the dissociations between the results of priming tests and the results of explicit memory tests are due to the fact that priming tests frequently focus on sensory items (and that performance on such tests is, therefore, enhanced by data-driven study conditions). In the terminology developed in the previous section, proponents of the TAP hypothesis essentially argue that what appears to be a general implicit memory effect (i.e., an effect mediated by the implicit memory system) is, in fact, due to the format of the experimental stimuli (namely, sensory stimuli) and the retrieval conditions. For my purposes, the interesting thing about this is that the existence of two specific competing explanations of the same data provides an opportunity for further experimental work to narrow down the question of which explanation is on the right track (and, thereby, to shape and develop the construct of implicit memory).

Such an experimental design was, in fact, devised (Blaxton 1989). The basic reasoning employed was that implicit memory had been operationally defined by way of tests that required perceptual processing (see table 7.2, cell B) and that explicit memory had been operationally defined by way of tests that required conceptual processes (see table 7.2, cell C). Blaxton supplemented these tests with ones that fit into cells A and D of table 7.2 (in condition A, subjects were presented with words that look and sound like words from the study phase and explicitly instructed to remember the words from the study phase). Now, if performance on implicit memory tests was, indeed, due to an implicit memory system, one would expect a dissociation between the columns. If the dissociation was due to processing requirements, one would expect to see a dissociation between the rows.

Overall, Blaxton took her results to confirm her hypothesis (i.e., "that dissociations among memory tasks are better explained in terms of the degree of overlap between mental operations at study and test than in terms of various memory systems underlying different tasks" [1989, 657]). However, two things need to be emphasized here. First, while it is true that since Blaxton's study appeared a lot of evidence of dissociations within types of memory tasks and between different modalities has accumulated, there is also evidence to the effect that, for cross-modal (or cross-format) priming, retrieval in one modality or format can be enhanced by a prime in a different modality or format (for references, see Graf and Ryan [1990]). Second, rather than confirming the TAP approach, any additional dissociation can also be taken to indicate the existence of additional systems. This has, indeed, sometimes been the tendency in cognitive neuropsychology.[12] All this is to say that, even though Blaxton's experiment is a beautiful application of the converging operations approach, it certainly cannot be regarded as a crucial experiment, one that decisively settles the matter. It is, however, a tool of conceptual development and data evaluation in that it targets specific auxiliary assumptions that are required to treat the data as allowing for a specific inference.

We can construe Blaxton's reasoning as being based on a suspicion that the dissociation between the data generated by implicit memory tests and that generated by explicit memory tests was the result of the stimulus materials used, both in the intervention phase and the measurement phase of the tests, rather than by distinct systems. On this logic, advocates of the systems hypothesis would have mistakenly thought that their data were indicative of how their experimental stimuli had differentially affected two distinct systems. They were, however, in fact an artifact of a failure to realize that the stimulus material and (in particular) the experimental instructions were not identical between the two testing conditions. From the perspective of the TAP approach, this allowed for an alternative explanation of the experimental data, one that held the experimental intervention to have differentially affected not two systems but two modes of processing. The latter interpretation can, in turn, be read as either revealing something about implicit memory or revealing that implicit memory, understood as a memory system, does not exist. By leaving this intentionally open, I take up an argument from the previous chapter, that is, that there are no mechanistic facts that settle whether there really is implicit memory (and what it really is). This is completely compatible with there being facts of the matter

---

12. See the discussion of the hypothesis that the existence of a domain-specific perceptual representation system accounts for dissociations between various perceptual tasks in Schacter (1994, 234). See also sec. 3.4.2.

about the mechanisms in play in any specific experimental effect. The crucial point here, however, is that, whichever mechanistic story ends up being told about whether implicit memory exists and what its underlying mechanisms are, converging operations experiments like Blaxton's are likely to play an important role in the conceptual dynamics leading up to this choice.

7.5. So What Do Converging Operations Converge On?

Given my claim that there is a certain degree of indeterminacy regarding the lessons learned from converging operations, one might ask what converging operations converge on? In this section, I approach this question from several angles. First (in sec. 7.5.1), I explore the relationship between converging operations and traditional notions of *robustness*. Second (in secs. 7.5.2 and 7.5.3), I address the question of how ideas about convergence are complicated by the claim that objects of psychological research are clusters of phenomena that are capable of realizing complex (behavioral, cognitive, and experiential) whole-organism capacities.[13]

7.5.1. Converging Operations and Robustness Analysis

Let me begin by briefly reminding readers of the discussion in chapter 2 in which I laid out that mid-twentieth-century American psychology saw a debate about how to validate psychometric tests, a debate that has to be placed in the context of the increasing importance of psychometric tests in American society, giving rise to the question of how to judge the adequacy of such tests. One answer was simply to judge tests by how well they predicted some outcome of interest (e.g., school or job success). However, this (criterion-oriented) way of assessing validity does not get at the heart of the question of whether a test measures what it purports to measure, a desideratum that became known by the name "construct validity." This issue is, of course, related to the question of whether (and under what description) the corresponding to-be-measured object or phenomenon exists. These questions (which were explicitly positioned against the notion that psychological concepts are simply to be operationally defined in

---

13. While this section discusses converging operations in relation to the literature on robustness, there are also parallels with the literature on the problem of coordination. For a recent account, see Ohnesorge (2022), which introduces the term "operational pluralism" to capture how researchers have tackled this problem in physical geodesy.

terms of specific tests) gave rise to the idea that, on the one hand, the reality of a mental entity could be established by measuring it in more than one way and, on the other, the validity of a psychometric test could be established by showing that its measurements of one entity were positively correlated with those of a different test purporting to measure the same entity while its measurements of different entities were not correlated, thereby ruling out method artifacts (Campbell and Fiske 1959).[14]

The issue of the construct validity of psychometric tests remains important, but for our purposes we can glean only limited insights from it since it is—in its traditional treatment—not really concerned with conceptual development (though note that Cronbach and Meehl [1955] did address this issue as well). Nor does it take into view the possibility that the entity under investigation might not be an atomistic unit but rather a cluster of phenomena (as I argued in chaps. 6 and 7). The same, I argue here, is true of traditional work on robustness and triangulation. As already indicated in chapter 2, one origin of the notion of *robustness* as discussed in philosophy of science can be traced back to the work of Donald Campbell, a mid-twentieth-century methodologist. In what follows, I give a quick rundown of some common themes in philosophical discussions of robustness leading up to recent work, which bears a surprising similarity to the notion of *converging operations* as explained in this chapter.[15]

Let me begin with Wimsatt's canonical formulation, according to which: "All the variants and uses of robustness have a common theme in the distinguishing of the real from the illusory; the reliable from the unreliable; the objective from the subjective; and, in general, that which is regarded as ontologically and epistemologically trustworthy and valuable from that which is unreliable, ungeneralizable, worthless, and fleeting" (1981/2012, 63). To establish such robustness, Wimsatt argued, one has to engage in robustness analysis, which requires demonstrating the invariance of results across several at least partially independent

---

14. There is a rich literature on test validity (e.g., Borsboom 2005; Borsboom, Mellenbergh, and van Heerden 2004) that I cannot cover here. I am also neglecting the wider literature about measurement in psychology and beyond (Alexandrova 2017; McClimans 2017; Tal 2020). I return to this briefly in my concluding remarks.

15. To my knowledge, the relevance of philosophical debates about robustness and triangulation to psychological research has been largely overlooked, with one noteworthy exception: Haig (2022). There are important overlaps and differences between Haig's analysis and my own that I cannot explore here.

procedures.[16] Writing at about the same time, and arguing along similar lines, Wesley Salmon (1984) formulated what has become a classic case study of robustness analysis—Jean Perrin's discovery of Avogadro's number—to illustrate his point. While not using the term *robustness*, Salmon cites the fact that Perrin used thirteen different experimental approaches to detect Avogadro's number, which provides evidence for the reality of "micro-entities [such] as atoms, molecules, and ions" (1984, 220). In turn, Sylvia Culp (1994, 1995) has drawn on Salmon's analysis to argue that robustness analysis is a form of triangulation that proceeds along the lines of Reichenbach's principle of the common cause. She attempts to back this up with a case study of the bacterial mesosomes that were mistakenly thought to be cellular organelles even though they were an artifact of the preparation procedures needed for microscopy.[17] Contrary to Rasmussen's (1993) analysis (which follows in the footsteps of Collins's [1985] experimenter's regress to argue that there were no decisive epistemic factors), she argues that the case was, ultimately, resolved by robustness reasoning in her sense. Robert Hudson (1999) pushed back, arguing that the case was decided not by robustness reasoning but by what he called "reliable process reasoning," that is, by way of a process of what in a different publication he called a calibration process (Hudson 2020). Marcel Weber (2004) has added to this argument by suggesting that the case was resolved simply by virtue of growing chemical knowledge that made it possible to understand which preparation technique was best suited for a particular investigative purpose.

On Hudson and Weber's analyses, some cases that initially seemed to be case studies of robustness analysis turned out not to be examples of robustness analysis after all. They were, rather, examples of trying to establish that specific experimental methods give rise to reliable data with respect to a specific subject matter. If this is a correct characterization, we might say that in scientific practice robustness analysis is not as prominent as philosophers have claimed. But one can also take this in a different direction and argue that perhaps robustness analysis just does not work the way it was previously assumed to. Versions of this latter

16. In philosophy of science, the notion of *robustness* has been taken up in two different contexts, namely, the epistemology of modeling, on the one hand, and the philosophy of experimentation, on the other (see Coko 2015). The focus here is on the latter and is constrained by my aim of drawing a connection between robustness and data reliability. But, for an illuminating discussion of the various meanings of the word "independent" invoked in standard accounts of robustness, see Coko (2020b).

17. According to the analysis in the previous section of artifacts produced, the experimental result indicating the existence of mesosomes was due to an unrecognized failure to meet the measurement challenge.

kind of argument are made by Alessandra Basso and Jonah Schupbach. Basso (2017), in particular, argues for a notion of *robustness* that can be used to establish measurement reliability. Schupbach argues that traditional notions of diverse evidence as probabilistic independence do not adequately capture the kind of diversity at work in robustness analysis, highlighting instead the "central notions explanation and elimination" (2018, 286). When we look at what he has in mind, the method he outlines looks surprisingly similar to the method of converging operations explained above (Garner, Hake, and Eriksen 1956) in that the aim is not to find statistically independent data that triangulate on the same object but rather to devise experiments that test competing explanations of the data from a previous experiment. Intriguingly, Schupbach also uses the famous example of Perrin's discovery of Avogadro's number, arguing that it was, indeed, a case of robustness analysis, but not for the reasons stated by Salmon and Culp. At the center of his argument is the question of why Perrin conducted exactly the number of experiments he did conduct. Why not stop after five methods, and why not continue past thirteen? As he puts it: "In any case, we can ask: What motivates us to seek and cite additional means of detection in an RA [robustness analysis]? What work are these additional means doing for us? The general account I am proposing responds that, at each increment of an RA, one cites an additional means of detection that has the power to discriminate between the target explanation of the result, H, and some competing potential explanation(s), Ho" (Schupbach 2018, 288).

What Schupbach is offering here is, of course, essentially the method of converging operations. Let me emphasize that, even though I am sympathetic to Schupbach's analysis, I am not committed to it being historically correct when it comes to the Perrin case (though it may well be). Nor am I committed to considering converging operations as a form of robustness reasoning. (For an excellent overview of the ways in which the Perrin case has been interpreted by philosophers, see Coko [2020a].) Let me also highlight that, even though there are similarities between Schupbach's analysis of the Perrin case and the converging operations approach as formulated by Garner, Hake, and Eriksen (1956), I want to push the analysis beyond a mere focus on data reliability and draw attention to the dynamics of conceptual development outlined above. Schupbach leaves intact the notion that the hypothesis under investigation and the central concepts that figure in it essentially remain unchanged. He also leaves intact the idea that the main point of robustness reasoning (as he constructs it) is to provide additional confirmation for the existence of an entity described by an already fully articulated concept. While this may be correct for the case of Avogadro's number,

I argue that it does not capture the research dynamics in psychology. There may well be cases in which the extension and descriptive content of the central concepts remain the same throughout the investigative process, especially when there are strong theoretical reasons for a given hypothesis. However, picking up my analysis from section 7.4 above, I argue that, in psychology, the elimination of alternative explanations for experimental data goes hand in hand with conceptual work concerning not only the identification of specific mechanisms but also the very shape of the object of research. I therefore agree with Thompson (2022) that, whereas traditional accounts of robustness reasoning have focused on its potential to give a confirmatory boost, it can have exploratory functions as well.

7.5.2. Converging on Clusters of Phenomena That Instantiate Whole-Organism Capacities

On the analysis just presented, converging operations do not follow the logic of robustness analysis as traditionally understood (i.e., as converging by independent means on the same object or result). Rather, the convergence in question is on a more and more nuanced description of the object, where this is achieved by successive and iterative steps of differential hypothesis testing. But there is more. I argue that, when we consider the ways in which the converging operations approach plays out in psychological research, we need to keep in mind my claims (a) that such objects are complex whole-organism capacities and (b) that questions about the shape of such objects cannot be neglected in the investigative efforts of psychologists. In this and the following subsection, I explain the implications of these two points for an adequate analysis of conceptual development in psychology.

As laid out above, one line of reasoning in the literature about robustness (going back to Wesley Salmon's work) treats it as a strategy of triangulation based on the common-cause principle. The basic idea is that, if we are trying to gain evidence for the existence of an unobservable entity, then it will help to have independent evidence for it, that is, evidence that does not presuppose the same theories. Versions of this line of reasoning are also found in Hacking's famous argument for entity realism, though it remains a contested question what the argument is precisely (Hacking 1983; see also Miller 2016). This line of reasoning may seem plausible if we are dealing with an unobservable or hidden entity such as the electron (Arabatzis 2011). However, I have argued in this book that this is not the right way to think about psychological kinds. Psychological kinds like memory, for example, are not "micro-entities [like] atoms, molecules, and ions" (Salmon 1984, 220). They are not atomic entities that reside inside the human or the animal skull but rather complex cognitive, behavioral, and sensory capacities that are individuated at the whole-organism level. This means that, even though researchers make inferences from their experimental data, this process is not adequately characterized as one of making inferences from independently generated data to some hidden entity. It is better characterized as one in which researchers make inferences about specific aspects of a given kind and the kind as a whole is already understood (albeit in an epistemically blurry fashion) to be composed of a cluster of phenomena, some of which (such as encoding and retrieval mechanisms or phenomenological features of recognizing something) are hidden and others (such as specific patterns of reported recall or recognition) instantiated in a directly observable fashion.[18]

Now, if objects of psychological research are clusters of phenomena, then there is, presumably, a sense in which one can triangulate on them. However, this is not by virtue of having independent access to one phenomenon but rather by virtue of accessing different phenomena that are associated with the same object and making the case that they are, indeed, integrated in a way that makes it epistemically and practically fruitful to refer to them as a "kind." Given my contention that objects of research are epistemically blurry, it is obvious that the very questions of what these phenomena are and how they jointly constitute the psychological kind are precisely what must be determined (Thompson 2022). This once again drives home the point that the issue to be attacked is really one of hypothesis-driven exploration and conceptual development. The key question is which phenomena jointly constitute any given psychological kind and how to determine the shape of that kind.

To avoid misunderstandings, let me emphasize that, when we refer to objects of psychological research as "clusters of phenomena," I mean what I call "core phenomena" as opposed to "supporting phenomena" (relative to a given object). The latter are phenomena that would not intuitively be classified as belonging conceptually to the kind (examples that come to mind are blood flow and respiration, which are necessary in

---

18. This insight, which was explained in the previous chapter, fits well with the network approach in recent psychometrics (e.g., Borsboom 2023), according to which data measuring psychological attributes (like memory) are not well understood as effects of a unitary common cause since psychological attributes are typically networks of interrelated phenomena. Following Sijtsma (2006), Borsboom (2023) refers to psychological kind concepts as "organizing principles." Large parts of this book can be read as analyzing how such organizing principles figure in the epistemic practices of psychologists.

the sense of being vital to the functioning of the organism more generally). I therefore pass over them here.[19] With regard to core phenomena, I pick up the idea—already developed in chapter 6—that there are three types of phenomena relevant to psychological kinds, namely, phenomena having to do with cognitive, behavioral, and experiential attributes (see also Feest 2023). For example, spatial memory is often described in terms of the capacity to build, store, and retrieve mental maps (all of which are cognitive phenomena), to exhibit search behavior (a behavioral phenomenon), and to have access to specific kinds of experiences such as the experience of a mental map or a feeling of familiarity (experiential phenomenon). This should make it clear that within each of the three core categories there can be multiple phenomena that can, simultaneously or consecutively, play a role in the instantiation of a psychological kind. It is, therefore, apt to refer to the corresponding concepts as *ballung* concepts (Basso 2017; Thompson 2022). This notion has recently been introduced to describe concepts that are characterized by a "fuzzy cluster of features" (Basso 2017, 63; Cartwright and Runhardt 2014).

The discussion offered above has important consequences when it comes to empirical efforts to converge on a given object of research. First, depending on the question pursued, any one of the core phenomena associated with a research object can be targeted. Second, once a specific core phenomenon has been singled out, it figures as a proxy for the object.[20] The choice of proxy clearly has an impact on how the concept gets operationally defined and, thus, on how the object is measured. Morgan Thompson (2023) has rightly pointed out that, especially if it becomes the standard way of measuring the object, the choice of a given phenomenon and/or operational definition can give rise to a problematic path dependence in the research that is subsequently pursued. The worry is that, if researchers settle on an operational definition for a given research object, this not only opens up certain avenues for conceptual progress (as I have argued in this book) but also closes off others that might be equally (or more) promising. I accept this as an important challenge to my thesis but cannot pursue it here. I would like to suggest, however, that my notion of *psychological kinds* as complex whole-organism capacities highlights the need for conceptual and

---

19. I realize that, while the distinction between core phenomena and supporting phenomena is intuitively plausible, it is not trivial. In particular, a central issue concerns the ways in which noncore phenomena can moderate core phenomena. Developing this more fully will have to wait for another moment.

20. I am borrowing the notion of a *proxy* from Philipp Eichenberger (2023).

methodological work that integrates the totality of phenomena pertaining to them.

Let me also emphasize that, even though I think of psychological kinds as complex whole-organism capacities (rather than individual phenomena), it does not follow that psychological research cannot target individual phenomena that are thought to be relevant to the kind. Quite on the contrary. The illustrations presented above of the converging operations approach show that, within the context of exploring specific objects of research (perception and implicit memory), it is possible to test competing hypotheses differentially in order to determine the exact nature of the phenomena that are thought to be descriptively tied to a given kind. In other words, individuating an object of psychological research at the whole-organism level does not take away from the possibility of testing hypotheses pertaining to its component phenomena. This is one of the places where the method of converging operations is useful and appropriate. It can aid the process of developing the concept, both in the sense of fine-tuning it (adding descriptive detail) and in the sense of contracting or shifting its extension (ruling certain phenomena irrelevant to the object).

7.5.3. Converging on Context-Sensitive Kinds: The New Nomic Problem

There is an additional complication, however, one having to do with the ways in which the occurrence of effects can potentially be modified by external conditions. Consider the possibility of an experimental effect being moderated by a change in the experimental setting. In such a case, we might say that the effect (and by extension any object to which it is tied) is context sensitive. However, earlier I defined the notion of the *shape of a psychological object* in terms of the specific ways in which that object's behavior is modified by variations in contextual parameters. Given this phrasing, it might seem that to say of an object that it is context sensitive is simply to say that it has a specific shape.[21]

I want to distinguish this somewhat trivial notion of *context sensitivity* from a more interesting observation, namely, that (some of) the phenomena associated with a given psychological kind are more stable than others. By this I mean that some phenomena are more likely to be affected by contextual differences than others. There is a suggestion in the literature, for example, that phenomena related to social cognition are more sensitive to contextual variation than are phenomena related

---

21. I thank Quill Kukla for drawing my attention to this issue.

to more basic cognitive functions (Van Bavel et al. 2016a, 2016b). (I return to this briefly in my concluding remarks.) If this is the case (and the point is purely hypothetical right now), this means that there is a good chance that the kinds of social psychology are going to be sensitive to the conditions under which they are studied. For example, let us assume that prejudice as a psychological kind is a cluster of cognitive, experiential, and cognitive capacities that require specific contexts for their actualization. Let us also assume that there are contextual variations that can affect people's performance on a standard measure of prejudice that asks people to rank their attitude toward a specific social group or product in such a way that people who score high under one condition (e.g., when the test is not administered by a member of that social group) score low on another (e.g., when the test is administered by a member of that group). This hypothetical case fits the description of a small contextual variation that might be indicative of a variable that moderates the effect and (by extension) tells us something about the shape of the object *prejudice*. We might, then, speculate that behavioral expression of this kind is moderated by specific contextual variables.

But how should we read this hypothetical possibility? Should we say that under the moderating condition the person in question is not prejudiced? Or should we say that under the moderating condition prejudice might be expressed differently and that a different test is needed? More generally speaking, the problem is that, if we reject the notion of *converging operation* as understood under the common-cause conception (as I have argued we should), and if, in addition, we understand psychological kinds as context sensitive, we are in danger of losing a stable referent that would allow us to draw a clear distinction between the following two cases. The first is the case in which failure to measure a given kind under specific contextual variations means it is not present. The second is the case in which failure to measure a given kind means that different contexts call for different measures.

A similar question has been raised by Wajnerman-Paz and Rojas-Líbano, who ask what might be criteria for individuating phenomena in cognitive neurobiology across different contexts. As they put it: "How do we know whether contextual variation changes the phenomenon elicited in different contexts or whether it only modulates how the same phenomenon is manifested?" (2022, 7). They clarify that by "phenomenon," they mean the "behavioral manifestation of a neurocognitive capacity" (8). They argue that, if contextual factors are changed and, as a result, the behavioral phenomenon changes, then this new behavioral phenomenon is of a different kind if and only if the components, activities, and/or organizational properties (CAOs) are the same. In other words, two behavioral regularities are modulations of the same kind if they share the same underlying neural/cognitive mechanisms.

While the issue cannot be treated in detail here, I have two preliminary responses to Wajnerman-Paz and Rojas-Líbano's suggestion. The first response is that this is a plausible idea, one that also carries over to psychology. However, we need to keep in mind my argument in chapter 6 that CAOs are in no way metaphysically tied to specific kinds. Therefore, there is inevitably an element of convention involved when researchers settle on a specific CAO as defining the kind. Consequently, there can be disagreements and shifts regarding the very question of what the relevant CAOs are and even whether they are always going to be decisive. This gives rise to my second response, which grows out of my relational account of psychological kinds, articulated in chapter 6. If we conceptualize psychological kinds as clusters of phenomena that are not tightly held together by one underlying mechanism (as the homeostatic property cluster theory of natural kinds would have it), then the decision when specific phenomena are expressions of the kind is going to be determined by context. This has potentially alarming consequences if we also hold that the shapes of the kinds themselves are moderated by context. The first of these two points implies that the very notion of a *context-transcendent measurement instrument* is problematic. But, if that is the case, we have no neutral instrument to which to appeal when trying to settle questions about the shape of psychological objects.

The issue at hand is similar to Chang's (2004) problem of nomic measurement, that is, the problem of establishing a functional connection between a hypothetical entity and an observable event when we have no independent access to the entity. This is why I am calling the problem outlined above the "new problem of nomic measurement." The problem, in short, is that, in order to investigate the ways in which psychological objects are moderated by contexts, one might want to rely on neutral or acontextual experiments. But there is no such thing. Every experiment instantiates a context even if (as I argued above) the question which context is instantiated is typically underdetermined by the data. It follows that it is not clear why the experimental context should be treated as a neutral standard that settles facts about a given (presumed) psychological kind. This has implications not only for the study of the shape of a given (presumed) kind but also for the study of the CAOs that are assumed to be distinctive of the kind.

I should emphasize that I do not wish to claim that the new nomic problem creates an insurmountable obstacle for psychological research going forward. I do, however, argue for the importance of recognizing the shaky epistemic grounds on which psychological science must proceed. Kellen and colleagues (2022), who also analyze the predicament of psychological research in terms of Chang's nomic problem, suggest that one way of addressing these issues is by using theoretical principles to determine the appropriate functions that coordinate experimental data with the things we want to investigate. While I think theory will certainly play a role, I hope that this book has made it clear that I see the root of the problem as being on a deeper and more conceptual level, as having to do with the question of how to capture what psychological theories are about, that is, how the objects of psychological research can be individuated and described empirically. This book offers a specific answer to this question, one according to which objects of psychological research are often individuated at the whole-organism level by means of a folk-psychological vocabulary. This focus on the whole organism brings to the fore questions about the shape and scope of psychological kinds and, thus, draws attention to the context sensitivity of the psychological subject matter and the context dependence of experimental data in psychology.

7.6. Conclusion

In this chapter, I have drawn together several sets of themes that have come up throughout the book. One such theme, in particular, was the notion that operationism requires not just operational definitions of objects of research but also what Bridgman called "operational analyses," that is, the systematic exploration of the experimental context. I have suggested that such explorations are conducted by way of the method of converging operations, which we first encountered in chapter 2. This method targets the background assumptions about objects of psychological research that are (implicitly or explicitly) built into experimental designs and required for sound inferences from the data.

I highlighted three sets of background assumptions, namely, (*a*) that the individuation challenge has been met, (*b*) that the manipulation challenge has been met, and (*c*) that the measurement challenge has been met. And I argued that critically examining these assumptions can be an important driver of conceptual development insofar as they can prompt researchers to design new experiments that address ambiguities in the way previous data were interpreted. In this way, I argued, converging operations do more than mere error probing, instead having the potential to contribute to conceptual refinement. While traditional accounts of operationism in experimental psychology focus on the measurement challenge (something that results from the fact that there is no guarantee that any given measurement procedure in fact measures the object of interest), I have emphasized that in the course of operational analysis the individuation challenge and the manipulation challenge (i.e., the challenge of ensuring that a given experimental manipulation in fact affects the intended target) need to be addressed, too.

Picking up the argument made in previous chapters that objects of psychological research are clusters of phenomena that are epistemically blurry as well as the claim that psychological kinds are whole-organism capacities, I highlighted the importance of individuating objects of psychological research at the level of the whole organism. This led me to emphasize that conceptual development should focus not only on the ways in which converging operations can fine-tune our understanding of features internal to the organism but also on questions of shape and scope. Addressing this from the epistemic perspective of the researcher requires addressing the individuation challenge and, in particular, applying the method of converging operations to the question of what environments and populations are sampled by a given experiment.

Concluding Remarks

1. Introduction

This book has provided an analysis of the methodological position of operationism in psychology, focusing in particular on the practice of formulating operational definitions of key concepts, that is, of concepts pertaining to objects of psychological research. With the expression *object of research*, I tried to capture the notion that psychologists direct their attention to specific kinds of entities with the aim of investigating and exploring them. In this brief final chapter, I summarize some of my main points, highlight some ways in which they connect with current debates, and gesture at future directions.

2. Main Points

The suggestion that operationism should be construed as a methodology for exploring objects of psychological research captures three central contentions of this book, namely, (a) that much psychological research is concerned with objects that are epistemically blurry (in the sense that they are not fully known or understood), (b) that the corresponding concepts are characterized by their semantic openness (i.e., they are not fully articulated), and (c) that operational definitions of the concepts in question are crucial tools for the (interrelated) tasks of experimentation and conceptual development.

While operationism has long been (mis)understood as an attempt to reduce the meanings of concepts to measurement operations, this book has argued that—quite on the contrary—operational definitions in psychology importantly draw on the rich preexisting meanings of folk-psychological concepts. For example, it is only by virtue of understanding the meaning of the term "hunger" that an operational definition as *time since last feeding* seems reasonable (see chap. 1). Likewise, it is only because we understand spatial memory to be connected to the capacity to remember previously learned spatial information that it makes sense operationally to define the expression "spatial memory" in terms of the speed with which rats can find their way in a maze that they have previously explored (see chap. 5).

Psychological experiments are specifically designed with the aim of generating behavioral data that speak to questions about a given object of research. I have argued in this book that the very choice of behavioral data as relevant to a specific object of research is already deeply laden with conceptual assumptions about that object, assumptions that are ingrained in experiments via operational definitions. This gave rise (in chap. 3) to my claim that operational definitions are tools of research. A different way of putting my point about the choice of experimental data is to say that behavioral data are deemed relevant to a given object of research because they instantiate observable aspects—or empirical contours—of those very objects. As I laid out in this book, this has two significant and interrelated implications: one concerning the empirical contours of objects of psychological research and one concerning the status and reliability of the data produced to test claims about the object of research.

With regard to the first implication, the point (first developed in chap. 4) is that behavioral indicators serve the function of providing epistemic access to objects of psychological research by virtue of instantiating observational features that are parts of the objects. Given my claim that objects of psychological research are epistemically blurry, it follows that their very contours are characterized by epistemic blurriness. This consideration informed my thesis that psychological research cannot be viewed as (merely) trying to uncover the explanatory mechanisms that are responsible for specific behaviors. Rather, and in addition, psychological research aims (or ought to aim) at the delineation and description of its objects of research. I have referred to this task as one of "exploration." Regarding the second implication, the point is that the epistemic uncertainty regarding the nature and empirical contours of psychological objects directly results in uncertainty regarding the quality of the experimental data that are used to test hypotheses about those objects. Consequently, chapter 7 argued that the experimental exploration of objects of psychological research is deeply intertwined with that of investigating the suitability of experimental data (and vice versa).

Chapters 5 and 6 tied these two points together by providing some additional analyses regarding (*a*) the notion of an *object of psychological research* and (*b*) the corresponding notion of a *psychological kind*. In Chapter 5, I argued that objects of psychological research are typically clusters of phenomena. I took phenomena to be empirical regularities that can either be found at the behavioral (surface) level or be more hidden from direct view (e.g., in the form of mechanisms). Chapter 6, in turn, put forth a modestly realist analysis of psychological kinds that allows us to articulate the idea that, even though psychological kinds are historically situated and relational, we can still capture the intuition that researchers can be right or wrong about their objects. I argued for a general conception of psychological kinds as complex cognitive, experiential, and behavioral capacities of organisms that are instantiated by multiple phenomena.

Chapter 7 returned to my analysis of the epistemic function of operationism to discuss further the interplay between conceptual and methodological issues in the experimental study of psychological objects. I combined an original account of experimental inferences with a specific analysis of what we should mean by the *shape* of psychological objects (and, correspondingly, by the *scope* of the concepts used to describe the objects). My analysis of experimental inference made clear (*a*) that the validity of such inferences hangs on the availability of high-quality data and (*b*) that questions about data quality are deeply entangled with conceptual questions about the objects about which we want to make inferences. I argued that conceptual assumptions built into an experimental design concern not just the choice of operational definition (i.e., how the effect of an intervention on an object is measured) but also the choice of the experimental subjects, the experimental context, and the stimuli. In each of these instances, the question is what broader features of the object these experimental components are taken to instantiate. I argued that, while experimental inferences rely on answers to this question, these answers require constant probing. This, I suggested, is the epistemic role of the method of converging operations.

3. Current Relevance and Future Directions

This book started out as an examination of experimental practices in psychology. It soon became clear that the history of these practices was deeply entangled with the history of debates in philosophy of science. Discussions about the epistemology of experimental practices in psychology touch on past and present debates about robustness, measurement, validity, mechanisms, natural kinds in general, and the nature and boundaries of psychological kinds in particular. I believe that my analyses offer some impulses for these broader philosophical discussions.

Equally importantly, the account developed here touches on a variety of lively recent discussions in psychology that were prompted by, among other things, the replication crisis in psychology. In this final section, I outline a few implications of my analysis and make some suggestions as to their relevance for current theoretical and methodological debates within psychology and philosophy of psychology.

3.1. Individuating Psychological Objects: An Ecological Perspective

While this book has emphasized the epistemic blurriness of objects of psychological research, one underlying theme has been that researchers inevitably carry ontological assumptions into their investigative designs. These assumptions are, I argued, ingrained in the language researchers use, and they make their way into operational definitions, which delineate the objects of research and play a crucial role in determining what kinds of data are deemed relevant. I have emphasized the importance of such definitions as research tools, but I have also acknowledged that their adequacy should not be taken for granted. This raises the question by what standards they should be judged.

To a large extent, my response in this book has been that operational definitions should be judged by their utility as tools, that is, by the productive role they play in an ongoing and iterative research process. But this kind of bottom-up perspective is not the whole story. This was highlighted in chapters 5–7, in which I argued that objects of psychological research are pitched at a particular level, namely, that of the cognizing, sensing, and behaving individual organism. While this is an empirical assertion, it raises normative questions, that is, whether the whole-organism individuation of psychological kinds is correct and whether there is a fact of the matter that might settle how we should individuate psychological objects. Chapter 6 offered a perspective on this question. I acknowledged that there are conventional components involved in the ways in which we individuate psychological kinds, but I argued that this does not stand in the way of their being real. Nor does it stand in the way of providing a normative justification for individuating psychological objects on the whole-organism level for scientific purposes.

I have provided three such justifications. First, individuating psychological objects at the whole-organism level makes psychology responsive to the kinds of questions that arise in real-life contexts where macro-level folk-psychological concepts are used. This is how it should be since people turn to scientific psychology in the hopes of receiving answers to questions that arise in real life. Second, a focus on the whole organism makes us aware of the fact that such (sensing, cognizing, and behaving) units are always embedded in environments. Third, such a focus is underwritten by research suggesting that humans take environmental features into account when classifying observable organismic phenomena as belonging to a specific psychological kind. All three of these reasons—but specifically the last two—encourage an ecological perspective in psychology.

Ecological approaches to psychology are not new, of course, having roots that go back to Darwinian psychology, pragmatist approaches (in particular William James and John Dewey), and early behaviorism. They are, perhaps, most prominently associated with the work of James Gibson (1904–79). Of the historical figures whose work I have analyzed in this book, Edward Tolman's approach bears clear marks of an ecological orientation, not least by virtue of his close association with Egon Brunswik. The ecological approach did not gain widespread traction in mid- to late twentieth-century psychology, though some more recent philosophical work has pushed for its revival (e.g., Chemero 2009). I have shown in this book that, despite its relatively marginal status in mainstream academic psychology, the Brunswikian version of ecological psychology with its emphasis on the whole organism left its mark on a pervasive methodological practice, namely, that of operationally defining a given concept in terms of behavioral data generated by canonical experiments. Looking ahead, I argue that these ecological roots should be explored and developed both for their theoretical insights and for their methodological insights.

3.2. Representative Design and the Importance of Context

With respect to the methodological implications of an ecologically oriented psychology, one important aspect to explore more deeply concerns that of experimental design. One good starting point is the ideas of Egon Brunswik as laid out in publications such as his last (posthumously published) book, *Perception and the Representative Design of Psychological Experiments* (1947/1956).

Brunswik is sometimes credited with having coined the concept of *ecological validity*, which many authors take to be synonymous with *external validity*, that is, to be referring to questions about the extent to which inferences from experimental results to the real world are warranted. However, as Araújo, Davids, and Passos (2007) argue convincingly, for Brunswik the expression referred to a different issue, namely, that of whether a perceptual variable is indicative of an environmental feature. That being said, Brunswik was also keenly interested in the question of how to ensure that psychological results were informative of psychological phenomena outside the experimental context. This prompted him to focus on the issue of experimental design, highlighting in particular the question of what it takes for an experimental design to be representative of the circumstances and phenomena of interest. Consequently, he argued that questions about generalizability should be asked before, not after, an experiment is conducted (Araújo, Davids, and Passos 2007, 72).

Another way of putting this is to say that the experimental space cannot be treated as acontextual. This idea dovetails with the contextual theory of evidence, which I introduced in chapter 5 and elaborated further in chapter 7, where I argued that experimental inferences rely for their soundness (relative to a given intended scope) on specific contextual assumptions. We can read Brunswik as saying that those contextual assumptions, which often remain implicit, need to be reflected on and explicitly considered in the design of any given experiment. What I added to this insight in chapter 7 was the thesis that questions about the scope of a given experimental inference are directly mirrored by questions about the shape of the object researchers are making an inference about. More broadly, I suggested that the issue of an object's shape concerns not just the question of what context is instantiated in any given experiment but also how the object would behave in other contexts. However, given the epistemic blurriness surrounding objects of psychological research, this question gives rise to a predicament that I called the "new problem of nomic measurement." In order to establish differences in the ways in which an object behaves under different circumstances, one needs to have already settled on an acontextual way of measuring it. But the contextual theory of evidence requires evidence to be adequate to the specific shape the object might have, raising the question of how to evaluate the relevant data prior to knowing the shape of the object.

What might at first glance look like a stalemate ceases to be one if we keep in mind that no single experiment, no matter how well designed, can answer questions about the shape of an object conclusively. This is where the method of converging operations comes in. This method is not simply about determining the generalizability of a given experimental effect or the reliability of data. It is rather about exploring the shape of a given object of research, thereby playing a crucial role in the process of conceptual articulation and development, where the latter must clearly be regarded as an iterative process (Bringmann, Elmer, and Eronen 2022). Converging operations are, I have argued in this book, an integral part of this iterative process insofar as they involve multiple well-designed (yet necessarily conceptually open) experiments that build on each other. I therefore suggest that research needs to focus on issues such as what are worthwhile objects, what are theoretically informed, fruitful ways of conceptualizing them, and what are alternative explanations of the resulting empirical data.

This gets me to my second response to the new problem of nomic measurement, namely, that the explorations that are conducted with the help of the method of converging operations need not take place in a theory-free space. With regard to experimental work in ecological psychology, this gives rise to three questions that deserve further attention, that is, (*a*) what kind of theorizing to engage in, (*b*) how best to implement such theorizing in experimental designs, and (*c*) how to use experimental results to inform a cycle of renewed theorizing. With regard to the theoretical underpinnings of experimental designs, for example, ecological psychologists will likely attempt to use stimuli that are informed by theoretical preconceptions about environmental complexity as it presents itself to the organism. They might also call for designs that enable experimental subjects to move around (Araújo, Davids, and Passos 2007). Future philosophical and methodological work might explore midcentury ecological approaches (e.g., Barker 1969; Gibson 1979) in relation to philosophical analyses of embedded and embodied cognition (for a foray in this direction, see Isaac and Ward [2021]). Picking up on Barker (1969) in particular, such work might also look into the relationship between experimental research and fieldwork.

3.3. Responses to the Credibility Crisis in Psychology

The discussion in the previous section opens up several avenues to topics that have been discussed in experimental psychology and the relatively new field of metascience. The debates were initially fueled by two facts: (*a*) that questionable research practices are widespread and (*b*) that many formerly well-established experimental results fail to replicate. Many researchers therefore found it to plausible to infer that at least some of the replication failures (perhaps even a large part) were in fact due to shoddy research practices in the original studies. The debate has advanced significantly in the last few years, gaining in nuance and depth. In this section, I explain some of the issues at stake as they present themselves from the vantage point of the analysis developed in this book.

3.3.1. *Replication vs. Converging Operations*

There seems to be a consensus among many psychologists (and philosophers of psychology) that psychology faces a crisis. However, there are widely divergent views about the nature, severity, and root cause of the crisis. Some debates have focused on explicating the very notion of *replication* and tied their explications to specific accounts of the epistemic function of replication studies. In this context, one target of debate has been the distinction between direct replication and conceptual replication, where the former tries to redo a previous experiment as faithfully as possible and the latter tries to test the same hypothesis with different methods (Schmidt 2009). As Rubin clarifies, since there can be no such thing as an exact replication, direct replications should be understood as recreating only "the theoretically essential" elements of a study (Rubin 2021, 5814).

Advocates of direct replication argue that it is needed as a way to make psychological claims falsifiable (e.g., LeBel et al. 2017). In turn, conceptual replication is sometimes portrayed as providing stronger evidence for a hypothesis by virtue of triangulating on the subject matter in an independent way (Stroebe and Strack 2014). Now, on the face of it, both of these functions (generating high-quality data and generating independent data for the same hypothesis) seem important and worthwhile. However, the way the debate has played out, they are often pitched against each other. Critics of direct replication point out that the mere repeatability of an experimental effect does not say anything about its meaning or relevance for a given hypothesis (e.g., Devezer and Buzbas 2022). And critics of conceptual replication argue that a failed conceptual replication is equally uninformative (e.g., Nosek and Errington 2020).

Sidestepping questions about the meaning and value of conceptual replication, my account of psychological experiments makes it clear that advocates of conceptual replication are right to point out that experimental effects are generated not for their own sake but rather as a means to obtain data that test specific hypotheses. This draws attention to the importance of distinguishing clearly between the replication of experiments, the replication of experimental effects, and the replication of experimental results. (As I argued in chap. 7, experimental results are the claims that are inferred from the experimental effects.) These distinctions clarify that advocates of conceptual replication are more focused on replicating experimental results and that advocates of direct replication are more focused on replicating experimental effects (i.e., data). Consequently, advocates of direct replication might say that replicable effects are a mark of high-quality data and, thus, a necessary condition for any experimental inference one might wish to draw.[1] In addition,

---

1. For a version of this argument (though he argues this of replication in general, rejecting the distinction between direct and conceptual replication), see Machery (2021).

they might say that close replications of experiments are a way to ensure that this condition is met.

On the basis of the accounts offered in chapters 7, I do not believe that questions about the quality of data (sometimes referred to as "data reliability") can, ultimately, be separated from questions about the validity and soundness of the inferences we want to draw from them. According to the contextual theory of evidence, to treat data as good is to claim that they are good for the purpose they were generated for. Hence, questions of data quality cannot be separated from conceptual issues. This concerns not only conceptual replication since (as remarked above) even direct replications rely on judgments about theoretically relevant features of the experiments (see also Feest 2019). To put it differently, all experiments that want to draw any conclusion (even if it is just to the robustness of a given effect) are in some sense conceptual. Conceptual judgments concern the subject matter under investigation and the auxiliaries required for their experimental investigation. This is why failed replications can stimulate conceptual work (Devezer and Penders 2023).

My analysis of the method of converging operations elucidates how forward-looking conceptual development can proceed despite, or rather because of, the undeniable underdetermination characteristic of the experimental situation. Importantly, this method does not itself appeal to the notion of *replication*. It differs in this way from, for example, Nosek and Errington's analysis, which holds that replications play the diagnostic function of allowing researchers to determine "generalizability across the conditions that inevitably differ from the original study" (2020, 4). This latter suggestion is intriguing because it acknowledges as important the descriptive question of how effects might be similar or different across experimental contexts. My analysis embeds this idea in a systematic and general account of experimental inferences pertaining to the shapes and other descriptive features of the objects of psychological research. Importantly, within the converging operations approach, the follow-up experiments are not replications because they are not directed at the same questions as the original experiment was. This is compatible with the idea that they can contribute to the development of a concept about a specific object of research. My account of converging operations thus provides us with a rationale of how experiments might build up on each other in the dynamic and iterative process of exploring a given object of research.[2]

---

2. Rubin (2021) integrates a similar intuition with a proposal about the statistical underpinnings of such exploratory hypothesis testing. Devezer and Buzbas (2023) investigate the role that formal models can play in such an iterative process.

3.3.2. The Replication Crisis and the Call for Theory

My remarks in the previous section resonate with another current in the crisis literature, namely, one that has emphasized the need to think systematically about the nature of theory in psychology as well as about the question of how theories should be turned into testable hypotheses.

An important contribution to this debate was Muthukrishna and Henrich (2017), which argued that, even though many had responded to the replication crisis with methodological reforms, it was equally important to think more about theoretical frameworks. Pointing out that "psychology textbooks are largely a potpourri of disconnected empirical findings," Muthukrishna and Henrich argues: "With decades of data now under suspicion, without a general and unifying theory of human behaviour, we have no principled way to navigate this morass" (2017, 1). I am sympathetic to the spirit of this pronouncement, but I would like to qualify the formulation by pointing out that an integrated psychological theory need not (and probably should not) be restricted to behavior and certainly not to human behavior. On my account of psychological objects/kinds, they are experiential, cognitive, and behavioral capacities that are exhibited by whole organisms (see chap. 6). This means that the capacity to behave in certain ways is only one of three aspects that can make up a psychological object of research. In other words, I propose a slightly richer understanding of the targets of psychological theorizing. Going hand in hand with this proposal I also suggest that our existing concepts of those targets should serve as the starting point of theory construction since they provide us with an (albeit epistemically blurry) notion of what the theories are supposed to be about.

With this out of the way, Muthukrishna and Henrich (2017) do offer an instructive proposal concerning the relationship between theories/theoretical frameworks, on the one hand, and what I call "objects of research," on the other. Their main point is that an approach that focuses on exploratory research at the level of specific objects (e.g., social influence) requires a rationale that constrains the choice of worthwhile hypotheses to pursue. Such constraints, they argue, can be provided by theoretical frameworks, and they define the concept of *theoretical frameworks* as "broad bodies of connected theories" (2017, 2). Agreeing with this, I would like to suggest here that the ecological approaches to psychology that I have highlighted above provide such a framework. Such approaches not only provide a rationale for my claim that psychological kinds should be individuated at the level of whole-organism capacities but also explain why it is reasonable to individuate them in the ways in which they present themselves to organisms (something reflected in our folk psychology).

Now, apart from a broad theoretical framework that governs the general outlook, additional questions concern (*a*) the kind of theorizing (e.g., mathematical vs. mechanistic) that is appropriate for objects of psychological research and (*b*) how best to construct theories/models of psychological research. Participants in these discussions have proposed methods to make concept formation and theory construction more rigorous (e.g., Borsboom et al. 2021; Devezer et al. 2021; and Fried 2020). Here, too, however, the underdetermination worry remains. As long as theoretical formulations are imprecise and vague, the inference chain between theoretical hypotheses and experimental data remains too unspecific to allow for unambiguous interpretations of the data (Oude Maatman 2021; Scheel et al. 2021). The account I have developed in this book touches on many of these issues. So let me briefly seize the opportunity to draw connections, point out parallels, and disagreement, and flag issues for further debate.

First, as I argued in chapter 5, psychological theories deal with capacities (Cummins 1983), and this calls—at least in part—for mechanistic theorizing (see also van Rooij and Baggio 2020), that is, theories that spell out the mechanistic underpinnings of capacities to instantiate cognitive, experiential, and behavioral phenomena. However, I have argued that theoretical work cannot be reduced to the task of explaining previously identified explananda. Rather, the (descriptive and taxonomic) process of exploring objects of psychological research should be regarded as theoretical as well, albeit, perhaps, in a bottom-up fashion. I am, therefore, sympathetic to recent interest in exploratory experimentation in psychology, though I am skeptical of the strict distinction between exploratory experimentation and confirmatory experimentation (see also Devezer and Buzbas 2023; and Rubin and Donkin 2022). As I laid out in chapter 4, I prefer the concept *exploratory research* to the more narrow *exploratory experimentation*. This is because I am less invested in the idea of a specific kind of experimentation and more interested in the general focus on exploring specific objects of research. Converging operations are, as I construe them, hypothesis guided and exploratory at the same time.

3.3.3. Description and Generalization

The previous section has argued that, even though theories can serve important orienting, explanatory, and integrative functions for research, it is important also to keep in mind the question of what the targets of psychology's theories are. While scholars in the history of psychology have at various moments preferred different answers to this question (ranging from consciousness to behavior and cognition), I have argued in this book that many objects of psychological research are best understood as bundles of capacities that are displayed by organisms in their entirety. It is precisely those organisms in their entirety that are typically subjected to psychological experiments. It is also those organisms in their entirety as they feel, think, and behave in the real world that psychological research ought to elucidate (see also Gozli 2019).

But psychological research happens under local conditions, using specific concepts and specific operational definitions, highlighting specific aspects of a given object of research, and making specific assumptions about relevant background conditions. This brings me to the issue of how to think about generality, that is, the question of what inferences we can draw from experiments conducted with individual organisms in specific lab settings. Tal Yarkoni (2022) recently raised a similar question, arguing that replication failures are symptomatic of a much more severe crisis, that is, a crisis of generalizability. This crisis is, he argues, prompted in part by the vagueness and informality of psychological concepts compared to the rigor of statistical inferences, where the latter provide psychological results with an illusory and unwarranted appearance of precision. I would add to this emphasis on the vagueness and informality of concepts the blurriness of the corresponding objects. Yarkoni himself concludes with a variety of recommendations, one of which I want to highlight here, namely, to "take descriptive research more seriously" (Yarkoni 2022). By this he means that researchers should engage in the project of describing the variability of experimental effects before jumping to conclusions about real-world effects and causal mechanisms.

This diagnosis resonates with several points I have made in this book, though my analysis gives it a different twist insofar as I would argue (*a*) that thinking carefully about descriptions requires us to think about the objects of descriptions and (*b*) thinking about objects commits us to categories that inevitably carry a certain degree of generality. This latter point leads me to argue that worries about the generality of psychological inferences should not be understood as suggesting that researchers typically start out with nongeneral facts about objects. In other words, as long as we describe only isolated effects, we are not describing capacities of macro-level psychological objects at all. For example, consider the case in which the effect of multiple different experimental treatments on performance on a memory test are investigated. It is, of course, possible to use such tests and avoid any commitment to what they are testing. However, I argue that it is much more natural (and common) to treat such effects as *memory* effects, and thus as pertaining to specific object of research.

My claim is, thus, that, instead of aiming for some kind of sanitized internal validity of experiments, we have to endorse the fact that, when describing the result of an experiment, researchers already make use of general concepts that (naturally) transcend the specifics of the experiment. Those concepts, I argued in chapter 3, underlie operational definitions that function as research tools. I have argued that such tools are used to delineate and describe the objects in question (chaps. 4 and 5). In chapter 7, I spelled out more specifically how such descriptions will aim, at least in part, to say something about the shape of an object. Questions about the shape of objects involve questions about whether and how objects behave in specific contexts. In this regard, my focus on the whole organism implies a broader understanding of what the description of psychological objects amounts to. Moreover, I have argued that experimental designs necessarily transport descriptive assumptions about the object's shape.

Summing up, I agree with Yarkoni (2022) that we ought to take descriptive work seriously, and I have a specific reading of what this means, that is, the targeted probing of general assumptions about the shapes of specific objects. Such targeted probing can be described as a form of exploratory experimentation, Steinle-style (chap. 4), understood to be the systematic varying of experimental parameters. However, it should be clear now that this kind of experimentation stands in the service of a larger project, that is, the project of exploring objects of psychological research. What is required here is the systematic variation of experimental parameters with the aim of figuring out which real-life features of the world are sampled by any given aspect of an experiment. This raises the question of how we should individuate aspects of experiments. I turn to this question next.

3.3.4. *Replicability, Mismatch Explanations, and Context Sensitivity*

Given my claim that questions about the shapes of objects are the flip side of questions about real-life contexts in which features of the objects instantiate capacities, we need to get the notion of *context* into clearer view. This is relevant also because, in recent debates about replication and replicability, it has occasionally been suggested that replication failures had resulted from the context sensitivity of the phenomenon under investigation and that replication failures might, thereby, reveal limiting conditions for a specific effect (e.g., Strack 2017, 702). The basic idea is that, when we deal with context-sensitive phenomena, small differences in the replication experiment might affect the subject matter differentially. In such a case, the nonreplication of the effect would be due to the fact that the two experiments were, in fact, different, that is, that the second experiment is not really a replication of the first. Colaço, Bickle, and Walters (2022) refer to this kind of reasoning as a "mismatch-explanation."[3]

Mismatch explanations are intriguing because they can be appealed to by both sides of any given experimental disagreement. On the one hand, as just noted, if a replication of a previous experiment does not yield the same effects, it can be argued that the second experiment did not, in fact, replicate the first (and, hence, cannot refute the results of the first). On the other hand, it could be replied that such a high degree of context sensitivity must mean that the phenomenon is not very robust, that is, perhaps not of significant scientific interest. In this section, I want to steer a middle ground, arguing that context-sensitive effects are of scientific interest precisely because it is ultimately an empirical question just how context sensitive a given object of psychological research is. Investigating this empirical question requires experimental designs that allow researchers to vary the contexts. This raises the question what kinds of experimental contexts can be varied and (with an eye on the generality question addressed in the previous subsection) what experiment-transcendent contexts can be inferred to.

In chapter 7, I introduced Cronbach's (1982) contention that experiments can be subdivided into several components: the treatment, the units (research subjects), the measurement operations, and the settings. It is typically assumed that each of the specific components in a given experiment are representative of some context of interest. In any given experiment, only one context gets varied (the so-called treatment), while the others are held constant. Mismatch explanations, then, claim either that the treatments of the two experiments were not the same or that one of the other three experimental components did not match. In their analysis of mismatch explanations, Colaço, Bickle, and Walters (2022) draw on Machery's (2021) resampling account of replication, which relies on a very similar understanding of experimental components as Cronbach's. Machery argues that an experiment is a replication of a previous one if all its components sample the same populations (or, as I am putting it, instantiate or are representative of the same contexts). Hence, a mismatch between two experiments occurs when the components of the two experiments do not sample from the same populations.

---

3. Notice that Collins's (1985) argument about the experimenters' regress essentially claims that it is impossible to resolve whether any given mismatch explanation is true. I have a slightly less pessimistic reading (see Feest 2016).

An interesting consequence of Machery's (2021) analysis is that an experiment can look quite different from a previous one and still be a replication (e.g., if it uses two different measurement procedures for the same object). Conversely, an experiment might look quite similar to a previous one and still not be a replication (e.g., the second one may have introduced a minor variation that ends up making a huge difference). While I find this to be a convincing account of the notion of *replication*, it serves only to highlight even more the difficulty, in practical terms, of figuring out when a mismatch explanation is adequate. My analysis suggests that, in order to explore this question, one needs specific hypotheses about where a mismatch might have occurred. And this, again, amounts to using the method of converging operations, which aims at the exploration of an object of research.

So, when an experimental effect is not replicated, this gives rise to the question of whether the two experiments differed in some way and whether this difference might be pointing to a contextual variable that makes a difference to the effect in question. In psychology, such variables are referred to as "moderator variables" (e.g., Fritz and Artur 2017). The suggestion I am making here is that converging operations can be used to test whether specific variables moderate experimental effects, pointing to specific ways in which an object of research might be context sensitive. A version of this question was addressed by Van Bavel and colleagues, though they approached this from the opposite angle. Starting out with the hunch that some objects of psychological research are more context sensitive than others, they conducted a metastudy about the question of whether experiments about traits that are judged (by independent raters) to be more context sensitive are less likely to replicate. They found that, indeed, "contextual sensitivity was negatively correlated with the success of the replication attempt" (Van Bavel et al. 2016a, 4656). Even though this conclusion was challenged (see Inbar 2016; and Van Bavel et al. 2016b), the question of context sensitivity, with its implications for data quality, replicability, and the shape of psychological objects, remains important and understudied. My analysis of experimental inferences in psychology (see sec. 7.3.1) offers a schematic account of relevant contexts to consider.

3.3.5. WEIRDness and the Shape of Objects

Notice that investigations of the shapes of psychological objects are not restricted to environmental features of the experiment (stimulus, manipulation, physical background, etc.). Descriptive features of objects can be moderated by physical features of the experiment, but they can also be moderated by differences in the composition of the group of research participants that we may never learn about if we always sample from the same population. This is also implied by the taxonomy of experimental components, one of which is the sample of research participants.

I argue similarly that, from the perspective of trying to determine the shape of an object of research, demographic attributes of research participants are one of the contexts to consider. To put this differently, if there are experimental effects associated with a given object of research that are moderated by specific demographic features of the study participants, then this can provide us with important descriptive information about the shape of the object. Such demographic features can include age, gender, ethnicity, class, geographic location, culture, etc. Notice that this is a hypothetical claim, that is, that I am not suggesting that the shape of psychological objects is invariably going to be affected by such features. My claim is only that this is one of several contexts that are potentially relevant when investigating the context sensitivity (and, hence, shape) of a given object.

The recommendation to diversify the populations from which psychologists sample came into prominence some years ago when Henrich, Heine, and Norenzayan (2010) pointed to the problem of WEIRDness. The acronym refers to the fact that most psychological research is conducted with white, educated participants that come from industrialized, rich, and democratic nations. The latter issue was met with interest by experimental philosophers, who realized that philosophy had long relied on the assumption that many cognitive features are widely shared across history and cultures (Machery, Knobe, and Stich 2023). Note that, if this assumption were true, the demographic composition of the research participants would be irrelevant since, in that case, any group of participants would sample the population of interest (i.e., humans). What Henrich, Heine, and Norenzayan pointed out, however, was that this is an empirical question and, thus, cannot be taken for granted. Group membership might be a moderator of specific experimental effects, and this question is worthy of investigation.

Rather than pointing out that there may be culture-specific variations in cognitive traits, I prefer to think about the matter in terms of the shape of psychological kinds themselves. In doing so, I hope to get away from the idea that researchers already have a good (decontextualized) understanding of the kinds in question and are merely interested in similarities and difference between kinds of peoples with respect to that kind. With my own approach, by contrast, I question that we can ever start with a decontextualized (pure, internal) and secure understanding of a psychological kind. To study a psychological kind is to study it in specific contexts, and, in order to determine facts about the shape of the kind, one has to vary contexts. In this book, I have suggested, however, that such variations need not (and for the most part should not) be random. My analysis of the method of converging operations suggests that the variations in question can be implemented in accordance with specific hypotheses regarding possible moderators for effects associated with a kind.

This way of thinking about contextual variability is, thus, in keeping with my commitment to the idea that objects of psychological research are epistemically blurry. Determining the shape of psychological objects includes the task of investigating the ways in which its articulations might be moderated by demographic factors without the comfort of knowing what an unmoderated or shapeless psychological kind might look like. Further probing into questions like this may well involve deconstructing the very WEIRD/non-WEIRD dichotomy into a much more nuanced understanding of the variability of psychological traits (Ghai 2021). It may also involve recognizing that WEIRDness extends to the unrepresentativeness of researchers themselves and to the choice of research topics (Barrett 2020; Sanches de Oliveira and Baggs 2023).

Let me quickly mention two more points. First, as already hinted at in chapter 7, one obvious response to my claim that there can be no decontextualized way of studying psychological kinds is that kinds might be individuated by specific underlying mechanisms that provide identity conditions independent of context. I argued against this notion in chapter 6, but I recognize that my argument hangs on a specific notion of *psychological kinds* as complex whole-organism capacities. This certainly deserves further attention, as does the suggestion that some objects are more context sensitive than others. Second, my argument that we focus on exploring the shapes of our objects of interest fits well with the ecological perspective articulated above, which suggests that the ways in which the shape of a given kind is moderated by specific populations may itself be a function of the specific habitats of those populations. Something to this effect was already recognized by early commentators on the concept of *WEIRDness* who noted that the task was not merely to figure out how cognitive traits are modulated by cultural context but also to gain explanatory knowledge of the ways in which cultural context shapes and conditions those very traits (Kesebir, Shigehiro, and Spellman 2010).

As also pointed out above, there is a sense in which the issue of how much cross-cultural variation we are likely to find with respect to specific kinds can be regarded as an empirical question. I am certainly not prejudging this question, to which the current literature offers competing answers (see the debate between Knobe and Machery as referenced in Machery, Knobe, and Stich 2023). My interest is at a more foundational level. In order even to begin to be able to ask and investigate empirical psychological questions, we already have to presuppose concepts and background assumptions. They are necessary tools, but they also need to be continually questioned and probed.

3.3.6. Validity of Measures and of Inferences

Since operational definitions are tied to measurement operations, there are obvious questions to be asked here concerning validity. In psychology, validity is typically regarded as a feature of a measurement tool (Cronbach and Meehl 1955) or of an experimental inference (Shadish, Cook, and Campbell 2002). With regard to the former (i.e., psychometric measurement), a common distinction that is drawn in the literature is that between predictive validity (or, more generally, criterion-oriented validity) and construct validity. Predictive validity concerns the extent to which a test result can predict a behavior of interest. Construct validity, in turn, has two interrelated aspects, namely, (*a*) whether a given instrument does a good job measuring the referent of a particular construct/concept and (*b*) whether the construct gives an adequate representation of the object (and, indeed, whether there even is a corresponding object) (Feest 2020a). (For a thorough recent overview and discussion of the history and philosophy of measurement in psychology, see Briggs 2021.)

In this book, I have spent relatively little time discussing validity in the psychometric sense, apart from the discussion of convergent and discriminant validity in section 2.5. The main reason for this is that many of the writings about validity take place in the context of needing to validate specific psychometric tools to give them the stamp of approval for specific realms of application. For example, if we base career decisions on how we do on an aptitude test, we would like to be sure that that test has been validated by some standard, such as, for example, that it has been cross-validated with other aptitude tests or that it is a good predictor of job performance. My focus in this book, by contrast, has been not on how to validate an existing instrument of psychometric measurement but rather how to use experiments to investigate objects of research. As such, it was concerned with questions about the validity of experimental inferences, given a good deal of uncertainty about the validational status of specific instruments.

It bears stressing that I have used ideas of the validity and soundness of experimental inferences mainly because they express a normative ideal that can push conceptual development by means of converging operations, not because they are something that can be achieved once and for all. Future work should delve more deeply into the ways in which this proposal fits with other recent work on validity coming out of the philosophy of measurement (Larroulet Philippi 2020; Tal 2020), the philosophy of economics (e.g., Jimenez-Buedo and Russo 2021), psychometrics (Flake and Fried 2020; Slaney 2017), and general experimental methodology (Vazire, Schiavone, and Bottesini 2022).

3.3.7. *The Objects and Phenomena That We Value*

In chapter 6, I acknowledged the worry that, by relying on existing concepts and operationally defining them in specific ways, researchers can create a kind of path dependence in subsequent research that is based on what becomes a canonical way of conceptualizing and measuring the subject matter (Thompson 2023). Given my contention that objects of psychological research are best understood as clusters of phenomena, it is easy to see why this worry is legitimate: there might be different salient properties associated with a given object. Which one is picked in an operational definition of the corresponding concept may be subject to pragmatic criteria, for example, the criterion of how straightforwardly or easily it can be measured. When those criteria are picked, other relevant phenomena get neglected or forgotten. Moreover, the choice of criteria can obviously be attached to values that are not equally shared by everybody.

I think that this is a very plausible version of the common concern that operationism is reductionist. While I agree with the worry, I would argue that the problem it points to is not inherent to operationism. Operationism, as I have construed it in this book, does not make the semantic claim that the meaning of a concept is nothing but a sum of operations. Rather, it makes the methodological claim that we cannot investigate an object unless we have first defined it operationally. This does not mean that we have operationally defined it well. It also does not mean that any given operational definition can do justice to the object as a whole even if it captures a salient phenomenon associated with an object. Finally, it does not mean that the object in question is even a worthy object of study.

These issues are important and deserve additional attention, touching on, among other things, an insight that has long been central to the literature about science and values. For example, Helen Longino pointed out already in 1990 that we study objects "under a description" (Longino 1990, 99) and that the ways in which we describe objects are likely informed by particular perspectives, values, and aims. They also touch on work from the philosophy of social science, where it has been argued that many of the concepts we use to describe objects of research contain ineliminably normative components (e.g., Alexandrova 2017). Future philosophical work on the role of operational definitions, especially when societally sensitive topics are concerned, will likely pay close attention to the ways in which normative components of concepts can be made invisible by operational definitions. While these discussions are important, I would like to suggest that they are not about whether to define central concepts operationally. The question, rather, is which research objects to focus on, what description to study them under, and how to define operationally the concepts central to such descriptions.

3.3.8. Giving Researchers' Agency Its Due

Let me conclude by highlighting an undercurrent of this entire book, namely, the central status of human researchers in the research process. Concepts are operationally defined by humans, experiments are designed by humans, experimental interventions are conducted by humans, experimental inferences are made by humans, etc.

As has been pointed out by recent commentators, the crisis of confidence in psychology has cast a lot of suspicion on the trustworthiness of human researchers, resulting in reflections on how better to regulate scientific practices and debias the investigative process (Flis 2019). Relatedly, there is also a lot of interest in the prospects of research assisted by artificial intelligence, research involving both the use of automatically generated (big) data and the use of algorithms to search for patterns in those data. I am firmly convinced that the open science movement has done valuable work in the direction of increasing the credibility of psychology. I am also firmly convinced that artificial intelligence is going to play an increasingly important role in scientific research.

However, I agree with those who have pointed out that (insofar as they are merely procedural) methodological innovations can do only part of the work required for high-quality research and that big data and machine-learning methods cannot be relied on to eliminate biases (Leonelli 2023). It is in this way that my book connects with other recent work in philosophy of psychology that has emphasized the epistemic agency of researchers (e.g., Gozli 2019; and Osbeck 2018). To put this in the terminology developed here, if we understand the objects of psychological research to be epistemically blurry, this implies the existence of epistemic agents from whose perspective they are blurry. If it were not for this perspective, we would not be doing science at all. And it is relative to this perspective that we ultimately judge whether specific scientific answers are satisfactory. This obviously does not mean that researchers can do no wrong or that we should trust them blindly. My book has tried to articulate some ways in which researchers who are making a good-faith effort to explore their objects can (and often do) proceed.

ACKNOWLEDGMENTS

This book builds on many years of research, and even though I wish I had been able to write it faster, I am immensely grateful that I had the time and the resources to take the time it took. While a very distant ancestor of this project started at the University of Pittsburgh, many of the ideas in this book elaborate on arguments that were developed in papers that I published over the last twenty years or so. As a result, the present book synthesizes a number of larger conversations that started unfolding after I left Pittsburgh: conversations about how to think philosophically about experimental practice, about the role of error in science, about how to integrate historical research into philosophical accounts of scientific rationality, and (most recently) about the specific problems that psychology faces as a scientific discipline. I am grateful to have been a small part of these conversations and grateful to everyone who has engaged with my analyses. It is my hope that this has allowed the book to become more mature and well-rounded than it would otherwise have been.

With a book this long in the making, it becomes a daunting task to properly acknowledge and thank all the people (family, friends, colleagues, travel companions) who have helped make it possible. So I figure it will be best to start at the beginning and extend my enormous gratitude to my parents, Christa Feest and Johannes Feest, whose intellectual curiosity, openness, resilience, and ongoing critical engagement with a rapidly changing world remain models of how to live one's life. I am grateful for the privilege of having parents who have always supported me and believed in me, and I am more than grateful for having these particular two individuals be my parents. I would also like to thank my brothers, Caspar and David, for being the best siblings one could ask for.

Next in line to thank are my friends, fellow students, and professors at the University of Frankfurt (Germany), where I studied psychology several lifetimes ago and also dabbled in philosophy. In particular, I am grateful to one of my first mentors, Wolfgang Detel, who recognized that I might find the Department of History and Philosophy of Science at the University of Pittsburgh congenial to my interests. I am also grateful to Martin Carrier, who wrote a letter of recommendation even though he barely knew me at the time.

It is hard to overstate how formative my years as a graduate student in the Department of History and Philosophy of Science (HPS) were for me. I am grateful to all members of the Pitt HPS faculty at the time for providing me with an intellectual home and laying the foundations for my academic identity. In particular, I thank Merrilee Salmon for her warm welcome into the profession.

During my time at Pitt, Paul Griffiths and Mark Wilson joined the HPS and Philosophy Departments, respectively, and both ended up becoming important teachers. I owe them both thanks for teaching graduate seminars that helped me think creatively about scientific concepts. I am also grateful to Bob Olby, another member of my committee, for supervising my very first attempt to say something about operationism. Jim Bogen moved to Pittsburgh during my time there and became an incredibly supportive informal adviser. Jim early on expressed enthusiasm about my work in a way that was both baffling and uplifting. I hope that it is apparent on many pages of this book that the enthusiasm is mutual. Most importantly, Peter Machamer gave me just the right mix of encouragement and critical feedback, pushing me to look closely at scientific practices (past and present), to take seriously my hunches about their philosophical significance, and to always question philosophical platitudes. It is my hope that he would have approved of this book. Last, but certainly not least, I was incredibly lucky to share my early research with an amazing group of peers, who made my time in Pittsburgh intellectually rich and stimulating in ways that went far beyond the academy.

Between 2003 and 2006, I had the good fortune to be a postdoc at the Max Planck Institute for the History of Science in Berlin. I was a member of Department III, led by Hans-Jörg Rheinberger, whose trust in me and hands-off yet caring style of mentoring were exactly what I needed. I cannot thank him enough for giving me the mental space as well as the physical and intellectual resources to pursue the project that ultimately culminated in this book (as well as a bunch of other projects). During my time at the Max Planck Institute, I was able to connect with many scholars who shared my interest in the history and philosophy of experimentation, the intertwined histories of psychology and philosophy of science, and the role of concepts in scientific research. The institute was an extraordinarily vibrant place, and I am extremely grateful for all the friends I made while there, of whom Jutta Schickore, Friedrich Steinle, and Thomas Sturm were particularly closely aligned with my interests. It was also during that time that I first met Hasok Chang, who has been a steadfast supporter of my work on operationism ever since, and Gary Hatfield, whose work in the history and philosophy of psychology continues to inspire me.

After my time at the Max Planck Institute, I was an assistant professor of philosophy at the Technical University of Berlin (2006–12). I would like to thank the Stifterverband für die deutsche Wissenschaft for a generous grant that funded that position, providing me with several more years of financial stability to pursue my research. In 2011–12, I was a visiting researcher at the Center for Philosophy of Science at the University of Pittsburgh. I would like to thank the center, its director at the time, John Norton, and all my fellow "fellows" but especially Dana Tulodziecki for their input and many helpful discussions. Between the fall of 2012 and February 2014, I was a visiting scholar/professor at the University of Michigan and a visiting researcher at the Max Planck Institute for Human Development (Berlin). Many heartfelt thanks go to Gerd Gigerenzer for inviting me to the latter.

In March 2014, I took up my current position as a philosophy professor at the Leibniz University in Hannover. My ten years at Hannover have been a whirlwind of travel (including a transatlantic commute), teaching, and joint academic adventures of all sorts. I count myself extremely lucky to have landed a job in a department that not only specializes in philosophy of science but also has a particular focus on epistemological and ethical issues that arise in scientific practice. I am grateful to Paul Hoyningen-Huene for his role in founding this department and to my Hannover colleagues for the ways in which they not only have helped make it grow but also continue to fill it with life, ideas, and a general spirit of support and camaraderie.

An earlier version of this book (which at the time was still missing the last two chapters) was discussed at a manuscript workshop that I held at Leibniz University in the summer of 2019. I am hugely grateful to the following friends and colleagues for making the time to read the manuscript and gather for two days of fun and helpful discussion: Corinne Bloch-Mullins, Hasok Chang, David Colaço, Philipp Haueis, Katie Kendig, and Dana Tulodziecki. We should all write more books if it means that we can hang out and talk about the history and philosophy of science! I would also like to thank Karen Darling from the University of Chicago Press for not giving up on this book. Huge thanks go, in addition, to two reviewers for the University of Chicago Press. They pushed me to clarify, qualify, and rethink some of the main ideas of the book, particularly in chapters 6 and 7. While I cannot speak for them, I certainly think that their critical feedback has resulted in substantial improvements.

I am pretty certain that very few of the things I have done over the last nineteen years would have been possible without the unwavering love and support of my partner, John Carson. But, of course, I cannot be sure of this because I cannot imagine my life without him. I am so very lucky that he seems to feel the same. I am also grateful to Arlo, who takes me for walks. However, this book could not be dedicated to anyone other than our son, Robinson. He grew into his own person much faster than my manuscript grew into a book, and at the end of the day, that is significantly more important than any book could ever be.

REFERENCES

Abraham, Tara. 2016. *Rebel Genius: Warren S. McCulloch's Transdisciplinary Life in Science.* Cambridge, MA: MIT Press.

Ackermann, Robert. 1989. "The New Experimentalism." *The British Journal for the Philosophy of Science* 40 (2): 185–90.

Addis, Laird. 1999. "Gustav Bergmann (4 May 1906–21 April 1987)." *American National Biography*, vol. 2, 639–41. Oxford University Press.

Alexandrova, Anna. 2017. *A Philosophy for the Science of Well-Being.* New York: Oxford University Press.

American Psychological Association. 1954. "Technical Recommendations for Psychological Tests and Diagnostic Techniques." *Psychological Bulletin* 51 (2, pt. 2): 1–38.

Amsel, Abram, and Michael Rashotte. 1984. *Mechanisms of Adaptive Behavior: Clark L. Hull's Theoretical Papers, with Commentary.* New York: Columbia University Press.

Anastasi, Anne. 1950. "The Concept of Validity in the Interpretation of Test Scores." *Educational and Psychological Measurement* 10 (1): 67–78.

Andersen, Hanne, Peter Barker, and Xiang Chen. 2006. *The Cognitive Structure of Scientific Revolutions.* Cambridge: Cambridge University Press.

Anderson, Michael. 2014. *After Phrenology: Neural Reuse and the Interactive Brain.* Cambridge, MA: MIT Press.

———. 2016. "Précis of *After Phrenology: Neural Reuse and the Interactive Brain.*" *Behavioral and Brain Sciences* 39:e120. https://doi.org/10.1017/S0140525X15000631.

Arabatzis, Theodore. 2011. "On the Historicity of Scientific Objects." *Erkenntnis* 75:377–90. https://doi.org/10.1007/s10670-011-9344-5.

———. 2012. "Experimentation and the Meaning of Scientific Concepts." In *Scientific Concepts and Investigative Practice*, ed. Uljana Feest and Friedrich Steinle, 149–66. Berlin: de Gruyter.

Arabatzis, Theodore, and Vasso Kindi. 2008. "The Problem of Conceptual Change in the Philosophy and History of Science." In *International Handbook of Research on Conceptual Change*, ed. Stella Vosniadou, 247–372. New York: Routledge.

Araújo, Duarte, Keith Davids, and Pedro Passos. 2007. "Ecological Validity, Representative Design, and Correspondence between Experimental Task Constraints and Behavioral Setting: Comment on Rogers, Kadar, and Costall (2005)." *Ecological Psychology* 19 (1): 69–78.

Ash, Mitchell. 1995. *Gestalt Psychology and German Culture, 1890–1967*. Cambridge: Cambridge University Press.

Atkinson, Richard, and Richard Shifrin. 1968. "Human Memory: A Proposed System and Its Control Processes." In *The Psychology of Learning and Motivation*, vol. 2, ed. K. W. Spence and J. T. Spence, 89–195. New York: Academic Press.

Baddeley, Alan, and Graham Hitch. 1974. "Working Memory." In *The Psychology of Learning and Motivation*, vol. 8, ed. G. A. Bower, 47–89. New York: Academic Press.

Baddeley, Alan, and Elizabeth Warrington. 1970. "Amnesia and the Distinction between Long- and Short-Term Memory." *Journal of Verbal Learning and Verbal Behavior* 9 (2): 176–89.

Barker, Peter. 2011. "The Cognitive Structure of Scientific Revolutions." *Erkenntnis* 75 (3): 445–65. https://doi.org/10.1007/s10670-011-9333-8.

Barker, Roger. 1969. *Ecological Psychology: Concepts and Methods for Studying the Environment of Human Behavior*. Stanford, CA: Stanford University Press.

Barsalou, Lawrence. 1987. "The Instability of Graded Structure: Implications for the Nature of Concepts." In *Concepts and Conceptual Development: Ecological and Intellectual Factors in Categorization*, ed. Ulric Neisser, 101–40. Cambridge: Cambridge University Press.

Barrett, Clark. 2020. "Towards a Cognitive Science of the Human: Cross-Cultural Approaches and Their Urgency." *Trends in Cognitive Science* 24 (8): 620–38. https://doi.org/10.1016/j.tics.2020.05.007.

Basso, Alessandra. 2017. "The Appeal to Robustness in Measurement Practice." *Studies in History and Philosophy of Science: Part A* (65–66): 57–66.

Beaman, Philip. 2001. "The Size and Nature of a Chunk." *Behavioral and Brain Sciences* 24 (1): 118.

Bechtel, William. 2008a. "Mechanisms in Cognitive Psychology: What Are the Operations?" *Philosophy of Science* 75 (5): 983–94.

———. 2008b. *Mental Mechanisms: Philosophical Perspectives on Cognitive Neuroscience*. New York: Routledge.

Bechtel, William, and Jennifer Mundale. 1999. "Multiple Realizability Revisited: Linking Cognitive and Neural States." *Philosophy of Science* 66 (2): 175–207.

Bechtel, William, and Robert Richardson. 1993. *Discovering Complexity: Decomposition and Localization as Strategies in Scientific Research*. Princeton, NJ: Princeton University Press.

Bergmann, Gustav, and Kenneth W. Spence. 1941. "Operationism and Theory in Psychology." *Psychological Review* 48:1–14.

Bickle, John. 2016. "Revolutions in Neuroscience: Tool Development." *Frontiers in Systems Neuroscience* 10 (24). https://doi.org/10.3389/fnsys.2016.00024.

Bird, Alexander, and Emma Tobin. 2018. "Natural Kinds." In *The Stanford Encyclopedia of Philosophy* (Spring 2018 ed.), ed. Edward N. Zalta. https://plato.stanford.edu/archives/spr2018/entries/natural-kinds.

Blaxton, Teresa. 1989. "Investigating Dissociations among Memory Measures: Support for a Transfer-Appropriate Processing Framework." *Journal of Experimental Psychology: Learning, Memory, and Cognition* 15:657–68.

Bloch-Mullins, Corinne. 2012. "Early Concepts in Investigative Practice—the Case of the Virus." In *Scientific Concepts and Investigative Practice*, ed. Uljana Feest and Friedrich Steinle, 191–218. Berlin: de Gruyter.

———. 2018. "Bridging the Gap between Similarity and Causality: An Integrated Approach to Concepts." *The British Journal for the Philosophy of Science* 69 (1): 605–32.

———. 2019. "Similarity Reimagined (with Implications for a Theory of Concepts)." *Theoria* 87:31–68. https://doi.org/10.1111/theo.12197.

———. 2020. "Scientific Concepts as Forward-Looking: How Taxonomic Structure Facilitates Conceptual Development." *Journal of the Philosophy of History* 14:1–27.

Bogen, James, and James Woodward. 1988. "Saving the Phenomena." *Philosophical Review* 97 (3): 303–52.

Boon, Mieke. 2012. "Scientific Concepts in the Engineering Sciences: Epistemic Tools for Creating and Intervening with Phenomena." In *Scientific Concepts and Investigative Practice*, ed. Uljana Feest and Friedrich Steinle, 219–44. Berlin: de Gruyter.

Boone, Worth, and Gualtiero Piccinini. 2016. "The Cognitive Neuroscience Revolution." *Synthese* 193 (5): 1509–34.

Boring, Edwin G. 1923. "Intelligence as the Test Tests It." *New Republic* 36:35–37.

———. 1927. "Edward Bradford Titchener: 1867–1927." *American Journal of Psychology* 38:489–506.

———. 1929. *A History of Experimental Psychology*. New York: Appleton-Century-Crofts.

———. 1933. *The Physical Dimensions of Consciousness*. New York: The Century.

———. 1935. "The Relation of the Attributes of Sensation to the Dimensions of the Stimulus." *Philosophy of Science* 2 (2): 236–45.

———. 1942. *Sensation and Perception in the History of Experimental Psychology*. New York: Appleton-Century-Crofts.

———. 1945. "The Use of Operational Definitions in Science." *Psychological Review* 52 (5): 243–45.

———. 1961. "Psychologist at Large." In *Psychologist at Large: An Autobiography and Selected Essays*, ed. Edwin Boring, 3–83. New York: Basic Books.

Borsboom, Denny. 2005. *Measuring the Mind: Conceptual Issues in Contemporary Psychometrics*. Cambridge: Cambridge University Press.

———. 2023. "Psychological Constructs as Organizing Principles." In *Essays on Contemporary Psychometrics, Methodology of Educational Measurement and Assessment*, ed. L. Andries van der Ark, Wilco H. M. Emons, and Rob R. Meijer, 89–108. Springer. https://doi.org/10.1007/978-3-031-10370-4_5.

Borsboom, Denny, Gideon Mellenbergh, and Jaap van Heerden. 2004. "The Concept of Validity." *Psychological Review* 111 (4): 1061–71.

Borsboom, Denny, Han van der Maas, Jonas Dalege, Rogier Kievit, and Brian Haig. 2021. "Theory Construction Methodology: A Practical Framework for Building Theories in Psychology." *Perspectives on Psychological Science* 16 (4): 756–66. https://doi.org/10.1177/1745691620969647.

Bowers, Jeffrey S., and Sid Kouider. 2003. "Developing Theories of Priming with an Eye on Function." In *Rethinking Implicit Memory*, ed. Jeffrey S. Bowers and Chad S. Marsolek, 19–40. Oxford: Oxford University Press.

Boyd, Richard. 1991. "Realism, Anti-Foundationalism and the Enthusiasm for Natural Kinds." *Philosophical Studies* 61 (1–2): 127–48.

———. 1999. "Kinds, Complexity and Multiple Realization." *Philosophical Studies* 95:67–98.

———. 2000. "Kinds as the 'Workmanship of Men': Realism, Constructivism, and Natural Kinds." In *Rationalität, Realismus, Revision/Rationality, Realism, Revision*, ed. Julian Nida-Rümelin, 52–89. Berlin: de Gruyter.

Brandom, Robert. 1998. *Making It Explicit: Reasoning, Representing, and Discursive Commitment*. Cambridge, MA: Harvard University Press.

Bridgman, Percy Williams. 1927. *The Logic of Modern Physics*. New York: Macmillan.
———. 1938. "Operational Analysis." *Philosophy of Science* 5 (2): 114–31.
———. 1945. "Some General Problems of Operational Analysis." *Psychological Review* 52 (2): 246–49.
Brigandt, Ingo. 2010. "The Epistemic Goal of a Concept: Accounting for the Rationality of Semantic Change and Variation." *Synthese* 177:19–40. https://doi.org/10.1007/s11229-009-9623-8.
———. 2012. "The Dynamics of Scientific Concepts: The Relevance of Epistemic Aims and Values." In *Scientific Concepts and Investigative Practice*, ed. Uljana Feest and Friedrich Steinle, 75–104. Berlin: de Gruyter.
———. 2022. "How to Philosophically Tackle Kinds without Talking about 'Natural Kinds.'" In "Engaging with Science, Values, and Society," ed. Ingo Brigandt, special issue, *Canadian Journal of Philosophy* 52 (3): 356–79. https://doi.org/10.1017/can.2020.29.
Briggs, Derek. 2021. *Historical and Conceptual Foundations of Measurement in the Human Sciences: Credos and Controversies*. New York: Routledge.
Bringmann, Laura F., Timon Elmer, and Markus Eronen. 2022. "Back to Basics: The Importance of Conceptual Clarification in Psychological Science." *Current Directions in Psychological Science* 31 (4): 340–46. https://doi.org/10.1177/09637214221096485.
Brun, Georg. 2016. "Explication as a Method of Conceptual Re-Engineering." *Erkenntnis* 81:1211–41.
———. 2020. "Conceptual Re-Engineering: From Explication to Reflective Equilibrium." *Synthese* 197 (3): 925–54. https://doi.org/10.1007/s11229-017-1596-4.
Brunswik, Egon. 1934. *Wahrnehmung und Gegenstandswelt: Grundlegung einer Psychologie vom Gegenstand her*. Leipzig: Deutike.
———. 1943. "Organismic Achievement and Environmental Probability." *Psychological Review* 50 (3): 255–72. https://doi.org/10.1037/h0060889.
———. 1947/1956. *Perception and the Representative Design of Psychological Experiments*. 2nd ed. Berkeley: University of California Press.
Burian, Richard. 1997. "Exploratory Experimentation and the Role of Histochemical Techniques in the Work of Jean Brachet, 1938–1952." In "Research Programs of the Rouge-Cloitre Group," ed. Richard Burian and Denis Thierry, special issue, *History and Philosophy of the Life Sciences* 19:27–45.
———. 2013. "Exploratory Experimentation." In *Encyclopedia of Systems Biology*, ed. Werner Dubitzky, Olaf Wolkenhauer, Kwang-Hyun Cho, and Hiroki Yokota, 720–23. New York: Springer. https://doi.org/10.1007/978-1-4419-9863-7_60.
Burnston, Daniel. 2016. "A Contextualist Approach to Functional Localization in the Brain." *Biology and Philosophy* 31 (4): 527–50.
———. 2021. "Getting over Atomism: Functional Decomposition in Complex Neural Systems." *The British Journal for the Philosophy of Science* 72 (3). https://doi.org/10.1093/bjps/axz039.
———. 2022. "Cognitive Ontologies, Task Ontologies, and Explanation in Cognitive Neuroscience." In *Neuroscience Experiment: Philosophical and Scientific Perspectives*, ed. John Bickle, Carl F. Craver, and Ann Sophie Barwich, 259–83. New York: Routledge.
Bursten, Julia. 2018. "Smaller Than a Breadbox: Scale and Natural Kinds." *The British Journal for the Philosophy of Science* 69 (1): 1–23.
Butler, Laure, and Diane Berry. 2001. "Implicit Memory: Intention and Awareness Revisited." *Trends in Cognitive Sciences* 5 (5): 192–97.

Campbell, Donald T. 1969/1988. "Definitional versus Multiple Operationalism." In *Methodology and Epistemology for Social Science*, ed. E. S. Overman, 31–36. Chicago: University of Chicago Press.

Campbell, Donald T., and Donald W. Fiske. 1959. "Convergent and Discriminant Validation by the Multitrait-Multimethod Matrix." *Psychological Bulletin* 56 (2): 81–105. https://doi.org/10.1037/h0046016.

Canali, Stefano. 2020. "Towards a Contextual Approach to Data Quality." *Data* 5 (4): 90. https://doi.org/10.3390/data5040090.

Carnap, Rudolf. 1928/1966. *Der logische Aufbau der Welt*. Hamburg: Meiner.

———. 1931a. "Die physikalische Sprache als Universalsprache der Wissenschaft." *Erkenntnis* 2:432–65. Reprinted as "The Physical Language as the Universal Language of Science," in *Readings in Twentieth-Century Philosophy*, ed. William P. Alston and George Nakhnikian (New York: Free Press, 1963), 393–424.

———. 1931b. "Überwindung der Metaphysik durch logische Analyse der Sprache." *Erkenntnis* 2:219–41.

———. 1932. "Psychologie in physikalischer Sprache." *Erkenntnis* 3:107–42. Translated by George Schick as "Psychology in Physical Language," in *Logical Positivism*, ed. Alfred J. Ayer (Glencoe, IL: Free Press, 1959), 165–96.

———. 1934. "On the Character of Philosophical Problems." *Philosophy of Science* 1:5–19.

———. 1936–37. "Testability and Meaning, Parts 1 and 2." *Philosophy of Science* 3 (4): 419–71, and 4 (1): 1–40.

———. 1950a. "Empiricism, Semantics, and Ontology." *Revue internationale de philosophie* 4 (2): 20–40.

———. 1950b/1962. *Logical Foundations of Probability*. 2nd ed. Chicago: University of Chicago Press; London: Routledge & Kegan Paul.

———. 1956. "The Methodological Character of Theoretical Concepts." In *Foundations of Science and the Concepts of Psychology and Psychoanalysis* (Minnesota Studies in the Philosophy of Science 1), ed. Herbert Feigl and Michael Scriven, 38–76. Minneapolis: University of Minnesota Press.

Carroll, David. 2017. *Purpose and Cognition: Edward Tolman and the Transformation of American Psychology*. Cambridge: Cambridge University Press.

Cartwright, Nancy. 2012. "Presidential Address: Will This Policy Work for You? Predicting Effectiveness Better: How Philosophy Helps." *Philosophy of Science* 79:973–89.

Cartwright, Nancy, and Rosa Runhardt. 2014. "Measurement." In *Philosophy of Social Science: A New Introduction*, ed. Nancy Cartwright and Eleonora Montuschi, 265–287. New York: Oxford University Press.

Carus, A. W. 2007. *Carnap and Twentieth-Century Thought: Explication as Enlightenment*. Cambridge: Cambridge University Press.

Chang, Hasok. 2004. *Inventing Temperature: Measurement and Scientific Progress*. Oxford: Oxford University Press.

———. 2012a. "Beyond Case-Studies: History as Philosophy." In *Integrating History and Philosophy of Science: Problems and Prospects* (Boston Studies in the Philosophy of Science 263), ed. Seymour Mauskopf and Tad Schmaltz, 109–24. Heidelberg: Springer.

———. 2012b. *Is Water H_2O? Evidence, Realism, and Pluralism*. Heidelberg: Springer.

———. 2014. "Epistemic Activities and Systems of Practice: Units of Analysis in Philosophy of Science after the Practice Turn." In *Science after the Practice Turn in the Philosophy, History, and Social Studies of Science*, ed. Léna Soler, Sjoerd Zwart, Michael Lynch, and Vincent Israel-Jost, 67–79. London: Routledge.

———. 2016. "The Rising of Chemical Natural Kinds through Epistemic Iteration." In *Natural Kinds and Classification in Scientific Practice*, ed. Catherine Kendig, 33–46. London: Routledge.

———. 2019. "Operationalism." In *The Stanford Encyclopedia of Philosophy* (Winter 2019 ed.), ed. Edward N. Zalta. https://plato.stanford.edu/archives/win2019/entries/operationalism.

Chemero, Anthony. 2009. *Radical Embodied Cognitive Science*. Cambridge, MA: MIT Press.

Cheon, Hyundeuk, and Edouard Machery. 2010. Review of *Creating Scientific Concepts*, by Nancy J. Nersessian. *Mind* 119 (475): 838–44.

Churchland, Paul. 1981. "Eliminative Materialism and the Propositional Attitudes." *Journal of Philosophy* 78 (2): 67–90.

Cohen, Neil, and Larry Squire. 1980. "Preserved Learning and Retention of Pattern Analyzing Skill in Amnesia: Dissociation of Knowing How and Knowing That." *Science* 210:207–10.

Coko, Klodian. 2015. "The Structure and Epistemic Import of Multiple Determination in Scientific Practice." PhD diss., Indiana University.

———. 2020a. "Jean Perrin and the Philosophers' Stories: The Role of Multiple Determination in Determining Avogadro's Number." *HOPOS: The Journal of the International Society for the History of Philosophy of Science* 10 (1):143–93.

———. 2020b. "The Multiple Dimensions of Multiple Determination." *Perspectives on Science* 28 (4): 505–41.

Colaço, David. 2018a. "Rethinking the Role of Theory in Exploratory Experimentation." *Biology and Philosophy* 33 (5–6): article 38. https://doi.org/10.1007/s10539-018-9648-9.

———. 2018b. "Rip It Up and Start Again: The Rejection of a Characterization of a Phenomenon." *Studies in History and Philosophy of Science: Part A* 72:32–40.

———. 2020. "Recharacterizing Scientific Phenomena." *European Journal for Philosophy of Science* 10: article 14. https://doi.org/10.1007/s13194-020-0279-z.

———. 2022. "What Counts as a Memory? Definitions, Hypotheses, and 'Kinding in Progress.'" *Philosophy of Science* 89 (1): 89–106.

Colaço, David, John Bickle, and Bradley Walters. 2022. "When Should Researchers Cite Study Differences in Response to a Failure to Replicate?" *Biology and Philosophy* 37: article 39. https://doi.org/10.1007/s10539-022-09873-y.

Collins, H. M. 1976. "The Seven Sexes: A Study in the Sociology of a Phenomenon; or, The Replication of Experiments in Physics." *Sociology* 6:141–84.

———. 1985. *Changing Order: Replication and Induction in Scientific Practice*. London: Sage.

Conway, Andrew, Michael Kane, Michael Bunting, D. Zach Hambrick, Oliver Wilhelm, and Randall Engle. 2005. "Working Memory Span Tasks: A Methodological Review and User's Guide." *Psychonomic Bulletin Review* 12 (5): 769–86.

Conway, Andrew, Michael Kane, and Randall Engle. 2003. "Working Memory Capacity and Its Relation to General Intelligence." *Trends in Cognitive Sciences* 7 (12): 547–52.

Cowan, Nelson. 2001. "The Magical Number 4 in Short-Term Memory: A Reconsideration of Mental Storage Capacity." *Behavioral and Brain Sciences* 24 (1): 87–114.

———. 2005. "Working-Memory Capacity Limits in a Theoretical Context." In *Human Learning and Memory: Advances in Theory and Application: The 4th Tsukuba International Conference on Memory*, ed. C. Izawa and N. Ohta, 155–75. Lawrence Erlbaum.

———. 2008. "What Are the Differences between Long-Term, Short-Term, and Working Memory?" In *Essence of Memory* (Progress in Brain Research 169), ed. Wayne S. Sossin, Jean-Claude Lacaille, Vincent F. Castellucci, and Sylvie Belleville, 323–38. Amsterdam: Elsevier.

Craver, Carl. 2004. "Dissociable Realization and Kind Splitting." *Philosophy of Science* 71 (5): 960–71.

———. 2007. *Explaining the Brain: Mechanisms and the Mosaic Unity of Neuroscience*. Oxford: Oxford University Press.

———. 2009. "Mechanisms and Natural Kinds." *Philosophical Psychology* 22 (5): 575–94.

Craver, Carl, and Talia Dan-Cohen. 2024. "Experimental Artefacts." *The British Journal for the Philosophy of Science* 75 (1): 253–74. https://doi.org/10.1086/715202.

Craver, Carl, and Lindley Darden. 2001. "Discovering Mechanisms in Neurobiology: The Case of Spatial Memory." In *Theory and Method in the Neurosciences*, ed. Peter Machamer, Rick Grush, and Peter McLaughlin, 112–37. Pittsburgh, PA: University of Pittsburgh Press.

———. 2013. *In Search of Mechanisms: Discovery across the Life Sciences*. Chicago: University of Chicago Press.

Crawford, Sean. 2014. "On the Logical Positivists' Philosophy of Psychology: Laying a Legend to Rest." In *New Directions in the Philosophy of Science* (Philosophy of Science in a European Perspective 5), ed. Maria Carla Galavotti, Denis Dieks, Wenceslao J. Gonzalez, Stephan Hartmann, Thomas Uebel, and Marcel Weber, 711–26. Cham: Springer.

Cronbach, Lee. 1982. *Designing Evaluations of Educational and Social Programs*. San Francisco: Jossey-Bass.

Cronbach, Lee, and Paul Meehl. 1955. "Construct Validity in Psychological Tests." *Psychological Bulletin* 52:281–302.

Culp, Sylvia. 1994. "Defending Robustness: The Bacterial Mesosome as a Test Case." In *PSA: Proceedings of the Biennial Meeting of the Philosophy of Science Association*, vol. 1, Contributed Papers, 46–57. Cambridge: Cambridge University Press.

———. 1995. "Objectivity and Experimental Inquiry: Breaking Data-Technique Circles." *Philosophy of Science* 62:430–50.

Cummins, Robert. 1983. *Psychological Explanation*. Cambridge, MA: MIT Press.

———. 2000. "'How Does It Work?' versus 'What Are the Laws?': Two Conceptions of Psychological Explanation." In *Explanation and Cognition*, ed. F. Keil and R. Wilson, 117–44. Cambridge, MA: MIT Press.

Daneman, Meredyth, and Patricia Carpenter. 1980. "Individual Differences in Working Memory and Reading." *Journal of Verbal Learning and Verbal Behavior* 19 (4): 450–66.

Danks, David. 2015. "Goal-Dependence in (Scientific) Ontology." *Synthese* 192: 3601–16.

Danks, David, and Frederick Eberhardt. 2009. "Conceptual Problems in Statistics, Testing and Experimentation." In *The Routledge Companion to Philosophy of Psychology*, ed. John Symons and Paco Calvo, 214–30. Routledge/Taylor & Francis.

Danziger, Kurt. 1980. "The History of Introspection Reconsidered." *Journal for the History of the Behavioral Sciences* 16:241–62.

———. 1997. *Naming the Mind: How Psychology Found Its Language*. London: Sage.

———. 2002. "How Old Is Psychology, Particularly Concepts of Memory?" *History and Philosophy of Psychology* 4 (1): 1–12.

———. 2003. "Where Theory, History and Philosophy Meet: The Biography of Psychological Objects." In *About Psychology: Essays at the Crossroads of History, Theory and Philosophy*, ed. D. B. Hill and M. J. Kral, 19–33. Albany, NY: SUNY Press.

Darden, Lindley. 1991. *Theory Change in Science: Strategies from Mendelian Genetics*. Oxford: Oxford University Press.

———. 2006. *Reasoning in Biological Discoveries: Essays on Mechanisms, Interfield Relations, and Anomaly Resolution*. Cambridge: Cambridge University Press.

———. 2009. "Discovering Mechanisms in Molecular Biology: Finding and Fixing Incompleteness and Incorrectness." In *Models of Discovery and Creativity* (Origins: Studies in the Philosophy of Scientific Creativity 3), ed. Joke Meheus and Thomas Nickles, 43–55. Dordrecht: Springer.

Devezer, Berna, and Erkan Buzbas. 2022. "Minimum Viable Experiment to Replicate." Preprint, submitted November 26. https://philsci-archive.pitt.edu/21475.

———. 2023. "Rigorous Exploration in a Model-Centric Science via Epistemic Iteration." *Journal of Applied Research in Memory and Cognition* 12 (2): 189–94. https://doi.org/10.1037/mac0000121.

Devezer, Berna, Danielle Navarro, Joachim Vandekerckhove, and Erkan Ozge Buzbas. 2021. "The Case for Formal Methodology in Scientific Reform." *Royal Society Open Science* 8:200805. https://doi.org/10.1098/rsos.200805.

Devezer, Berna, and Bart Penders. 2023. "Scientific Reform, Citation Politics and the Bureaucracy of Oblivion." *Quantitative Science Studies* 4 (4): 857–59. https://doi.org/10.1162/qss_c_00274.

Dewhurst, Joseph. 2019. "Context Sensitive Ontologies for a Non-Reductionist Cognitive Neuroscience." *Australasian Philosophical Review* 2 (2): 224–28.

———. 2021. "Folk Psychological and Neurocognitive Ontologies." In *Neural Mechanisms: New Challenges in the Philosophy of Neuroscience* (Studies in Mind and Brain 17), ed. Fabrizio Calzavarini and Marco Viola, 311–34. Dordrecht: Springer. https://doi.org/10.1007/978-3-030-54092-0_14.

Dubova, Marina, and Robert Goldstein. 2022. "Carving Joints into Nature: Reengineering Scientific Concepts in Light of Concept-Laden Evidence." *Trends in Cognitive Sciences* 27 (7). https://doi.org/10.1016/j.tics.2023.04.006.

Dunn, John, and Kim Kirsner. 1989. "Implicit Memory: Task or Process?" In *Implicit Memory: Theoretical Issues*, ed. Stephan Lewandowsky, John C. Dunn, and Kim Kirsner, 17–31. Hillsdale, NJ: Lawrence Erlbaum.

Dupre, John. 1999. "Are Whales Fish?" In *Folkbiology*, ed. Douglass Medin and Scott Atran, 461–76. Cambridge, MA: MIT Press.

———. 2002. *Humans and Other Animals*. Oxford: Oxford University Press.

Ebbinghaus, Hermann. 1885. *Über das Gedächtnis*. Leipzig: Verlag von Duncker und Humblot.

Ebbs, Gary. 2000. "The Very Idea of Sameness of Extension across Time." *American Philosophical Quarterly* 37 (3): 245–68.

———. 2003. "Denotation and Discovery." In *Socializing Metaphysics: The Nature of Social Reality*, ed. Frederick F. Schmitt, 247–68. London: Rowman & Littlefield.

Eichenberger, Philipp. 2023. "History and Analysis of Constructs and Instruments in Positive Psychology." PhD diss., Leibniz University Hannover.

Eigner, Kai. 2010. "Understanding Psychologists' Understanding: The Application of Intelligible Models to Phenomena." PhD diss., Vrije Universiteit University Amsterdam.

Elliott, Kevin. 2012. "Epistemic and Methodological Iteration in Scientific Research." *Studies in History and Philosophy of Science: Part A* 43:376–82.

Ereshefsky, Mark, and Thomas Reydon. 2015. "Scientific Kinds." *Philosophical Studies* 172:969–86.

Fechner, Gustav Theodor. 1860. *Elemente der Psychophysik*. Leipzig: Breitkopf & Haertel.

Feest, Uljana. 2005a. "Giving Up Instincts in Psychology—or Not?" In *Recent Contributions to the History of the Human Sciences*, ed. B. Gómez-Zúñiga and A. Mülberger, 242–59. Munich: Profil.

———. 2005b. "Operationism in Psychology—What the Debate Is About, What the Debate Should Be About." *Journal for the History of the Behavioral Sciences* 41 (2): 131–50.

———. 2010. "Concepts as Tools in the Experimental Generation of Knowledge in Cognitive Neuropsychology." *Spontaneous Generations: A Journal for the History and Philosophy of Science* 4 (1): 173–90.

———. 2011a. "Remembering (Short-Term) Memory: Oscillations of an Epistemic Thing." *Erkenntnis* 75:391–411.

———. 2011b. "What Exactly Is Stabilized When Phenomena Are Stabilized?" *Synthese* 182 (1): 57–71.

———. 2012a. "Exploratory Experiments, Concept Formation and Theory Construction in Psychology." In *Scientific Concepts and Investigative Practice*, ed. Uljana Feest and Friedrich Steinle, 167–89. Berlin: De Gruyter.

———. 2012b. "Introspection as a Method and Introspection as a Feature of Consciousness." *Inquiry* 55 (1): 1–16.

———. 2014a. "The Continuing Relevance of 19th-Century Philosophy of Psychology: Brentano and the Autonomy of Psychology." In *New Directions in the Philosophy of Science* (Philosophy of Science in a European Perspective 5), ed. Maria Carla Galavotti, Denis Dieks, Wenceslao J. Gonzalez, Stephan Hartmann, Thomas Uebel, and Marcel Weber, 693–709. Cham: Springer.

———. 2014b. "Phenomenal Experiences, First-Person Methods, and the Artificiality of Experimental Data." *Philosophy of Science* 81:927–39.

———. 2016. "The Experimenters' Regress Reconsidered: Replication, Tacit Knowledge, and the Dynamics of Knowledge Generation." *Studies in History and Philosophy of Science: Part A* 58:34–45.

———. 2017a. "Phenomena and Objects of Research in the Cognitive and Behavioral Sciences." *Philosophy of Science* 84 (5): 1165–76.

———. 2017b. "Physicalism, Introspection, and Psychophysics: The Carnap/Duncker Exchange." In *Eppur Si Muove: Doing History and Philosophy of Science with Peter Machamer: A Collection of Essays in Honor or Peter Machamer* (Western Ontario Series in Philosophy of Science 81), ed. Marcus P. Adams, Zvi Biener, Uljana Feest, and Jacqueline A. Sullivan, 113–25. Dordrecht: Springer.

———. 2019. "Why Replication Is Overrated." *Philosophy of Science* 86 (5): 895–905.

———. 2020a. "Construct Validity in Psychological Tests—the Case of Implicit Social Cognition." *European Journal for the Philosophy of Science* 12: article 13. https://doi.org/10.1007/s13194-021-00443-9.

———. 2020b. "Eigenpsychisches und Fremdpsychisches: Carnaps Verhältnis zur Psychologie." In *Der junge Carnap im historischen Kontext: 1918–1935/Young Carnap in an Historical Context: 1918–1935* (Veröffentlichungen des Instituts Wiener Kreis), ed. Christian Damböck and Gereon Wolters, 169–86. Dordrecht: Springer.

———. 2021. "Gestalt Psychology, Frontloading Phenomenology, and Psychophysics." *Synthese* 198 (suppl. 9): 2153–73. https://doi.org/10.1007/s11229-019-02211-y.

———. 2022. "Data Quality, Experimental Artifacts, and the Reactivity of the Psychological Subject Matter." *European Journal for the Philosophy of Science* 12: article 13. https://doi.org/10.1007/s13194-021-00443-9.

———. 2023. "Progress in Psychology." In *New Philosophical Perspectives on Scientific Progress*, ed. Yafent Shan, 184–203. New York: Routledge.

Feest, Uljana, and Friedrich Steinle. 2012. "Scientific Concepts and Investigative Practice: Introduction." In *Scientific Concepts and Investigative Practice*, ed. Uljana Feest and Friedrich Steinle, 1–23. Berlin: De Gruyter.

———. 2016. "Experiment." In *Oxford Handbook of Philosophy of Science*, ed. Paul Humphreys, 274–95. Oxford: Oxford University Press.

Feest, Uljana, and Thomas Sturm. 2011. "What (Good) Is Historical Epistemology? Editors' Introduction." In "What (Good) Is Historical Epistemology?," ed. Thomas Sturm and Uljana Feest, special issue, *Erkenntnis* 75:285–302.

Feigl, Herbert. 1945. "Operationism and Scientific Method." *Psychological Review* 52 (5): 250–59.

Fidler, Fiona, and John Wilcox. 2018. "Reproducibility of Scientific Results." In *The Stanford Encyclopedia of Philosophy* (Winter 2018 ed.), ed. Edward N. Zalta. https://plato.stanford.edu/archives/win2018/entries/scientific-reproducibility.

Figdor, Carrie. 2011. "Semantics and Metaphysics in Informatics: Toward an Ontology of Tasks." *Topics in Cognitive Science* 3 (2): 222–26.

Fiske, Donald, and Donald Campbell. 1992. "Citations Do Not Solve Problems." *Psychological Bulletin* 112:393–95.

Flake, Jessica, and Eiko Fried. 2020. "Measurement Schmeasurement: Questionable Measurement Practices and How to Avoid Them." *Advances in Methods and Practices in Psychological Science* 3 (4): 456–65.

Flis, Ivan. 2019. "Psychologists Psychologizing Scientific Psychology: An Epistemological Reading of the Replication Crisis." *Theory and Psychology* 29 (2): 158–81.

Francken, Jolien, and Marc Slors. 2014. "From Commonsense to Science, and Back: The Use of Cognitive Concepts in Neuroscience." *Cognition* 29:248–58.

———. 2018. "Neuroscience and Everyday Life: Facing the Translation Problem." *Brain and Cognition* 120:67–74.

Franklin, Allan. 1986. *The Neglect of Experiment*. Cambridge: Cambridge University Press.

———. 1999. *Can That Be Right? Essays on Experiment, Evidence, and Science*. Dordrecht: Kluwer.

Franklin, Laura. 2005. "Exploratory Experiments." *Philosophy of Science* 72:888–99.

Freeman, Karen. 1996. "Sigmund Koch, Psychologist And Philosopher, Dies at 79." *New York Times*, April 14. https://www.nytimes.com/1996/08/14/us/sigmund-koch-psychologist-and-philosopher-dies-at-79.html.

Frege, Gottlob. 1892/1980. "Über Sinn und Bedeutung." *Zeitschrift für Philosophie und philosophische Kritik* 100:25–50. Translated by M. Black as "On Sense and Reference" in *Translations from the Philosophical Writings of Gottlob Frege*, ed. and trans. P. Geach and M. Black (Oxford: Basil Blackwell, 1952), 56–78.

Fried, Eiko. 2020. "Lack of Theory Building and Testing Impedes Progress in the Factor and Network Literature." *Psychological Inquiry* 31 (4): 271–88. https://doi.org/10.1080/1047840X.2020.1853461.

Fritz, Matthew, and Ann Artur. 2017. "Moderator Variables." In *Oxford Research Encyclopedia of Psychology*. https://doi.org/10.1093/acrefore/9780190236557.013.86.

Galison, Peter. 1996. *Image and Logic*. Chicago: University of Chicago Press.

Garner, Wendell R., Harold Hake, and Charles W. Eriksen. 1956. "Operationism and the Concept of Perception." *Psychological Review* 63:149–59.

Gelman, Andrew, and Eric Loken. 2014. "The Statistical Crisis in Science." *American Scientist* 102 (6): 460–64.

Ghai, Sakshi. 2021. "It's Time to Reimagine Sample Diversity and Retire the WEIRD Dichotomy." *Nature Human Behaviour* 5:971–72. https://www.nature.com/articles/s41562-021-01175-9.

Gibson, James J. 1979. *The Ecological Approach to Visual Perception*. Boston: Houghton Mifflin.

Glennan, Stuart. 2017. *The New Mechanical Philosophy*. Oxford: Oxford University Press.

Glennan, Stuart, Phyllis Illari, and Erik Weber. 2022. "Six Theses on Mechanisms and Mechanistic Science." *Journal for General Philosophy of Science* 53:143–61. https://doi.org/10.1007/s10838-021-09587-x.

Godman, Marion. 2021. *The Epistemology and Morality of Human Kinds*. London: Routledge.

Goldman, Alvin. 1992. "Reliabilism." In *A Companion to Epistemology*, ed. J. Dancy and E. Sousa, 433–36. Oxford: Blackwell.

Gomez-Lavin, Javier. 2021. "Working Memory Is Not a Natural Kind and Cannot Explain Central Cognition." *Review of Philosophy and Psychology* 12:199–225. https://doi.org/10.1007/s13164-020-00507-4.

Goodwin, C. James. 1995. *Research in Psychology: Methods and Design*. New York: Wiley.

Gozli, Davood. 2019. *Experimental Psychology and Human Agency*. Springer.

Graf, Peter, and George Mandler. 1984. "Activation Makes Words More Accessible, but Not Necessarily More Retrievable." *Journal of Verbal Learning and Verbal Behavior* 23:553–68.

Graf, Peter, and Lee Ryan. 1990. "Transfer-Appropriate Processing for Implicit and Explicit Memory." *Journal of Experimental Psychology: Learning Memory, and Cognition* 16 (6): 978–92.

Graf, Peter, and Daniel Schacter. 1985. "Implicit and Explicit Memory for New Associations in Normal and Amnesic Subjects." *Journal of Experimental Psychology: Learning Memory, and Cognition* 11 (3): 501–18.

Graf, Peter, Larry Squire, and George Mandler. 1984. "The Information That Amnesic Patients Do Not Forget." *Journal of Experimental Psychology: Learning, Memory, and Cognition* 10 (1): 164–78.

Graham, George. 2019. "Behaviorism." In *The Stanford Encyclopedia of Philosophy* (Spring 2019 ed.), ed. Edward N. Zalta. https://plato.stanford.edu/archives/spr2019/entries/behaviorism.

Green, Christopher. 1992. "Of Immortal Mythological Beasts: Operationism in Psychology." *Theory and Psychology* 2:291–320.

———. 2009. "Darwinian Theory, Functionalism, and the First American Psychological Revolution." *American Psychologist* 64 (2): 75–83.

Griffiths, Paul. 1997. *What Emotions Really Are: The Problem of Psychological Categories*. Chicago: University of Chicago Press.

Grünbaum, Thor. 2021. "The Two Visual Systems Hypothesis and Contrastive Underdetermination." *Synthese* 198 (suppl. 17): 4045–68. https://doi.org/10.1007/s11229-018-01984-y.

Guala, Francesco. 2000. "Artefacts in Experimental Economics: Preference Reversals and the Becker-Degroot-Marschak Mechanism." *Economics and Philosophy* 16:47–75.
———. 2003. "Experimental Localism and External Validity." *Philosophy of Science* 70:1195–1205.
———. 2005. *The Methodology of Experimental Economics*. Cambridge University Press.
———. 2012. "Experimentation in Economics." In *Handbook of the Philosophy of Science*, vol. 13, *Philosophy of Economics*, ed. Uskali Mäki, 597–640. Boston: Elsevier/Academic Press.
Gundlach, Ralph, and Madison Bentley. 1930. "The Dependence of Tonal Attributes upon Phase." *American Journal of Psychiatry* 42:519–43.
Guttinger, Stephan. 2019. "A New Account of Replication in the Experimental Life Sciences." *Philosophy of Science* 86 (3): 453–71.
Hacking, Ian. 1983. *Representing and Intervening: Introductory Topics in Philosophy of Science*. Cambridge: Cambridge University Press.
———. 1992. "The Self-Vindication of the Laboratory Sciences." In *Science as Practice and Culture*, ed. Andrew Pickering, 29–64. Chicago: University of Chicago Press.
———. 1995. "The Looping Effects of Human Kinds." In *Causal Cognition: A Multidisciplinary Debate* (Symposia of the Fyssen Foundation), ed. Dan Sperber, David Premack, and Ann James Premack, 361–94. New York: Clarendon.
Haig, Brian. 2022. "Understanding Replication in a Way That Is True to Science." *Review of General Psychology* 26 (2): 224–40.
Halford, Graeme, Stephen Phillips, and William Wilson. 2000. "Processing Capacity Limits Are Not Explained by Storage Limits." *Behavioral and Brain Sciences* 24:123–24.
Halverson, H. M. 1924. "Tonal Volume as a Function of Intensity." *American Journal of Psychology* 35 (3): 360–67.
Hammond, Kenneth. 1966. "Probabilistic Functionalism: Egon Brunswik's Integration of the History, Theory, and Method of Psychology." In *The Psychology of Egon Brunswik*, ed. K. Hammond, 15–80. New York: Holt, Rinehart & Winston.
Hanson, Norwood R. 1958. *Patterns of Discovery: An Inquiry into the Conceptual Foundations of Science*. Cambridge: Cambridge University Press.
———. 1961. "Is There a Logic of Discovery?" In *Current Issues in Philosophy of Science*, ed. Herbert Feigl and Grover Maxwell, 20–41. New York: Holt, Rinehart & Winston.
Hardcastle, Gary. 1995. "S. S. Stevens and the Origins of Operationism." *Philosophy of Science* 62:404–24.
Haslanger, Sally. 2012. *Resisting Reality*. Oxford: Oxford University Press.
Hatfield, Gary. 2002a. "Behaviorism and Naturalism." In *Cambridge History of Philosophy: 1870–1945*, ed. T. Baldwin, 93–106. Cambridge: Cambridge University Press.
———. 2002b. "Psychology, Philosophy and Cognitive Science: Reflections on the History and Philosophy of Experimental Psychology." *Mind and Language* 17 (3): 207–32.
———. 2005. "Introspective Evidence in Psychology." In *Scientific Evidence: Philosophical Theories and Applications*, ed. Peter Achinstein, 259–86. Baltimore: Johns Hopkins University Press.
———. 2019. "Gibson and Gestalt: (Re)Presentation, Processing, and Construction." *Synthese* 198 (suppl. 9): 2213–41. https://doi.org/10.1007/s11229-019-02380-w.
———. 2020. "Modern Meanings of Subjectivity: Philosophical, Psychological, Physiological." In *Internationales Jahrbuch des Deutschen Idealismus: Psychologie/International Yearbook of German Idealism: Psychology*, vol. 15, ed. Dina Emundts and Sally Sedgwick, 77–104. Berlin: De Gruyter.

Haueis, Philipp. 2016. "The Life of the Cortical Column: Opening the Domain of Functional Architecture of the Cortex (1955–1981)." *History and Philosophy of the Life Sciences* 38: article 2. https://doi.org/10.1007/s40656-016-0103-4.

———. 2017. "Meeting the Brain on Its Own Terms: Exploratory Concept Formation and Noncognitive Functions in Neuroscience." PhD diss., Otto-von-Guericke University Magdeburg.

———. 2018. "Beyond Cognitive Myopia: A Patchwork Approach to the Concept of Neural Function." *Synthese* 195 (12): 5373–5402.

———. 2023. "Exploratory Concept Formation and Tool Development in Neuroscience." *Philosophy of Science* 90 (2): 354–75. https://doi.org/10.1017/psa.2022.79.

———. 2024. "A Generalized Patchwork Approach to Scientific Concepts." *British Journal for Philosophy of Science* (preprint). https://doi.org/10.1086/716179.

Haugeland, John. 1998. "Pattern and Being." In *Having Thought. Essays in the Metaphysics of Mind*, 267–90. Cambridge, MA: Harvard University Press.

Havstadt, Joyce. 2018. "Messy Chemical Kinds." *British Journal for Philosophy of Science* 69:719–43.

Heidbreder, Edna. 1933. *Seven Psychologies*. New York: Century.

Heidelberger, M. 1998. "Die Erweiterung der Wirklichkeit im Experiment." In *Experimental Essay/Versuche zum Experiment*, ed. Michael Heidelberger and Friedrich Steinle, 71–92. Baden-Baden: Nomos.

Hempel, Carl G. 1935/1980. "The Logical Analysis of Psychology." Translated by Wilfrid Sellars. In *Readings in the Philosophy of Psychology* (2vols.), ed. Ned Block, 1:14–23. Cambridge, MA: Harvard University Press. Originally appeared as "Analyse logique de la psychologie," *Revue de synthèse* 55:27–42.

———. 1950. "Problems and Changes in the Empiricist Criterion of Meaning." *Revue internationale de philosophie* 4 (11): 41–63.

———. 1952. *Fundamentals of Concept Formation in Empirical Science*. International Encyclopedia of Unified Science, Foundations of the Unity of Science, vol. 2, no. 7. Chicago: University of Chicago Press.

———. 1954/1956. "A Logical Appraisal of Operationism." In *The Validation of Scientific Theories*, ed. Philipp Frank, 56–59. Boston: Beacon. Originally appeared as "A Logical Appraisal of Operationism," *Scientific Monthly* 79 (4): 215–20.

Henrich, Joseph, Steven Heine, and Ara Norenzayan. 2010. "The Weirdest People in the World?" *Behavioral and Brain Sciences* 33:61–135. https://doi.org/10.1017/S0140525X0999152X.

Hibbert, Fiona. 2019. "What Is Scientific Definition?" *Journal of Mind and Behavior* 40 (1): 29–52.

Hirst, William. 1989. "On Consciousness, Recall, Recognition, and the Architecture of Memory." In *Implicit Memory: Theoretical Issues*, ed. Stephan Lewandowsky, John C. Dunn, and Kim Kirsner, 33–46. Hillsdale, NJ: Lawrence Erlbaum.

Hochstein, Eric. 2016. "Categorizing the Mental." *Philosophical Quarterly* 66 (265): 745–59.

———. 2017. "When Does 'Folk Psychology' Count as Folk Psychological?" *The British Journal for the Philosophy of Science* 68 (4): 1125–47.

Hogarth, Robin M. 2006. "The Challenge of Representative Design in Psychology and Economics." *Journal of Economic Methodology* 12:253–63.

Holt, Edwin B. 1912. *The New Realism*. New York: Macmillan.

———. 1915a. *The Freudian Wish and Its Place in Ethics*. New York: Henry Holt.

———. 1915b. "Response and Cognition." In *The Freudian Wish and Its Place in Ethics* (New York: Henry Holt), 153–208.

Holt, Edwin B., Walter T. Marvin, William P. Montague, Ralph B. Perry, Walter B. Pitkin, and Edward G. Spaulding. 1910. "The Program and First Platform of Six Realists." *Journal of Philosophy* 7:393–401.

Hon, Giora. 1989. "Towards an Epistemology of Experimental Errors: An Epistemological View." *Studies in History and Philosophy of Science: Part A* 20 (4): 469–504.

———. 1998. "'If This Be Error': Probing Experiment with Error." In *Experimental Essays/Versuche zum Experiment*, ed. Michael Heidelberger and Friedrich Steinle, 227–48. Baden: Nomos.

Hoyningen-Huene, Paul. 1987. "Context of Discovery and Context of Justification." *Studies in History and Philosophy of Science: Part A* 18 (4): 501–15.

———. 1993. *Reconstructing Scientific Revolutions: Thomas S. Kuhn's Philosophy of Science*. Chicago: University of Chicago Press.

Hudson, Robert. 1999. "A Study in the Nature of Experimental Reasoning." *Philosophy of Science* 66 (2): 289–309.

———. 2020. "The Reality of Jean Perrin's Atoms and Molecules." *The British Journal for the Philosophy of Science* 71 (1): 33–58.

Hull, Clark L. 1920. "Quantitative Aspects of the Evolution of Concepts: An Experimental Study." *Psychological Monographs* 28 (1). Princeton, NJ: Psychological Review.

———. 1928. *Aptitude Testing*. Yonkers, NY: World Books.

———. 1929. "A Functional Interpretation of the Conditioned Reflex." *Psychological Review* 36:498–511.

———. 1930a. "Simple Trial and Error Learning: A Study in Psychological Theory." *Psychological Review* 37 (3): 241–56.

———. 1930b. "Knowledge and Purpose as Habit Mechanisms." *Psychological Review* 37:511–25.

———. 1931a. "Goal Attraction and Directing Ideas Conceived as Habit Phenomena." *Psychological Review* 38:487–506.

———. 1931b. "A Mechanical Model of the Conditioned Reflex." *Journal of General Psychology* 5:99–106.

———. 1932. "The Goal Gradient Hypothesis and Maze Learning." *Psychological Review* 39:25–43.

———. 1933. *Hypnosis and Suggestibility: An Experimental Approach*. New York: Appleton-Century.

———. 1934. "The Concept of the Habit Family Hierarchy and Maze Learning, Parts 1 and 2." *Psychological Review* 41:33–52, 134–52.

———. 1935a. "The Conflicting Psychologies of Learning—a Way Out." *Psychological Review* 42:491–516.

———. 1935b. "The Mechanism of the Assembly of Behavior Segments in Novel Combinations Suitable for Problem Solution." *Psychological Review* 42:219–45.

———. 1937. "Mind, Mechanism, and Adaptive Behavior." *Psychological Review* 44:1–32.

———. 1938. "Logical Positivism as a Constructive Methodology in the Social Sciences." *Einheitswissenschaft* 6:35–38.

———. 1943a. *Principles of Behavior: An Introduction to Behavior Theory*. New York: Appleton-Century.

———. 1943b. "The Problem of Intervening Variables in Molar Behavior Theory." *Psychological Review* 50:273–91.

———. 1952a. *A Behavior System: An Introduction to Behavior Theory concerning the Individual Organism.* New Haven, CT: Yale University Press.

———. 1952b. "Clark L. Hull." In *A History of Psychology in Autobiography*, vol. 4, ed. Edwin G. Boring, H. Werner, Herbert Langfeldt, and Robert Yerkes, 143–62. Worcester, MA: Clark University Press.

———. 1962. "Psychology of the Scientist: IV, Passages from the Idea Book." *Perceptual and Motor Skills* 15:807–82.

Hull, Clark L., and H. D. Baernstein. 1929. "A Mechanical Parallel to the Conditioned Reflex." *Science* 70:14–15.

Hull, Clark L., Carl I. Hovland, Robert Ross, Marshall Hall, Donald T. Perkins, and Frederic B. Fitch. 1940. *Mathematico-Deductive Theory of Rote Learning: A Study in Scientific Methodology.* New Haven, CT: Yale University Press.

Humphreys, Paul. 2004. *Extending Ourselves: Computational Science, Empiricism, and Scientific Method.* Oxford: Oxford University Press.

Inbar, Yoel. 2016. "Association between Contextual Dependence and Replicability in Psychology May Be Spurious." *PNAS: Proceedings of the National Academy of Sciences* 113 (34): E4933–E4934. www.pnas.org/cgi/doi/10.1073/pnas.1608676113.

Irvine, Elizabeth. 2014. *Consciousness as a Scientific Concept.* Studies in Brain and Mind. Dordrecht: Springer.

Isaac, Alistair M. C. 2017. "Hubris to Humility: Tonal Volume and the Fundamentality of Psychophysical Quantities." *Studies in History and Philosophy of Science: Part A* 65–66: 99–111. https://doi.org/10.1016/j.shpsa.2017.06.003.

Isaac, Alistair M. C., and Dave Ward. 2021. "Introduction." In "Gestalt Phenomenology and Embodied Cognitive Science," ed. Alistair M. C. Isaac and Dave Ward, special issue, *Synthese* 198 (9): 2135–51. https://doi.org/10.1007/s11229-019-02391-7.

Israel, Harold E. 1945. "Two Difficulties in Operational Thinking." *Psychological Review* 52 (5): 273–91.

Israel, Harold E., and Bernard H. Goldstein. 1944. "Operationism in Psychology." *Psychological Review* 52 (5): 177–88.

Jacoby, Larry L. 1991. "A Process Dissociation Framework: Separating Automatic from Intentional Uses of Memory." *Journal of Memory and Language* 30:513–41.

Jacoby, Larry L., and Fergus I. M. Craik. 1979. "Effects of Elaboration of Processing at Encoding and Retrieval: Trace Distinctiveness and Recovery of Initial Context." In *Levels of Processing in Human Memory*, ed. Laird S. Cermak and Fergus I. M. Craik, 1–21. Hillsdale, NJ: Lawrence Erlbaum.

Jacoby, Larry L., and Mark Dallas. 1981. "On the Relationship between Autobiographical Memory and Perceptual Learning." *Journal of Experimental Psychology: General* 110:306–40.

Jimenez-Buedo, Maria. 2011. "Conceptual Tools for Assessing Experiments: Some Well-Entrenched Confusions regarding the Internal/External Validity Distinction." *Journal of Economic Methodology* 18 (3): 271–82.

Jimenez-Buedo, Maria, and Federica Russo. 2021. "Experimental Practices and Objectivity in the Social Sciences: Re-Embedding Construct Validity in the Internal-External Validity Distinction." *Synthese* 199 (3–4): 9549–79.

Jonides, John, Richard L. Lewis, Derek Evan Nee, Cindy A. Lustig, Marc G. Berman, and Katherine Sledge Moore. 2008. "The Mind and Brain of Short-Term Memory." *Annual Review of Psychology* 59:193–224.

Justus, James. 2012. "Carnap on Concept Determination: Methodology for Philosophy of Science." *European Journal for Philosophy of Science* 2:161–79.

Kane, Michael, Andrew Conway, Timothy Miura, and Gregory Colflesh. 2007. "Working Memory, Attention Control, and the N-Back Task: A Question of Construct Validity." *Journal of Experimental Psychology: Learning, Memory, and Cognition* 33:615–22.

Kästner, Lena. 2017. *Philosophy of Cognitive Neuroscience: Causal Explanations, Mechanisms and Experimental Manipulations*. De Gruyter.

———. 2018. "Integrating Mechanistic Explanations through Epistemic Perspectives." *Studies in History and Philosophy of Science: Part A* 68:68–79.

———. 2021. "Integration and the Mechanistic Triad: Producing, Underlying and Maintaining Mechanistic Explanations." In *Neural Mechanisms* (Studies in Brain and Mind 17), ed. Fabricio Calzavarini and Marco Viola, 337–61. Springer. https://doi.org/10.1007/978-3-030-54092-0_15.

Kästner, Lena, and Philipp Haueis. 2021. "Discovering Patterns: On the Norms of Mechanistic Inquiry." *Erkenntnis* 86:1635–60. https://doi.org/10.1007/s10670-019-00174-7.

Kellen, David, Clintin Davis-Stober, John Dunn, and Michael Kalish. 2022. "The Problem of Coordination and the Pursuit of Structural Constraints in Psychology." *Perspectives on Psychological Science* 16 (4): 767–78. https://journals.sagepub.com/doi/10.1177/1745691620974771.

Kendig, Catherine. 2016a. "Editor's Introduction: Activities of Kinding in Scientific Practice." In *Natural Kinds and Classification in Scientific Practice*, ed. Catherine Kendig, 1–13. New York: Routledge.

———. 2016b. "Homologizing as Kinding." In *Natural Kinds and Classification in Scientific Practice*, ed. Catherine Kendig, 106–25. New York: Routledge.

Kendig, Catherine, and John Grey. 2019. "Can the Epistemic Value of Natural Kinds Be Explained Independently of Their Metaphysics?" *The British Journal for the Philosophy of Science* 72 (2): 359–76. https://doi.org/10.1093/bjps/axz004.

Kesebir, Selin, Oishi Shigehiro, and Barbara Spellman. 2010. "The Socio-Ecological Approach Turns Variance among Populations from a Liability to an Asset." *Behavioral and Brain Sciences* 33 (2–3): 96–97. https://doi.org/10.1017/S0140525X10000129.

Khalidi, Muhammad Ali. 2013. *Natural Categories and Human Kinds: Classification in the Natural and Social Sciences*. Cambridge: Cambridge University Press.

———. 2022. *Cognitive Ontology: Taxonomic Practices in the Mind-Brain Sciences*. Cambridge: Cambridge University Press.

Kim, Jaegwon. 1992. "Multiple Realization and the Metaphysics of Reduction." *Philosophy and Phenomenological Research* 52 (1): 1–26.

Kinoshita, Sachiko. 2001. "The Role of Involuntary Aware Memory in the Implicit Stem and Fragment Completion Tasks: A Selective Review." *Psychonomic Bulletin and Review* 8 (1): 58–69.

Klein, Collin. 2012. "Cognitive Ontology and Region- versus Network-Oriented Analyses." *Philosophy of Science* 79 (5): 952–60.

Koch, Sigmund. 1941a. "The Logical Character of the Motivation Concept I." *Psychological Review* 48:15–38.

———. 1941b. "The Logical Character of the Motivation Concept II." *Psychological Review* 48:127–44.

Köhler, Wolfgang. 1917. *Intelligenzpruefungen an Anthropoiden*. Abhandlungen der Koeniglich-Preussischen Akademie der Wissenschaften, no. 1. Berlin: Koniglichen Akademie der Wissenschaften.

Krech, David. 1974. "David Krech." In *A History of Psychology in Autobiography*, vol. 6, ed. Gardner Lindzey, 221–50. Hoboken, NJ: Prentice Hall. Krech was formerly known as Isadore Krechevsky.

Krechevsky, Isadore. 1932a. "The Genesis of 'Hypothesis' in Rats." *University of California Publications in Psychology* 6:45–64.

———. 1932b. "'Hypothesis' versus 'Chance' in the Pre-Solution Period in Sensory Discrimination Learning." *University of California Publications in Psychology* 6:27–44.

Krickel, Beate. 2018. *The Mechanical World. The Metaphysical Commitments of the New Mechanistic Approach*. Springer. https://link.springer.com/book/10.1007/978-3-030-03629-4.

Kripke, Saul. 1980. *Naming and Necessity*. Cambridge, MA: Harvard University Press.

Kronfeldner, Maria. 2015. "Reconstituting Phenomena." In *Recent Developments in the Philosophy of Science: EPSA13 Helsinki* (European Studies in the Philosophy of Science), ed. Uskali Mäki, Iohannis Votsis, Stéphanie Ruphy, and Gerhard Schurz, 169–81. Cham: Springer. https://link.springer.com/chapter/10.1007/978-3-319-23015-3_13.

Krueger, Robert G., and Clark L. Hull. 1931. "An Electro-Chemical Parallel to the Conditioned Reflex." *Journal of General Psychology* 5:262–69.

Kuhn, Thomas S. 1970. *The Structure of Scientific Revolutions*. Enlarged ed. Chicago: University of Chicago Press.

Külpe, Oswald. 1893. *Grundriss der Psychologie*. Nuremburg: Engelman. Translated by Edward Bradford as *Outlines of Psychology, Based on the Results of Experimental Investigation* (New York: Macmillan, 1895).

Larroulet Philippi, Christian. 2020. "Valid for What? On the Very Idea of Unconditional Validity." *Philosophy of Social Science* 51 (2): 151–75. https://doi.org/10.1177/0048393120971169.

Laurence, Stephen, and Eric Margolis. 1999. "Concepts and Cognitive Science." In *Concepts: Core Readings*, ed. Eric Margolis and Stephen Laurence, 3–81. Cambridge, MA: MIT Press.

Leahey, Thomas. 1980. "The Myth of Operationism." *Journal of Mind and Behavior* 1 (2): 127–43.

LeBel, Etienne, Derek Berger, Lorne Campbell, and Timothy Loving. 2017. "Falsifiability Is Not Optional." *Journal of Personality and Social Psychology* 113 (2): 245–61.

Lenoir, Timothy. 1992. "Practical Reason and the Construction of Knowledge: The Lifeworld of Haber-Bosch." In *The Social Dimension of Science*, ed. Ernan McMullin, 158–97. Notre Dame, IN: University of Notre Dame Press.

Leonelli, Sabina. 2015. "What Counts as Scientific Data? A Relational Framework." *Philosophy of Science* 82:810–21.

———. 2016. *Data-Centric Biology. A Philosophical Study*. Chicago: University of Chicago Press.

———. 2020. "Scientific Research and Big Data." In *The Stanford Encyclopedia of Philosophy* (Summer 2020 ed.), ed. Edward N. Zalta. https://plato.stanford.edu/archives/sum2020/entries/science-big-data.

———. 2023. *The Philosophy of Open Science*. Cambridge: Cambridge University Press.

Lewin, Kurt. 1931. *Die psychologische Situation bei Lohn und Strafe*. Leipzig: Hirzel.

Lewis, Clarence Irving. 1927. *Mind and the World Order*. New York: Dover.

Lockhart, Robert S. 1989. "The Role of Theory in Understanding Implicit Memory." In *Implicit Memory: Theoretical Issues*, ed. Stephan Lewandowsky, John C. Dunn, and Kim Kirsner, 3–16. Hillsdale, NJ: Lawrence Erlbaum.

———. 2000. "Methods of Memory Research." In *The Oxford Handbook of Memory*, ed. Endel Tulving and Fergus Craik, 45–57. Oxford: Oxford University Press.

Logan, Gordon, Michael Coles, and Arthur Kramer. 1996. "Introduction." In *Converging Operation in the Study of Visual Selection*, ed. Arthur Kramer, Michael Coles, and Gordon Logan, xi–xxv. Washington, DC: American Psychological Association.

Longino, Helen. 1990. *Science as Social Knowledge: Values and Objectivity in Scientific Inquiry*. Princeton, NJ: Princeton University Press.

MacCorquodale, Kenneth, and Paul Meehl. 1948. "On a Distinction between Hypothetical Constructs and Intervening Variables." *Psychological Review* 55:95–107.

Mace, John. 2003. "Involuntary Aware Memory Enhances Priming on a Conceptual Implicit Memory Task." *American Journal of Psychology* 116 (2): 281–90.

———. 2005. "Experimentally Manipulating the Effects of Involuntary Conscious Memory on a Priming Task." *American Journal of Psychology* 118 (2):159–82.

Machamer, Peter, Lindley Darden, and Carl Craver. 2000. "Thinking about Mechanisms." *Philosophy of Science* 67 (1): 1–25.

Machery, Edouard. 2009. *Doing without Concepts*. New York: Oxford University Press.

———. 2021. "What Is a Replication?" *Philosophy of Science* 87:545–67.

Machery, Edouard, Joshua Knobe, and Stephen Stich. 2023. "Editorial: Cultural Variation and Cognition." *Review of Philosophy and Psychology* 14:339–47. https://doi.org/10.1007/s13164-023-00687-9.

Mallon, Ron. 2018. "Constructing Race: Racialization, Causal Effects, or Both?" *Philosophical Studies* 175:1039–56. https://doi.org/10.1007/s11098-018-1069-8.

Manski, Charles. 1999. *Identification Problems in the Social Sciences*. Cambridge, MA: Harvard University Press.

Markman, Arthur, and Dedre Gentner. 1993. "Splitting the Differences: A Structural Alignment View of Similarity." *Journal of Memory and Language* 32 (4): 517–35.

Massimi, Michela. 2022. *Perspectival Realism*. Oxford: Oxford University Press.

Mayo, Deborah G. 1996. *Error and the Growth of Experimental Knowledge*. Chicago: University of Chicago Press.

McBride, Dawn. 2007. "Methods for Measuring Conscious and Automatic Memory: A Brief Review." *Journal of Consciousness Studies* 14:198–215.

McCaffrey, Joseph. 2015. "The Brain's Heterogeneous Functional Landscape." *Philosophy of Science* 82 (5): 1010–22.

McCaffrey, Joseph, and Jessey Wright. 2022. "Neuroscience and Cognitive Ontology: A Case for Pluralism." In *Neuroscience and Philosophy*, ed. Felipe de Brigard and W. Sinnott-Armstrong, 427–66. London: MIT Press.

McClimans, Leah. 2017. "Psychological Measures, Risks, and Values." In *Measurement in Medicine: Philosophical Essays on Assessment and Evaluation*, ed. Leah McClimans, 89–105. London: Rowman & Littlefield.

McDougall, William. 1908/1914. *An Introduction to Social Psychology*. 8th ed. London: Menuhen.

Meyer, David, and David Kieras. 1997. "A Computational Theory of Executive Cognitive Processes and Multiple-Task Performance: Part 1, Basic Mechanisms." *Psychological Review* 104 (1): 3–65.

Meyer, David, and Roger Schvanefeldt. 1971. "Facilitation in Recognizing Pairs of Words: Evidence of a Dependence between Retrieval Operations." *Journal of Experimental Psychology: General* 90 (2): 227–34.

Michaelian, Kourken. 2011. "Is Memory a Natural Kind?" *Memory Studies* 4 (2): 170–89. https://doi.org/10.1177/1750698010374287.

Miller, Boaz. 2016. "What Is Hacking's Argument for Entity Realism?" *Synthese* 193:991–1006. https://doi.org/10.1007/s11229-015-0789-y.

Miller, George. 1956. "The Magical Number Seven, Plus or Minus Two: Some Limits on Our Capacity for Processing Information." *Psychological Review* 63:81–97.

Milner, Peter. 2000. "Magical Attention." *Behavioral and Brain Sciences* 24:131.

Mitchell, Sandra. 2003. *Biological Complexity and Integrative Pluralism*. Cambridge: Cambridge University Press.

———. 2020. "Perspectives, Representation, and Integration." In *Understanding Perspectivism: Scientific Challenges and Methodological Prospects*, ed. Michela Massimi and Casey D. McCoy, 78–193. New York: Routledge.

Morawski, Jill. 1986. "Organizing Knowledge and Behavior at Yale's Institute of Human Relations." *Isis* 77:219–43.

Morris, Richard G. M, Paul Garrud, John N. P. Rawlins, and John O'Keefe. 1982. "Place Navigation Impaired in Rats with Hippocampal Lesions." *Nature* 297:681–83.

Muthukrishna, Michael, and Joseph Henrich. 2017. "A Problem in Theory." *Nature Human Behavior*. https://doi.org/10.1038/s41562-018-0522-1.

Nagel, Ernest. 1961. *The Structure of Science: Problems in the Logic of Scientific Explanation*. New York: Harcourt, Brace & World.

Nersessian, Nancy. 1984. *Faraday to Einstein: Constructing Meaning in Scientific Theories*. Dordrecht: Kluwer.

———. 1992. "How Do Scientists Think? Capturing the Dynamics of Conceptual Change in Science." In *Cognitive Models of Science*, ed. Ron Giere, 3–45. Minneapolis: University of Minnesota Press.

———. 2008. *Creating Scientific Concepts*. Cambridge, MA: MIT Press.

———. 2012. "Modeling Practices in Conceptual Innovation: An Ethnographic Study of a Neural Engineering Research Laboratory." In *Scientific Concepts and Investigative Practice*, ed. Uljana Feest and Fridrich Steinle, 245–70. Berlin: De Gruyter.

Newen, Albert, Anna Welpinghus, and Georg Juckl. 2015. "Emotion Recognition as Pattern Recognition: The Relevance of Perception." *Mind and Language* 30 (2): 187–208.

Newman, E., Stanley Smith Stevens, and Paul H. Davis. 1937. "Factors in the Production of Aural Harmonics and Combination of Tones." *Journal of the Acoustical Society of America* 9:107–14.

Newton, Isaac. 1687. *Philosophiae naturalis principia mathematica* (Mathematical principles of natural philosophy). London.

Nickles, Thomas. 2009. "The Strange Story of Scientific Method." In *Models of Discovery and Creativity*, ed. Joke Meheus and Thomas Nickles, 167–207. Dordrecht: Springer.

Norton, John. 2003. "A Material Theory of Induction." *Philosophy of Science* 70: 647–70.

Nosek, Brian A., and Timothy M. Errington. 2020. "What Is Replication?" *PLoS Biology* 18 (3): e3000691. https://doi.org/10.1371/journal.pbio.3000691.

Novick, Rose, and Philipp Haueis. 2023. "Patchworks and Operations." *European Journal for Philosophy of Science* 13: article 15. https://doi.org/10.1007/s13194-023-00515-y.

O'Donnell, John. 1985. *The Origins of Behaviorism: American Psychology, 1870–1920*. New York: New York University Press.

Ohnesorge, Miguel. 2022. "Pluralizing Measurement: Physical Geodesy's Measurement Problem and Its Resolution." *Studies in History and Philosophy of Science: Part A* 96:51–67.

O'Malley, Maureen. 2007. "Exploratory Experimentation and Scientific Practice: Metagenomics and the Proteorhodopsin Case." *History and Philosophy of the Life Sciences* 29 (3): 335–58.

Osbeck, Lisa. 2018. *Values in Psychological Science: Re-Imagining Epistemic Priorities at a New Frontier*. Cambridge: Cambridge University Press.

Oude Maatman, Freek. 2021. "Psychology's Theory Crisis, and Why Formal Modelling Cannot Solve It." Preprint. https://psyarxiv.com/puqvs.

Pascqual-Leone, Juan. 2000. "If the Magical Number Is 4, How Does One Account for Operations within Working Memory?" *Behavioral and Brain Sciences* 24:136–38.

Pennington, L., and J. Finan. 1940. "Operational Usage in Psychology." *Psychological Review* 47:254–66.

Pepper, Stephen. 1942. *World Hypotheses: A Study in Evidence*. Berkeley: University of California Press.

Perry, Ralph Barton. 1917a. "Purpose as Systematic Unity." *Monist* 27:252–75.

———. 1917b. "Purpose as Tendency and Adaptation." *Philosophical Review* 26:477–95.

———. 1918. "Docility and Purposiveness." *Psychological Review* 25:1–20.

———. 1921a. "A Behavioristic View of Purpose." *Journal of Philosophy* 18:85–105.

———. 1921b. "The Cognitive Interest and Its Refinements." *Journal of Philosophy* 18 (14): 365–75.

———. 1921c. "The Independent Universality of Purpose and Belief." *Journal of Philosophy* 18 (7): 169–80.

———. 1926. *General Theory of Value: Its Meaning and Basic Principles Construed in Terms of Interest*. New York: Longman, Green.

Piccinini, Gualtiero, and Carl Craver. 2011. "Integrating Psychology and Neuroscience: Functional Analyses as Mechanism Sketches." *Synthese* 183:283–311.

Place, Ullin T. 1956. "Is Consciousness a Brain Process?" *British Journal of Psychology* 47:44–50.

Poldrack, Russ. 2010. "Mapping Mental Function to Brain Structure." *Perspectives on Psychological Science* 5 (6): 753–61.

Poldrack, Russ, and Tal Yarkoni. 2016. "From Brain Maps to Cognitive Ontologies." *Annual Review of Psychology* 67:587–612.

Polger, Thomas, and Lawrence Shapiro. 2016. *The Multiple Realization Book*. Oxford: Oxford University Press.

Popper, Karl. 1935/1992. *The Logic of Scientific Discovery*. New York: Routledge.

Pöyhönen, Samuli. 2016. "Memory as a Cognitive Kind: Brains, Remembering Dyads, and Exograms." In *Natural Kinds and Classification in Scientific Practice*, ed. Catherine Kendig, 145–56. New York: Routledge.

Pramling, Niklas. 2011. "Possibilities as Limitations: A Study of the Scientific Uptake and Molding of G. A. Miller's Metaphor of a Chunk." *Theory and Psychology* 21:277–97.

Pratt, Carroll C. 1939. *The Logic of Modern Psychology*. New York: Macmillan.

———. 1945. "Operationism in Psychology." *Psychological Review* 52 (5): 262–69.

Price, Cathy, and Karl Friston. 2005. "Functional Ontologies for Cognition: The Systematic Definition of Structure and Function." *Cognitive Neuropsychology* 22 (3): 262–75.

Putnam, Hilary. 1962/1975a. "The Analytic and the Synthetic." In *Philosophical Papers*, vol. 2, *Mind, Language and Reality*, 33–69. New York: Cambridge University Press. Originally appeared as "The Analytic and the Synthetic," in *Scientific Explanation, Space and Time* (Minnesota Studies in the Philosophy of Science 3), ed. Herbert Feigl and Grover Maxwell, 358–97 (Minneapolis: University of Minnesota Press).

———. 1975b. "The Meaning of 'Meaning.'" In *Philosophical Papers*, vol. 2, *Mind, Language and Reality: Philosophical Papers*, 215–71. New York: Cambridge University Press. Originally appeared as "The Meaning of 'Meaning,'" in *Mind, Language and Reality* (Minnesota Studies in the Philosophy of Science 7), ed. Keith Gunderson, 215–71 (Minneapolis: University of Minnesota Press, 1975).

Quine, Williard van Orman. 1951. "Two Dogmas of Empiricism." *Philosophical Review* 60:20–43.

———. 1969. "Natural Kinds." In *Ontological Relativity and Other Essays*, 114–38. New York: Columbia University Press.

Rashevsky, Nicolas. 1931. "Possible Brain Mechanisms and Their Physical Models." *Journal of General Psychology* 5:262–69.

Rasmussen, Nicolas. 1993. "Facts, Artifacts, and Mesosomes: Practicing Epistemology with the Electron Microscope." *Studies in History and Philosophy of Science: Part A* 24:227–65.

Redick, Thomas, and Dakota Lindsey. 2013. "Complex Span and *N*-Back Measures of Working Memory: A Meta-Analysis." *Psychonomic Bulletin and Review* 20:1102–13. https://doi.org/10.3758/s13423-013-0453-9.

Reichenbach, Hans. 1920/1965. *The Theory of Relativity and A Priori Knowledge*. Berkeley: University of California Press.

———. 1938. *Experience and Prediction: An Analysis of the Foundations and the Structure of Knowledge*. Chicago: University of Chicago Press.

———. 1956. *The Direction of Time*. Berkely: University of California Press.

Renn, Jürgen. 2008. "The Historical Epistemology of Mechanics." Foreword to *The English Galileo: Thomas Harriot's Work on Motion as an Example of Preclassical Mechanics*, by Matthias Schemmel, vii–x. Dordrecht: Springer.

Reydon, Thomas. 2016. "From a Zooming-in Model to a Co-Creation Model: Towards a More Dynamic Account of Classification and Kinds." In *Natural Kinds and Classification in Scientific Practice*, ed. Catherine Kendig, 79–93. New York: Routledge.

Rheinberger, Hans-Jörg. 1997. *Toward a History of Epistemic Things: Synthesizing Proteins in the Test Tube*. Stanford, CA: Stanford University Press.

———. 2005. "A Reply to Bloor: 'Toward a Sociology of Epistemic Things.'" *Perspectives on Science* 13 (3): 406–10.

———. 2006. *Experimentalsysteme und epistemische Dinge: Eine Geschichte der Proteinsynthese im Reagenzglas*. Frankfurt: Suhrkamp.

———. 2011. "Infra-Experimentality: From Traces to Data, from Data to Patterning Facts." *History of Science* 49 (3): 337–48. https://doi.org/10.1177/007327531104900306.

Rich, G. 1916. "A Preliminary Study of Tonal Volumes." *Journal of Experimental Psychology* 1:13–22.

———. 1919. "A Tonal Study of Attributes." *American Journal of Psychiatry* 30:121–64.

Richardson-Klavehn, Alan, John M. Gardiner, and Rosalind I. Java. 1994. "Involuntary Conscious Memory and the Method of Opposition." *Memory* 2:1–29.

Ricker, Timothy, Angela AuBuchon, and Nelson Cowan. 2010. "Working Memory." *Wiley Interdisciplinary Review of Cognitive Science* 1 (4): 573–85. https://doi.org/10.1002/wcs.50.

Roback, A. A. 1923/1937. *Behaviorism and Psychology*. Cambridge, MA: Sci-Art.

Robins, Sarah. 2022. "Implicit Memory." In *Routledge Handbook of Philosophy of Implicit Cognition*, ed. Robert Thompson, 353–61. New York: Routledge.

Robischon, Thomas. 1967. "New Realism." In *Encyclopedia of Philosophy*, ed. P. Edwards, 5:485–89. New York: Macmillan/Free Press.

Roediger, Henry L., III. 2003. "Reconsidering Implicit Memory." In *Rethinking Implicit Memory*, ed. Jeffrey S. Bowers and Chad S. Marsolek, 3–18. Oxford: Oxford University Press.

Roediger, Henry L., III, and Lyn M. Goff. 1998. "Memory." In *A Companion to Cognitive Science*, ed. William Bechtel and George Graham, 250–64. Oxford: Blackwell.

Roediger, Henry L., III, Susan Weldon, and Bradford Challis. 1989. "Explaining Dissociations between Implicit and Explicit Measures of Retention: A Processing Account." In *Varieties of Memory and Consciousness: Essays in Honour of Endel Tulving*, ed. Henry L. Roediger III and Fergus Craik, 3–41. Hillsdale, NJ: Lawrence Erlbaum.

Rogers, Tim. 1989. "Operationism in Psychology: A Discussion of Contextual Antecedents and an Historical Interpretation of its Longevity." *Journal of the History of the Behavioral Sciences* 25:139–53.

Rol, Menno, and Nancy Cartwright. 2012. "Warranting the Use of Causal Claims: A Non-Trivial Case for Interdisciplinarity." *Theoria* 27 (2): 189–202.

Romero, Felipe. 2019. "Philosophy of Science and the Replicability Crisis." *Philosophy Compass*. https://doi.org/10.1111/phc3.12633.

———. 2022. "Novelty vs. Replicability: Virtues and Vices in the Reward System of Science." *Philosophy of Science* 84 (5): 1031–43. https://doi.org/10.1086/694005.

Rosch, Eleanor. 1978. "Principles of Categorization." In *Cognition and Categorization*, ed. Eleanor Rosch and Barbara B. Lloyd, 27–48. Hillsdale, NJ: Lawrence Erlbaum.

Rouse, Joseph. 2002. *How Scientific Practices Matter: Reclaiming Philosophical Naturalism*. Chicago: University of Chicago Press.

———. 2006. "Practice Theory." In *Philosophy of Anthropology and Sociology: Handbook of the Philosophy of Science*, vol. 5, ed. Stephen Turner and Mark Risjord, 490–540. Amsterdam: Elsevier.

———. 2011. "Articulating the World: Experimental Systems and Conceptual Understanding." *International Studies in the Philosophy of Science* 25 (3): 243–54.

———. 2015. *Articulating the World: Conceptual Understanding and the Scientific Image*. Chicago: University of Chicago Press.

Rubin, Mark. 2021. "What Type of Type I Error? Contrasting the Neyman-Pearson and Fisherian Approaches in the Context of Exact and Direct Replications." *Synthese* 198:5809–34. https://doi.org/10.1007/s11229-019-02433-0.

Rubin, Mark, and Chris Donkin. 2022. "Exploratory Hypothesis Tests Can Be More Compelling Than Confirmatory Hypothesis Tests." *Philosophical Psychology*. https://doi.org/10.1080/09515089.2022.2113771.

Russell, Bertrand, and Alfred N. Whitehead. 1910–13/1963. *Principia Mathematica*. 3 vols. Cambridge: Cambridge University Press.

Ryle, Gilbert. 1949/1983. *The Concept of Mind*. Harmondsworth: Penguin.

Salmon, Wesley. 1984. *Scientific Explanations and the Causal Structure of the World*. Princeton, NJ: Princeton University Press.

———. 1985. "Empiricism: The Key Questions." In *The Heritage of Logical Positivism*, ed. Nicholas Rescher, 1–22. Lanham, MD: University Press of America.

Sanches de Oliveira, Guilherme, and Edward Baggs. 2023. *Psychology's Weird Problems*. Cambridge: Cambridge University Press.

Schacter, Daniel. 1987. "Implicit Memory: History and Current Status." *Journal of Experimental Psychology: Learning, Memory, and Cognition* 13:501–18.

———. 1989. "On the Relation between Memory and Consciousness: Dissociable Interactions and Conscious Experience." In *Varieties of Memory and Consciousness: Essays in Honour of Endel Tulving*, ed. Henry L. Roediger III and Fergus Craik, 355–90. Hillsdale, NJ: Lawrence Erlbaum.

———. 1990. "Introduction to 'Implicit Memory: Multiple Perspectives.'" *Bulletin of the Psychonomic Society* 28 (4): 338–40.

———. 1992. "Understanding Implicit Memory: A Cognitive Neuroscience Approach." *American Psychologist* 47:559–69.

———. 1994. "Priming and Multiple Memory Systems: Perceptual Mechanisms of Implicit Memory." In *Memory Systems*, ed. Daniel Schacter and Endel Tulving, 233–68. Cambridge, MA: MIT Press.

———. 1999. "Implicit vs. Explicit Memory." In *The MIT Encyclopedia of Cognitive Sciences*, ed. Robert Wilson and Frank Keil, 156–57. Cambridge, MA: MIT Press.

Schacter, Daniel, Jeffrey Bowers, and Jill Booker.1989. "Intention, Awareness, and Implicit Memory: The Retrieval Intentionality Criterion." In *Implicit Memory: Theoretical Issues*, ed. Stephan Lewandowsky, John C. Dunn, and Kim Kirsner, 47–65. Hillsdale, NJ: Lawrence Erlbaum.

Scheel, Anne, Leonid Tiokhin, Peter Isager, and Daniël Lakens. 2021. "Why Hypothesis Testers Should Spend Less Time Testing Hypotheses." *Perspectives on Psychological Science* 16 (4):744–755.

Schickore, Jutta. 2005. "'Through Thousands of Errors We Reach the Truth'—but How? On the Epistemic Roles of Error in Science." *Studies in History and Philosophy of Science: Part A* 36:539–56.

———. 2011. "More Thoughts on HPS: Another 20 Years Later." *Perspectives on Science* 19 (4): 453–81.

———. 2016. "'Exploratory Experimentation' as a Probe into the Relation between Historiography and Philosophy of Science." *Studies in History and Philosophy of Science: Part A* 55:20–26.

———. 2017. *About Method: Experimenters, Snake Venom, and the History of Writing Scientifically*. Chicago: University of Chicago Press.

———. 2019. "The Structure and Function of Experimental Control in the Life Sciences." *Philosophy of Science* 86:203–18.

Schickore, Jutta, and Klodian Coko. 2014. "Using Multiple Means of Determination." *International Studies in the Philosophy of Science* 27 (3): 295–313.

Schmidt, Stefan. 2009. "Shall We Really Do It Again? The Powerful Concept of Replication Is Neglected in the Social Sciences." *Review of General Psychology* 13 (2): 90–100.

Schubert, Torsten, and Peter Frensch. 2000. "How Unitary Is the Capacity-Limited Attentional Focus?" *Behavioral and Brain Sciences* 24:146–47.

Schupbach, Jonah. 2018. "Robustness Analysis as Explanatory Reasoning." *British Journal of the Philosophy of Science* 69:275–300.

Sellars, Wilfrid. 1948. "Concepts as Involving Laws and Inconceivable without Them." *Philosophy of Science* 15:287–815.

Shadish, William, Thomas Cook, and Donald Campbell. 2002. *Experimental and Quasi-Experimental Designs for Generalized Causal Inferences*. Boston: Houghton, Mifflin.

Shapere, Dudley. 1976. "The Influence of Knowledge on the Description of Facts." *PSA: Proceedings of the Biennial Meeting of the Philosophy of Science Association: Symposia and Invited Papers* 2:281–98.

———. 1982. "Reason, Reference, and the Quest for Knowledge." *Philosophy of Science* 49:1–23.

———. 1984. *Reasons and the Search for Knowledge: Investigations in the Philosophy of Science*. Boston Studies in the Philosophy of Science 78. Dordrecht: Reidel.

Sijtsma, Klaas. 2006. "Psychometrics in Psychological Research: Role Model or Partner in Science?" *Psychometrika* 71 (3): 451–55. https://doi.org/10.1007/s11336-006-1497-9.

Skinner, Burrhus Frederic. 1938. *The Behavior of Organisms*. New York: Appleton-Century-Crofts.

———. 1945/1972. "The Operational Analysis of Psychological Terms." In *Cumulative Record. A Selection of Papers* (3rd ed.), 370–87. New York: Appleton-Century-Crofts.

Slaney, Kathleen. 2017. *Validating Psychological Constructs: Historical, Philosophical, and Practical Dimensions*. Palgrave Macmillan.

Slater, Matthew. 2015. "Natural Kindness." *British Journal of the Philosophy of Science* 66:375–411.

Smart, J. J. C. 1959. "Sensations and Brain Processes." *Philosophical Review* 68 (2): 141–56.

Smith, Eden. 2018a. "Interdependent Concepts and Their Independent Uses: Mental Imagery and Hallucinations." *Perspectives on Science* 26 (3): 360–99. https://doi.org/10.1162/posc_a_00278.

———. 2018b. "The Structured Uses of Concepts as Tools Comparing fMRI Experiments That Investigate either Mental Imagery or Hallucinations." PhD diss., University of Melbourne.

———. 2019. "Examining the Structured Uses of Concepts as Tools: Converging Insights." *Filozofia Nauki* (The philosophy of science) 27 (4): 7–22. https://doi.org/10.14394/filnau.2019.0023.

———. 2020. "Examining Tensions in the Past and Present Uses of Concepts." *Studies in History and Philosophy of Science: Part C: Studies in History and Philosophy of Biological and Biomedical Sciences* 84:84–94.

Smith, Laurence. 1986. *Behaviorism and Logical Positivism: A Reassessment of Their Alliance*. Stanford, CA: Stanford University Press.

Smith, Roger. 1997. *The Norton History of the Human Sciences*. New York: Norton.

Soler, Léna. 2011. "Tacit Aspects of Experimental Practices: Analytical Tools and Epistemological Consequences." *European Journal of Philosophy of Science* 1:393–433.

Stadler, Friedrich. 2001. *The Vienna Circle: Studies in the Origins, Development, and Influence of Logical Empiricism*. Vienna: Springer.

Steel, Daniel. 2008. *Across the Boundaries: Extrapolation in Biology and Social Science*. Oxford: Oxford University Press.

Steinle, Friedrich. 1997. "Entering New Fields: Exploratory Uses of Experimentation." *Philosophy of Science* 64:S65–S74.

———. 2009. "How Experiments Make Concepts Fail: Faraday and Magnetic Curves." In *Going Amiss in Experimental Research* (Boston Studies in the Philosophy of Science 267), ed. Giora Hon, Jutta Schickore, and Friedrich Steinle, 119–35. Berlin: Springer.

———. 2012. "Goals and Fates of Concepts: The Case of Magnetic Poles." In *Scientific Concepts and Investigative Practice*, ed. Uljana Feest and Friedrich Steinle, 105–25. Berlin: De Gruyter.

———. 2016. *Exploratory Experiments: Ampere, Faraday, and the Origins of Electrodynamics*. Translated by Alex Levine. Pittsburgh: University of Pittsburgh Press. Originally appeared as *Explorative Experimente: Ampere, Faraday und die Ursprünge der Elektrodynamik* (Stuttgart: Steiner, 2005).

Sternberg, Robert J. 1992. "*Psychological Bulletin*'s Top 10 'Hit Parade.'" *Psychological Bulletin* 112:387–88.

Stevens, Stanley S. 1933. "Materialism." Harvard University Archives, HUG (FP)-2.45, box 2, folder "Seminar Papers."

———. 1934a. "The Attributes of Tones." *PNAS: Proceedings of the National Academy of Sciences* 20:457–59.

———. 1934b. "Tonal Density." *Journal of Experimental Psychology* 17:585–92.

———. 1934c. "The Volume and Intensity of Tones." *American Journal of Psychology* 46:397–408.

———. 1935a. "The Operational Basis of Psychology." *American Journal of Psychology* 47:323–30.

———. 1935b. "The Operational Definition of Psychological Concepts." *Psychological Review* 42:517–27.

———. 1935c. "The Relation of Pitch to Intensity." *Journal of the Acoustical Society of America* 6:150–54.

———. 1936a. "Psychology, the Propaedeutic Science." *Philosophy of Science* 3:90–103.

———. 1936b. "A Scale for the Measurement of a Psychological Magnitude: Loudness." *Psychological Review* 43:405–16.

———. 1937. "On Hearing by Electrical Stimulation." *Journal of the Acoustical Society of America* 8:191–95.

———. 1939a. "On the Problem of Scales for the Measurement of Psychological Magnitudes." *Unified Science* 9:94–99.

———. 1939b. "Psychology and the Science of Science." *Psychological Bulletin* 36:221–63.

———. 1946. "On the Theory of Scales of Measurement." *Science* 103:677–80.

———. 1951. "Mathematics, Measurements, and Psychophysics." In *Handbook of Experimental Psychology*, ed. S. S. Stevens, 1–49. New York: Wiley.

———. 1968. "Edwin Garrigues Boring: 1886–1968." *American Journal of Psychology* 81:589–606.

———. 1974a. *Psychophysics: Introduction to Its Perceptual, Neural, and Social Prospects*. Edited by Geraldine Stevens. New York: Wiley.

———. 1974b. "S. S. Stevens." In *A History of Psychology in Autobiography*, vol. 6, ed. Gardner Lindzey, 395–420. Hoboken, NJ: Prentice Hall.

Stevens, Stanley S., and Paul H. Davis. 1936. "Physiological Acoustics: Pitch and Loudness." *Journal of the Acoustical Society of America* 8:1–13.

———. 1938. *Hearing: Its Psychology and Physiology*. New York: Wiley.

Stevens, Stanley S., and R. Clark Jones. 1939. "The Mechanism of Hearing by Electrical Stimulation." *Journal of the Acoustical Society of America* 10:261–69.

Stevens, Stanley S., J. Volkmann, and E. B. Newman. 1937. "A Scale for the Measurement of the Psychological Magnitude Pitch." *Journal of the Acoustical Society of America* 8:185–90.

Stinson, Catherine. 2016. "Mechanisms in Psychology: Ripping Nature at Its Seams." *Synthese* 193:1585–614.

Stone, Caroline. 2019. "A Defense and Definition of Construct Validity in Psychology." *Philosophy of Science* 86 (5): 1250–61.

Strack, Fritz. 2017. "From Data to Truth in Psychological Science: A Personal Perspective." *Frontiers in Psychology* 8: article 702. https://doi.org/10.3389/fpsyg.2017.00702.

Strapasson, Bruno Angelo, and Saulo de Freitas Araujo. 2020. "Methodological Behaviorism: Historical Origins of a Problematic Concept (1923–1973)." *Perspectives on Behavior Science* 43:415–29.

Strawson, Peter F. 1963. "Carnap's Views on Constructed Systems versus Natural Languages in Analytic Philosophy." In *The Philosophy of Rudolf Carnap*, ed. Paul Arthur Schilpp, 503–18. Lasalle, IL: Open Court.

Stroebe, Wolfgang, and Fritz Strack. 2014. "The Alleged Crisis and the Illusion of Exact Replication." *Perspectives on Psychological Science* 9 (1): 59–71. https://doi.org/10.1177/1745691613514450.

Sturm, Thomas, and Annette Mülberger. 2012. "Crisis Discussions in Psychology—New Historical and Philosophical Perspectives." *Studies in History and Philosophy of Science: Part C: Studies in History and Philosophy of Biological and Biomedical Sciences* 43:425–33.

Sullivan, Jacqueline Anne. 2007. "Validity and Reliability of Experiment in the Neurobiology of Learning and Memory." PhD diss., University of Pittsburgh.

———. 2009. "The Multiplicity of Experimental Protocols: A Challenge to Reductionist and Non-Reductionist Models of the Unity of Neuroscience." *Synthese* 167:511–39.

———. 2010. "Reconsidering 'Spatial Memory' and the Morris Water Maze." *Synthese* 177:261–83.

———. 2014. "Is the Next Frontier in Neuroscience a 'Decade of the Mind'?" In *Brain Theory: Essays in Critical Neurophilosophy*, ed. Charles Wolfe, 46–67. New York: Palgrave Macmillan.

———. 2016a. "Construct Stabilization and the Unity of the Mind-Brain Sciences." *Philosophy of Science* 83 (5): 662–73.

———. 2016b. "Neuroscientific Kinds through the Lens of Scientific Practice." In *Natural Kinds and Classification in Scientific Practice*, ed. Catherine Kendig, 47–56. New York: Routledge.

———. 2017a. "Coordinated Pluralism as a Means to Facilitate Integrative Taxonomies of Cognition." *Philosophical Explorations* 20 (2): 129–45.

———. 2017b. "Long-Term Potentiation: One Kind or Many?" In *Eppur Si Muove: Doing History and Philosophy of Science with Peter Machamer: A Collection of Essays in Honor or Peter Machamer* (Western Ontario Series in Philosophy of Science 81), ed. Marcus Adams, Zvi Biener, Uljana Feest, and Jacqueline Sullivan, 113–25. Dordrecht: Springer.

Suppe, Frederick. 1974/1977. *The Structure of Scientific Theories*. 2nd ed. Urbana: University of Illinois Press.

Tal, Eran. 2020. "Measurement in Science." In *The Stanford Encyclopedia of Philosophy* (Fall 2020 ed.), ed. Edward N. Zalta. https://plato.stanford.edu/archives/fall2020/entries/measurement-science.

Thompson, Morgan. 2022. "Epistemic Risk in Methodological Triangulation: The Case of Implicit Attitudes." *Synthese* 201 (1): 1–22.

———. 2023. "Path-Dependence in Measurement: A Problem for Coherentism." *Philosophy of Science*: 1–11. https://doi.org/10.1017/psa.2023.147.

Titchener, Edward B. 1908. *Lectures on the Elementary Psychology of Feeling and Attention*. New York: Macmillan.

———. 1910. *A Text-Book of Psychology.* New York: Macmillan.
———. 1929. *Systematic Psychology: Prolegomena.* New York: Macmillan.
Tolman, Edward C. 1918. "Nerve Process and Cognition." *Psychological Review* 25:423–42.
———. 1920. "Instinct and Purpose." *Psychological Review* 27:217–33.
———. 1922a. "Can Instincts Be Given Up in Psychology?" *Journal of Abnormal Psychology and Social Psychology* 17 (2): 139–52. Reprinted in Edward C. Tolman, *Behavior and Psychological Man: Essays in Motivation and Learning* (Berkeley: University of California Press, 1958), 9–22.
———. 1922b. "Concerning the Sensation Quality: A Behavioristic Account." *Psychological Review* 29:140–45.
———. 1922c. "A New Formula for Behaviorism." *Psychological Review* 29 (1): 44–53. Reprinted in Edward C. Tolman, *Behavior and Psychological Man: Essays in Motivation and Learning* (Berkeley: University of California Press, 1958), 1–8.
———. 1923a. "A Behavioristic Account of the Emotions." *Psychological Review* 30 (3): 217–27. Reprinted in Edward C. Tolman, *Behavior and Psychological Man: Essays in Motivation and Learning* (Berkeley: University of California Press, 1958), 23–31.
———. 1923b. "The Nature of Instinct." *Psychological Bulletin* 20:206–16.
———. 1925a. "Behaviorism and Purpose." *Journal of Philosophy* 22 (2): 36–41. Reprinted in Edward C. Tolman, *Behavior and Psychological Man: Essays in Motivation and Learning* (Berkeley: University of California Press, 1958), 32–37.
———. 1925b. "Purpose and Cognition: The Determiners of Animal Learning." *Psychological Review* 32 (4): 285–97. Reprinted in Edward C. Tolman, *Behavior and Psychological Man: Essays in Motivation and Learning* (Berkeley: University of California Press, 1958), 38–47.
———. 1926a. "A Behavioristic Theory of Ideas." *Psychological Review* 33 (5): 352–69. Reprinted in Edward C. Tolman, *Behavior and Psychological Man: Essays in Motivation and Learning* (Berkeley: University of California Press, 1958), 48–62.
———. 1926b. "The Fundamental Drives." *Journal of Abnormal Psychology* 20:349–58.
———. 1927. "A Behaviorist's Definition of Consciousness." *Psychological Review* 34 (6): 433–39. Reprinted in Edward C. Tolman, *Behavior and Psychological Man: Essays in Motivation and Learning* (Berkeley: University of California Press, 1958), 63–68.
———. 1930. Review of *The Great Apes*, by Robert M. Yerkes and Ada W. Yerkes. *American Anthropologist* 32:313–16.
———. 1932a. "Lewin's Concept of Vectors." *Journal of General Psychology* 7:3–15.
———. 1932b. *Purposive Behavior in Animals and Men.* New York: Century.
———. 1933. "Gestalt and Sign Gestalt." *Psychological Review* 40 (5): 391–411. Reprinted in Edward C. Tolman, *Behavior and Psychological Man: Essays in Motivation and Learning* (Berkeley: University of California Press, 1958), 77–93.
———. 1935. "Psychology versus Immediate Experience." *Philosophy of Science* 2 (3): 356–80. Reprinted in Edward C. Tolman, *Behavior and Psychological Man: Essays in Motivation and Learning* (Berkeley: University of California Press, 1958), 94–114.
———. 1936. "Operational Behaviorism and Current Trends in Psychology." *Proceedings of the 25th Anniversary of the Celebration of the Inauguration of Graduate Studies, University of Southern California*, 89–103. Los Angeles: University of Southern California Press. Reprinted in Edward C. Tolman, *Behavior and Psychological Man: Essays in Motivation and Learning* (Berkeley: University of California Press, 1958), 115–29.
———. 1937. "An Operational Analysis of 'Demands.'" *Erkenntnis* 6:383–92.

———. 1938. "The Determiners of Behavior at a Choice Point." *Psychological Review* 45 (1): 1–41. Reprinted in Edward C. Tolman, *Behavior and Psychological Man: Essays in Motivation and Learning* (Berkeley: University of California Press, 1958), 144–78.

———. 1948. "Cognitive Maps in Rats and Men." *Psychological Review* 55 (4): 189–208.

———. 1952. "Edward Chace Tolman." In *A History of Psychology in Autobiography*, vol. 4, ed. Edwin G. Boring, H. Werner, Herbert Langfeld, and Robert Yerkes, 323–39. Worcester, MA: Clark University Press.

———. 1955. "Egon Brunswik, Psychologist and Philosopher of Science." *Science* 122:910.

———. 1956. "Egon Brunswik: 1903–1955." *American Journal of Psychology* 69:315–42.

———. 1958. *Behavior and Psychological Man: Essays in Motivation and Learning*. Berkeley: University of California Press.

———. 1959. "Principles of Purposive Behavior." In *Psychology: A Study of a Science*, vol. 2, ed. Sigmund Koch, 92–157. New York: McGraw Hill.

Tolman, Edward, and Egon Brunswik. 1935. "The Organism and the Causal Texture of the Environment." In *The Psychology of Egon Brunswik*, ed. Kenneth Hammond, 457–86. New York: Holt, Rinehart & Winston, 1966. Originally appeared as "The Organism and the Causal Texture of the Environment," *Psychological Review* 42:43–77.

Tolman, Edward, and Isadore Krechevsky. 1933. "Means-End-Readiness and Hypothesis—a Contribution to Comparative Psychology." *Psychological Review* 40:60–70.

Tulodziecki, Dana. 2013. "Shattering the Myth of Semmelweis." *Philosophy of Science* 80:1065–75.

Tulving, Endel. 1983. *Elements of Episodic Memory*. New York: Oxford University Press.

———. 1985. "How Many Memory Systems Are There?" *American Psychologist* 40:385–98.

———. 2000. "Concepts of Memory." In *The Oxford Handbook of Memory*, ed. Endel Tulving and Fergus Craik, 34–43. Oxford: Oxford University Press.

Tulving, Endel, and Daniel Schacter. 1990. "Priming and Human Memory Systems." *Science* 247:301–6.

Tulving, Endel, Daniel Schacter, and Heather Starck. 1982. "Priming Effects in Word-Fragment Completion Are Independent of Recognition Memory." *Journal of Experimental Psychology: Learning, Memory, and Cognition* 8:336–42.

Tversky, Barbara. 1993. "Cognitive Maps, Cognitive Collages, and Spatial Mental Models." In *Spatial Information Theory: A Theoretical Basis for GIS*, ed. Andrew U. Frank and Irene Campari, 14–24. Berlin: Springer.

Underwood, Benton J. 1957. *Psychological Research*. New York: Appleton-Century-Crofts.

Uygun Tunç, Duygu, and Mehmet Necip Tunç. 2023. "A Falsificationist Treatment of Auxiliary Hypotheses in Social and Behavioral Sciences: Systematic Replications Framework." *Metapsychology* 7. https://doi.org/10.15626/MP.2021.2756.

Van Bavel, Jay, Peter Mende-Siedleckia, William Bradya, and Diego Reineroa. 2016a. "Contextual Sensitivity in Scientific Reproducibility." *PNAS: Proceedings of the National Academy of Sciences* 113 (23): 6454–59. www.pnas.org/cgi/doi/10.1073/pnas.1521897113.

———. 2016b. "Contextual Sensitivity Helps Explain the Reproducibility Gap between Social and Cognitive Psychology." *PNAS: Proceedings of the National Academy of Sciences* 113 (34): E4935–E4936. www.pnas.org/cgi/doi/10.1073/pnas.1609700113.

van Rooij, Iris, and Giosuè Baggio. 2020. "Theory Development Requires an Epistemological Sea Change." *Psychological Inquiry* 31 (4): 321–25. https://doi.org/10.1080/1047840X.2020.1853477.

———. 2021. "Theory before the Test: How to Build High-Verisimilitude Explanatory Theories in Psychological Science." *Perspectives on Psychological Science*. 16 (4): 682–97. https://journals.sagepub.com/doi/full/10.1177/1745691620970604.

Vazire, Simine, Sarah Schiavone, and Julia Bottesini. 2022. "Credibility beyond Replicability: Improving the Four Validities in Psychological Science." *Current Directions in Psychological Science* 31 (2): 162–68.

Verhaegh, Sander. 2021. "Psychological Operationisms at Harvard: Skinner, Boring, and Stevens." *Journal for the History of the Behavioral Sciences* 57 (2): 194–212.

Vessonen, Eleni. 2020. "Respectful Operationalism." *Theory and Psychology* 31 (1): 84–105.

———. 2021. "Conceptual Engineering and Operationalism in Psychology." *Synthese* 199 (3–4): 10615–37.

Wajnerman-Paz, Abel, and Daniel Rojas-Líbano. 2022. "On the Role of Contextual Factors in Cognitive Neuroscience Experiments: A Mechanistic Approach." *Synthese* 200: article 402. https://doi.org/10.1007/s11229-022-03870-0.

Warrington, Elizabeth, and Lawrence Weiskrantz. 1968. "New Methods of Testing Long-Term Retention with Special Reference to Amnesic Patients." *Nature* 217:972–74.

———. 1970. "Amnesic Syndrome: Consolidation or Retrieval?" *Nature* 228:629–30.

———. 1982. "Amnesia: A Disconnected Syndrome?" *Neuropsychologia* 20:233–48.

Waters, R., and L. A. Pennington. 1938. "Operationism in Psychology." *Psychological Review* 45:414–23.

Watson, John B. 1908. "The Behavior of Noddy and Sooty Terns." *Carnegie Publications* 103:187–255.

———. 1913. "Psychology as the Behaviorist Views It." *Psychological Review* 20:158–77.

———. 1914. *Behavior: An Introduction to Comparative Psychology*. New York: Henry Holt.

Weber, Marcel. 2004. *Philosophy of Experimental Biology*. Cambridge: Cambridge University Press.

———. 2009. "The Crux of Crucial Experiments: Duhem's Problems and Inference to the Best Explanation." *British Journal of the Philosophy of Science* 60:19–49.

———. 2018. "Experiment in Biology." In *The Stanford Encyclopedia of Philosophy* (Summer 2018 ed.), ed. Edward N. Zalta. https://plato.stanford.edu/archives/sum2018/entries/biology-experiment.

Weisskopf, Daniel. 2011. "Models and Mechanisms in Psychological Explanation." *Synthese* 183:313–38. https://doi.org/10.1007/s11229-011-9958-9.

———. 2020. "Anthropic Concepts." *Noûs* 54 (2): 451–68.

Wilson, Mark. 1982. "Predicate Meets Property." *Philosophical Review* 91 (4): 549–89.

———. 2006. *Wandering Significance: An Essay in Conceptual Behavior*. Oxford: Oxford University Press.

Wimsatt, William C. 1981/2012. "Robustness, Reliability, and Overdetermination." In *Characterizing the Robustness of Science* (Boston Studies in the Philosophy of Science 292), ed. Léna Soler, Emiliano Trizio, Thomas Nickles, and William C. Wimsatt (Dordrecht: Springer, 2012), 61–87. Originally appeared as "Robustness, Reliability, and Overdetermination," in *Scientific Inquiry and the Social Sciences*, ed. Marilynn B. Brewer and Barry E. Collins (San Francisco: Jossey-Bass), 123–62.

———. 1994. "The Ontology of Complex Systems: Levels of Organization, Perspectives and Causal Thickets." *Canadian Journal of Philosophy of Science* 20 (suppl.): 207–70.

Woodger, J. H. 1937. *The Axiomatic Method in Biology*. Cambridge: Cambridge University Press.

Woodward, James. 1989. "Data and Phenomena." *Synthese* 79:393–472.

———. 2000. "Data, Phenomena, and Reliability." *Philosophy of Science* 67:S163–79.

———. 2005. *Making Things Happen: A Theory of Causal Explanation*. Oxford: Oxford University Press.

Woodworth, R. 1931. *Contemporary Schools of Psychology*. New York: Ronald Press.

Wundt, Wilhelm. 1874/1893. *Grundzüge der Physiologischen Psychologie*. Vol. 1. 4th ed. Leipzig: Wilhelm Engelmann.

Yarkoni, Tal. 2022. "The Generalizability Crisis." *Behavioral and Brain Sciences* 45:e1. https://doi.org/10.1017/S0140525X20001685.

Yarkoni, Tal, Russ A. Poldrack, Thomas E. Nichols, David C. van Essen, and Tor D. Wager. 2011. "Large-Scale Automated Synthesis of Human Functional Neuroimaging Data." *Nature Methods* 8 (8): 665–70. https://doi.org/10.1038/nmeth.1635.

Yerkes, Robert. 1916. *The Mental Life of Monkeys and Apes*. Behavior Monographs 3 (1). Cambridge, MA: Henry Holt.

———. 1932. "Robert Mearns Yerkes, Psychobiologist." In *A History of Psychology in Autobiography*, vol. 2, ed. C. Murchison, 381–407. New York: Russell & Russell.

Zoll, P. 1934. "The Relation of Tonal Volume, Intensity, and Pitch." *American Journal of Psychiatry* 46:99–106.

INDEX

American Psychological Association, 87
Amsel, Abram, 67
Andersen, Hanne, 130–31
Arabatzis, Theodore, 128n14
Araújo, Duarte, 32n2, 266
Araujo, Saulo de Freitas, 32n2
Artur, Ann, 276
Ash, Mitchell, 54n23, 212
Atkinson, Richard, 105
AuBuchon, Angela, 120

Baddeley, Alan, 105, 106
Baggio, Giosuè, 179n10, 272
Barker, Peter, 130–31
Barker, Roger, 268
Barsalou, Lawrence, 130, 131
Basso, Alessandra, 253
Beaman, Philip, 149, 151
Bechtel, William, 176, 186–88; and Mundale, 206; and Richardson, 2, 183–85
behavior, 23, 31; adaptive, 57, 66; classification of, 211–15, 220, 223; description of, 51, 52, 214, 221; discriminatory, 38, 72, 98, 105, 244, 246; molar, 51, 65; molecular, 51, 65; navigational, 4, 177; purposive, 51, 53, 59
behavioral sciences, 3, 17, 141
behaviorism, 32, 44, 58, 62, 70, 266; analytical, 31, 46; logical, 31, 32; methodological, 31; operant, 33; purposive, 47. *See also* Tolman, Edward C.

Bentley, Madison, 40n11
Bergmann, Gustav, 67, 86n10; and Spence, 64, 76, 80–81, 82n5, 240
Bickle, John, 275
bilateral reduction sentence, 83
Bird, Alexander, 201
Blaxton, Teresa, 248–50
Bloch-Mullins, Corinne, 131, 212–13
Bogen, James, 4, 27, 164–70, 172, 176, 189, 192
Booker, Jill, 143–44
Boone, Worth, 190
Boring, Edwin, 35–36, 36n7, 39–41, 44–45, 64n34, 69, 80
Borsboom, Denny, 89n13, 251n14, 255n18
Bowers, Jeffrey, 122, 123, 143–44
Boyd, Richard, 201–3, 205, 208n11
Brandom, Robert, 19n5
Bridgman, Percy, 5, 6, 22, 24, 30–31, 34n5, 45, 56, 62, 74, 76, 80, 83, 130, 243, 260
Brigandt, Ingo, 132n16, 204
Briggs, Derek, 279
Brun, Georg, 112
Brunswik, Egon, 47, 54–55, 54n24, 56, 214, 266–67
Burian, Richard, 2, 153–54, 153n11, 161
Burnston, Daniel, 191, 216, 220
Bursten, Julia, 19n5, 130
Buzbas, Erkan, 269, 270n2

Campbell, Donald, 91–93, 94, 96–97, 251
capacity/capacities: behavioral, 24, 28, 179, 180, 210, 271; cognitive, 24, 28, 210, 222; complex, 222, 223; experiential, 24, 28, 210; whole-organism, 24, 28, 220, 223, 256, 271
Carnap, Rudolf, 32, 32nn3–4, 46, 56, 72n2, 81–85, 111–12
Carpenter, Patricia, 147
Carroll, David, 46n17
Cartwright, Nancy, 230n2
Chang, Hasok, 13, 16, 22, 31, 37n9, 89, 103, 123, 129, 145, 198, 259–60
Chemero, Anthony, 266
Chen, Xiang, 130–31
Cheon, Hyundeuk, 132
chunk(s), 147–52, 158, 167, 208
chunking, 139, 147–52, 158, 167, 171–74, 181, 188–89, 228. *See also* working memory capacity
Churchland, Paul, 220
Cohen, Neil, 105
Coko, Klodian, 93, 252n16, 253
Colaço, David, 3n2, 104nn1–2, 157, 184n12, 275
Collins, Harry M., 23, 252, 275n3
concept formation, 3, 9, 23, 58–59, 94–95, 100, 171, 226–27, 272
concepts: classical theory of, 124–26; description theory of, 26, 124–25; patchwork account of, 19, 130; as tools, 26
conceptual: assumptions, 7, 17, 104–5, 161, 165, 170, 215, 263–64; change, 19, 43, 108, 111, 118–19, 124, 130–33; development, 23, 103, 112, 119, 130, 226, 236, 240, 243, 246, 251–55, 270, 280; openness, 1, 3, 25, 85, 95, 118, 138, 140, 152–54, 157
construct(s), 72, 90; empirical, 76, 86; formation, 93; hypothetical, 64–65, 134; psychological, 65, 82, 220, 221; theoretical, 83, 90; validity/validation, 88–91, 94, 250, 251, 279. *See also* intervening variables; MacCorquodale, Kenneth; Meehl, Paul
context: of discovery, 8–10; of justification, 8–10; sensitive, 216, 220, 223, 227, 257, 258, 274–76, 278; specificity, 9, 116, 156

contextual: theory of evidence, 235, 267, 270; variation, 257–58
conventionalism, 196, 200–203
converging operations, 29, 63, 74, 97–99, 160, 226, 228, 240, 243–61
Conway, Andrew, 107
Cowan, Nelson, 120, 148–52
Craver, Carl, 164n1, 179, 182, 184n12, 186, 201, 201n5, 205–6; and Dan-Cohen, 242; and Darden, 2, 16, 176–77, 182, 186; and Piccinini, 180
Crawford, Sean, 32
Cronbach, Lee, 231, 275; and Meehl, 87–91, 94–97, 251
Culp, Sylvia, 93, 252–53
Cummins, Robert, 179

Dallas, Mark, 122
Dan-Cohen, Talia, 242
Daneman, Meredyth, 147
Danks, David, 206, 234
Danziger, Kurt, 199, 208
Darden, Lindley, 2, 11, 16, 176–77, 182, 186
data, 6, 21–22; generation/production of, 28–29; inferences from, 103, 113, 227, 228–30, 232, 236, 255, 260; and phenomena, 27; quality of, 138, 171, 192, 227, 236–38, 270; reliability of, 29, 239, 240, 243, 246, 252–53, 263, 267, 270
Davids, Keith, 266
Davis, Paul H., 39n10
decomposition, 183, 184, 187; mechanistic, 187–88; phenomenal, 187–88. *See also* localization
description, 194, 206, 214; of experimental cause/manipulation, 227, 230; of experimental effect, 229; level of, 11, 209, 216, 221; of object of research, 126, 138–52, 154, 161, 163, 182, 254, 263, 273–74; phenomenological, 27, 182, 193; thick, 13
description theory of meaning. *See* concepts
descriptive, 10; features, 27, 37, 118, 126, 138, 139, 140–41, 151, 154–55, 159, 171. *See also* description: of object of research
Devezer, Berna, 269, 270n2
Dewhurst, Joseph, 220–21

discovery, 2, 7, 15–16, 19, 69, 183, 186–88, 191; and justification, 7–8, 10, 84; logic of, 8; of mechanisms, 165, 182, 193
Donkin, Chris, 272
Dupre, John, 201

Ebbinghaus, Hermann, 61, 107
Ebbs, Gary, 128–29, 133
Eberhardt, Frederick, 234
ecological: design, 266–67; psychology, 54, 211–15, 265–66
Eichenberger, Philipp, 256n20
Eigner, Kai, 84n8
eliminativism, 44, 78, 221
Elliott, Kevin, 16–17
epistemic: access, 87, 101, 114, 169, 186, 263; iteration, 16–17; perspective/standpoint of researchers, 26, 84, 101, 118, 135–38, 194, 261, 208; things, 2, 19–20, 136; tools, 117, 133; uncertainty, 2–4, 25, 73–75, 85, 95, 118, 137–39, 239, 263
epistemically blurry, 4, 19, 136–37, 160–61, 164, 173–74, 178–79, 262–63, 265, 278
Ereshefsky, Mark, 202
Eriksen, Charles W., 25, 97–100, 243–47, 253
evidence, relational theory of, 168–70
experiment: crucial, 246–47, 249; defining, 42, 49, 50, 70
experimental artifacts, 226–27, 240–43; data, 21. *See also* data
experimental design, 25–26, 76, 104, 114–16, 174, 232–33, 248, 260, 266–68, 274
experimental inferences, 29, 226–27, 230, 233, 236–37, 238, 241, 264, 267, 269, 270, 276, 279
experimental paradigm, 6, 19, 40, 41, 116, 161, 184
experimental results, 236–37
explanation: constitutive, 175, 177, 179, 181, 220; etiological, 199; mechanistic, 24, 27, 164, 165, 175, 176, 178, 179–82, 184
exploratory: experimentation, 2, 3, 27, 115, 152–60, 172, 174; research, 138, 152, 155–57, 159–60, 161, 171, 271, 272
extension of concept, 1, 37, 114, 123–30, 136, 154, 171, 173, 204, 207, 254, 257

Feigl, Herbert, 45, 67, 80
Fidler, Fiona, 20n6
Fiske, Donald W., 91–93, 94, 96–97
folk psychology, 4, 28, 103–4, 112, 141, 198, 207, 219–21, 223, 225, 271
Francken, Jolien, 219–20
Franklin, Allan, 2, 3, 240n8
Franklin-Hall, Laura, 156–57
Frege, Gottlob, 124
Frensch, Peter, 150–51
Fried, Eiko, 272
Friston, Karl, 215, 216
Fritz, Matthew, 276
functional dissociation, 1, 17, 106, 109, 124, 142–43, 248–49

Galison, Peter, 9
Gardiner, John M., 144, 145
Garner, Wendell R., 25, 97–100, 243–47, 253
Gelman, Andrew, 21
Gentner, Dedre, 212
Gibson, James, 54, 214, 266
Glennan, Stuart, 175
Gomez-Lavin, Javier, 113n9
Goodwin, C. James, 88
Gozli, Davood, 273, 281
Graf, Peter, 141, 249
Graham, George, 32n2
Green, Christopher, 34, 34n5, 45, 214
Grey, John, 208
Griffiths, Paul, 201, 210
Guala, Francesco, 226, 231–32, 241–42
Gundlach, Ralph, 40n11
Guttinger, Stephan, 240

Hacking, Ian, 2, 7, 10, 165, 169, 200n4, 254
Haig, Brian, 251n15
Hake, Harold, 25, 97–100, 243–47, 253
Halford, Graeme, 151
Halverson, H. M., 40
Hanson, Norwood Russell, 8, 191
Hardcastle, Gary, 36n8, 46
Hatfield, Gary, 44n13, 211n12, 214–15
Haueis, Philipp, 3n2, 19n5, 115, 130, 130n15, 153–54, 154n12, 191
Haugeland, John, 191
Havstadt, Joyce, 198

Heidbreder, Edna, 68
Heidelberger, Michael, 117n11
Heine, Steven, 277
Hempel, Carl G., 30, 32–33, 81–83
Henrich, Joseph, 271, 277
Hibbert, Fiona, 35n6
Hitch, Graham, 105, 106
Hochstein, Eric, 206
Hogarth, Robin M., 232
Holt, Edwin Bissel, 46, 51–52, 58
homeostatic property cluster (HPC) theory, 201–8
Hon, Giora, 242
Hoyningen-Huene, Paul, 10
Hudson, Robert, 252
Hull, Clark, 24–25, 32, 34, 34n5, 56–79
human kinds, 200
Humphreys, Paul, 31
hypothetical constructs, 64. *See also* Cronbach, Lee: and Meehl

Illari, Phyllis, 175
implicit memory, 1, 108–10, 122–23, 141–46, 187–88, 247–50
Inbar, Yoel, 276
individuation challenge, 29, 227, 229–30, 233–38, 241, 246
inference schema, 236–39
intervening variables, 47–51, 62–63, 100
Isaac, Alistair M. C., 40, 268
Israel, Harold E., 80

Jacoby, Larry L., 122, 145
Java, Rosalind I., 144, 145
Jimenez-Buedo, Maria, 230n2, 231–32, 280
Jonides, John, 121
Juckl, Georg, 213n13
Justus, James, 111n8

Kane, Michael, 107
Kästner, Lena, 164n2, 173, 174, 175, 177n9, 180n11, 182, 191
Kellen, David, 260
Kendig, Catherine, 104n1, 195n1, 200, 208n11
Khalidi, Muhammad Ali, 202, 208–10, 208n11, 216n15
Kieras, David, 147

Kim, Jaegwon, 209
Koch, Sigmund, 64, 86, 86n10
Koffka, Kurt, 46, 67
Köhler, Wolfgang, 54, 54n23
Kouider, Sid, 122, 123
Krech, David (Isadore Krechevsky), 55, 55n25
Krickel, Beate, 174, 175, 177
Kripke, Saul, 198
Kronfeldner, Maria, 185
Krueger, Robert G., 65
Kuhn, Thomas, 77, 115
Külpe, Oswald, 43

latent variables, 233–34, 236–37
Latour, Bruno, 118
Leahey, Thomas, 32–33, 56
Leonelli, Sabina, 21n7, 170, 192
Lewin, Kurt, 54, 67, 86n10
Lewis, C. I., 55, 115n10
localization, 182–89, 196, 215–20
logical positivism, 24, 30, 32, 81–84; and Stevens, 46
Loken, Eric, 21
Longino, Helen, 280

MacCorquodale, Kenneth, 48, 64
Mace, John, 144, 145n7
Machery, Edouard, 132, 269n1, 275–76, 279
manipulation challenge, 29, 233–38, 241, 244, 246, 260–61
Manski, Charles, 230n2
Markman, Arthur, 212
Massimi, Michela, 195, 195n2, 203–4, 203n7
Mayo, Deborah, 2, 3, 238, 242–43
McBride, Dawn, 145
McCaffrey, Joseph, 216, 222
McDougall, William, 53
meaning: of concepts/terms, 36–42, 71–79, 81–86, 124–32, 262–63; of experimental data, 21
measurement, challenge, 29, 227, 233–38, 241, 244, 246, 260
mechanisms, discovery of, 182–86
mechanistic: decomposition, 183, 187; explanation, 174–82

Meehl, Paul, 48, 64, 87–91, 94–97, 251
Mellenbergh, Gideon, 89n13, 251n14
memory: implicit, 1, 108–10, 122–23, 141–46, 187–88, 247–50; long-term, 105–6; short-term, 105–8; spatial, 176–81. *See also* working memory capacity
methodological: iteration, 16–17; maxim, 7, 12, 15; rules, 3; strategies, 15
Meyer, David, 109n7, 147, 229
Miller, Boaz, 254
Miller, George, 147–48
Milner, Peter, 151
Mitchell, Sandra, 173
Morawski, Jill, 66, 67
Morris Water Maze, 176–79
Mülberger, Annette, 20n6
multitrait-multimethod approach, 91–93
Mundale, Jennifer, 206
Muthukrishna, Michael, 271

Nagel, Ernest, 58
natural kinds, 197–211
Nersessian, Nancy, 7, 130–31
Newen, Albert, 213n13
Newton, Isaac, 60, 65, 66, 67, 75
Newtonian physics, 31
Nickles, Thomas, 8
nomological network, 83, 90, 94–95
Norenzayan, Ara, 277
Norton, John, 9
Nosek, Brian A., 269, 270
Novick, Rose, 130n15

objects of research, 2–4, 18; as clusters of phenomena, 27–28, 172–92; as epistemically blurry, 4, 26–27, 136–41
O'Donnell, John, 36, 214
Ohnesorge, Miguel, 250n13
O'Malley, Maureen, 159–60, 159n16
operational analysis, 6, 25, 76, 236–40, 243–50
operational definitions, 2, 5–6; as clarification, 110–12; and disposition terms, 83; and experiments, 7, 77, 114–16; as explication, 73, 77, 110–12; and logical positivism, 81–87; and measurement, 5–7, 27, 70; narrow, 33, 38, 47, 49, 70, 72, 96, 104, 106–7, 110, 114, 123; and objects of research, 18; rules of application, 112; as tools, 26; wide, 33, 47–48, 70, 72, 96, 104, 107, 112, 114
Osbeck, Lisa, 281

paradigm: experimental, 6, 19, 40–41, 77–79, 116, 161; notion of, 106, 115; recognition vs. recall, 142. *See also* Kuhn, Thomas
paradigmatic: conditions of application, 25, 42, 61, 70, 104, 106–12, 113–15; experimental conditions, 33; procedures, 6, 14, 26, 37, 110, 115–16, 132–33, 167
Passos, Pedro, 266
Pavlov, Ivan, 61, 62
Pennington, L. A., 77
Pepper, Stephen, 55
Perry, Ralph Barton, 46, 51–53, 55, 58
phenomena: clusters of, 27–28; vs. data, 165–68; as evidence, 168–71; as explananda, 174–79; hidden, 163–64, 169–71; surface, 164, 169–71
phenomenal: awareness, 143–46; decomposition (*see* decomposition); reconstitution, 182–84
Phillips, Stephen, 151
Piccinini, Gualtiero, 180, 190
Place, Ullin T., 221
Poldrack, Russ, 190–91, 215, 217
Polger, Thomas, 209, 221n18
Popper, Karl, 8
Pöyhönen, Samuli, 206–7
Pratt, Carroll C., 80
Price, Cathy, 215, 216
priming: effect, 4, 110; semantic, 122–23; sensory, 122–23; tests, 17, 109–10
psychological kinds, 28, 195–204; and cognitive ontology, 215–18; and kind splitting, 222; and realism, 218–21; as whole-organism kinds, 204–15
psychophysics, 35–38, 41–42
Putnam, Hilary, 125, 127, 198, 221n18

Quine, W. V., 84, 125–27, 140n5, 197–98

Rashevsky, Nicolas, 65
Rashotte, Michael, 67

Rasmussen, Nicolas, 252
reductionism, about natural kinds, 208–9
reference, 119, 120–39. *See also* meaning
Reichenbach, Hans, 10, 12, 115n10, 252
Reijula, Samuli. *See* Pöyhönen, Samuli
reliabilism, 88
reliability: of data, 236, 239–40, 270; of instruments, 88
Renn, Jürgen, 131
replication crisis, 20, 271
Reydon, Thomas, 202, 203n6
Rheinberger, Hans-Jörg, 4, 7, 19–20, 118, 136, 153n11
Rich, G., 40n11
Richardson, Robert, 2, 183–85
Richardson-Klavehn, Alan, 144, 145
Ricker, Timothy, 120
Roback, A. A., 47n18
Robins, Sarah, 146
robustness, 92–94, 250–54
Roediger, Henry L., III, 123
Rojas-Líbano, Daniel, 258, 259
Rol, Menno, 230n2
Romero, Felipe, 21
Rosch, Eleanor, 125
Rouse, Joseph, 126, 129
Rubin, Mark, 269, 270n2, 272
Russell, Bertrand, 66
Russo, Federica, 230n2, 280
Ryan, Lee, 249
Ryle, Gilbert, 32

Salmon, Wesley, 252–53, 254–55
Schacter, Daniel, 105, 110, 122–23, 126, 141, 143–44, 249n12
Schickore, Jutta, 13, 14, 93, 153, 240, 242
Schubert, Torsten, 150–51
Schupbach, Jonah, 253
Schvanefeldt, Roger, 109n7, 229
Sellars, Wilfrid, 83
Shapere, Dudley, 131, 140–41
Shapiro, Lawrence, 209, 221n18
Shifrin, Richard, 105
Sijtsma, Klaas, 255n18
Skinner, B. F., 32–34, 34n5, 45, 80
Slaney, Kathleen, 89
Slater, Matthew, 208n11
Slors, Marc, 219–20

Smart, J. J. C., 221
Smith, Eden, 119n13
Smith, Laurence, 52, 56, 57, 58, 63n33, 65, 66, 67
Soler, Léna, 230
Spence, Kenneth, 64, 65, 67, 76, 80–81, 82n5, 240
Squire, Larry, 105
Stadler, Friedrich, 46n16, 56
Starck, Heather, 110
Steel, Daniel, 171
Steinle, Friedrich, 2, 10, 16, 22, 27, 115, 153–57, 153n11, 161, 274
Sternberg, Robert J., 91n17
Stevens, Stanley Smith, 24, 32, 33, 34, 35–46, 47, 48, 49, 56, 58, 61, 62, 68, 69, 70, 72, 75, 78, 79, 80, 85, 95n18, 97, 98, 105
Stinson, Catherine, 188n13
Stone, Caroline, 89n14
storage capacity. *See* working memory capacity
Strack, Fritz, 274
Strapasson, Bruno Angelo, 32n2
Strawson, Peter F., 111n8
Sturm, Thomas, 20n6
Sullivan, Jacqueline, 3n2, 116, 178, 180n11, 200, 218, 239
Suppe, Frederick, 31

theory construction, 25, 56, 60–64, 73–75, 100, 271–72
Thompson, Morgan, 254, 256
Titchener, Edward, 36, 40, 43, 44, 214
Tobin, Emma, 201
Tolman, Edward C., 24, 32, 34, 36n7, 46–58, 60, 61, 62, 68, 69, 70, 72, 75, 79, 85, 95n18, 96, 97, 176, 180, 214, 266
Tulodziecki, Dana, 11
Tulving, Endel, 105, 110
Tunç, Mehmet Necip, 238

Underwood, Benton, 95–96
Uygun Tunç, Duygu, 238

validity: construct, 250–51, 279; external, 231–32, 266–67; internal, 231–32, 274; predictive, 279
Van Bavel, Jay, 276

van Heerden, Jaap, 89n13, 251n14
van Rooij, Iris, 179n10, 272
variables: independent and dependent, 7, 47–50, 114; intervening, 47–51, 62–63, 100; latent, 233–34, 236–37
Verhaegh, Sander, 45
verificationism, 82–85
Vessonen, Eleni, 31

Wajnerman-Paz, Abel, 258, 259
Walters, Bradley, 275
Ward, Dave, 268
Warrington, Elizabeth, 121
Waters, R., 77
Watson, John B., 32, 51, 53
Weber, Erik, 175
Weber, Ernst, 41
Weber, Marcel, 11, 241, 247n10, 252
Weisskopf, Daniel, 204

Welpinghus, Anna, 213n13
Whitehead, Alfred, 66
Wilcox, John, 20n6
Wilson, Mark, 19n5, 128, 129, 130
Wilson, William, 151
Wimsatt, William, 86n11, 92, 93, 251
Woodger, J. H., 67
Woodward, James, 4, 27, 164, 165–72, 176, 189, 192, 239
Woodworth, R., 68
working memory capacity, 120, 148–51
Wright, Jessey, 222
Wundt, Wilhelm, 36, 43

Yarkoni, Tal, 22, 190–91, 273, 274
Yerkes, Robert, 36n7, 46, 53–54, 54n23

Zoll, P., 40n11

www.ingramcontent.com/pod-product-compliance
Ingram Content Group UK Ltd.
Pitfield, Milton Keynes, MK11 3LW, UK
UKHW042114180325
456433UK00003B/196